Handbook of
Electronic Components

Handbook of
Electronic Components

Edited by Aidan Foley

*C*LANRYE
*I*NTERNATIONAL
www.clanryeinternational.com

Clanrye International,
750 Third Avenue, 9th Floor,
New York, NY 10017, USA

ISBN: 978-1-64726-129-0

Cataloging-in-Publication Data

Handbook of electronic components / edited by Aidan Foley.
  p. cm.
Includes bibliographical references and index.
ISBN 978-1-64726-129-0
1. Electronic apparatus and appliances. 2. Electronic instruments. 3. Electronics. I. Foley, Aidan.
TK7870 .H36 2022
621.381--dc23

For information on all Clanrye International publications
visit our website at www.clanryeinternational.com

CLANRYE
INTERNATIONAL

Contents

Preface

Every book is initially just a concept; it takes months of research and hard work to give it the final shape in which the readers receive it. In its early stages, this book also went through rigorous reviewing. The notable contributions made by experts from across the globe were first molded into patterned chapters and then arranged in a sensibly sequential manner to bring out the best results.

Electronic components are discrete devices in electronic systems that affect electrons and electric fields. They can be active, passive or electromechanical. Active components include triode vacuum tubes, transistors and tunnel diodes which rely on a source of energy, while passive components are resistors, inductors, transformers and capacitors. Electromechanical components, such as piezoelectric devices and crystals drive an electrical operation with the use of moving parts or electrical connections. Electronic components can have a number of electrical leads or terminals. These connect to an electrical circuit that delivers a particular function. This book outlines the varied components of an electronic circuit and their applications in detail. It consists of contributions made by international experts. Scientists and students actively engaged in electronic engineering will find this book full of crucial and unexplored concepts.

It has been my immense pleasure to be a part of this project and to contribute my years of learning in such a meaningful form. I would like to take this opportunity to thank all the people who have been associated with the completion of this book at any step.

Editor

On the Evaluation of Gate Dielectrics for 4H-SiC based Power MOSFETs

Muhammad Nawaz

ABB Corporate Research, Froskargränd 7, 724 78 Västerås, Sweden

Correspondence should be addressed to Muhammad Nawaz; muhammad.nawaz@se.abb.com

Academic Editor: Jiun-Wei Horng

This work deals with the assessment of gate dielectric for 4H-SiC MOSFETs using technology based two-dimensional numerical computer simulations. Results are studied for variety of gate dielectric candidates with varying thicknesses using well-known Fowler-Nordheim tunneling model. Compared to conventional SiO_2 as a gate dielectric for 4H-SiC MOSFETs, high-k gate dielectric such as HfO_2 reduces significantly the amount of electric field in the gate dielectric with equal gate dielectric thickness and hence the overall gate current density. High-k gate dielectric further reduces the shift in the threshold voltage with varying dielectric thicknesses, thus leading to better process margin and stable device operating behavior. For fixed dielectric thickness, a total shift in the threshold voltage of about 2.5 V has been observed with increasing dielectric constant from SiO_2 (k = 3.9) to HfO_2 (k = 25). This further results in higher transconductance of the device with the increase of the dielectric constant from SiO_2 to HfO_2. Furthermore, 4H-SiC MOSFETs are found to be more sensitive to the shift in the threshold voltage with conventional SiO_2 as gate dielectric than high-k dielectric with the presence of interface state charge density that is typically observed at the interface of dielectric and 4H-SiC MOS surface.

1. Introduction

Power semiconductors provide basic building block in almost all energy conversion, transmission, and distribution networks used today. From system design point of view, semiconductor devices that enable reduced power losses, provide higher power density, facilitate compact converter design, and simultaneously bring lower overall system cost will be considered key technological booster for very high power applications. With power electronics reliability consideration in mind, semiconductor devices [1–10] are also evaluated for high power applications that demand harsh environment [3, 4, 9]. Recently, silicon carbide (SiC) material has gained substantial interest as a promising candidate for high power and high temperature applications. Compared to conventional Si material (used today for standard CMOS and for variety of high power applications ranging from 25 to 125°C), silicon carbide is a wide bandgap material (i.e., 3 times than that of Si) having larger thermal conductivity values (i.e., 3 times than that of Si) and larger breakdown field strength (i.e., 10 times than that of Si) and offers larger carrier saturation

velocity (i.e., 2 times than that of Si). Because of these unique features, SiC material is well blessed for high power devices [10–20] for applications ranging from −75 up to 550°C [4].

Silicon carbide based semiconductor devices such as Schottky diodes [3], junction field effect transistors (JFETs) [8, 9], bipolar junction transistors (BJTs) [4–7], metal oxide semiconductor field effect transistors (MOSFETs) [11–15], insulated gate field effect transistors (IGBTs) [16–20], and integrated gate commutated thyristors (IGCTs) [10] are explored today as potential candidates to meet the growing demand from the point of view of power system compactness that offer increased power density and simultaneously reduced overall system losses. While significant progress at material and device level research has been made and resulted in commercialization of some of these devices, reliability concern has been raised either in the passivation process for BJTs/JFETs [7, 8] or in gate dielectric process for MOSFETs/IGBTs [11–20] that may present one limiting factor for reliable performance of these devices in real applications. Note that most of these devices use silicon dioxide (SiO_2) as a passivation layer protecting the device surface or as gate dielectric process. While SiC can be thermally oxidized to yield SiO_2 over its surface, a fundamental inherent drawback of SiO_2 is its low dielectric constant, which is about 2.5 times lower than that of the SiC material and also possesses poor interface properties at the SiO_2/SiC junction. This leads to proportionally a larger electric field enhancement in the dielectric medium compared to that in the semiconductor layer underneath, which is a reason why new dielectrics with dielectric constant at least similar to that of the SiC material and with lower interface state densities are required for device applications. Note that this inequality of dielectric constant often requires device operation at an electric field far below the SiC material breakdown field in order to avoid premature SiO_2 breakdown at the device surface. The performance of these SiO_2/SiC interfaces for MOSFETs (and for BJTs as well) has been improved by various oxidation [21, 22] or nitridation methods [21, 22], but their realistic commercialization potential still remains limited by a low channel mobility due to a high interface state density near the conduction band. Generally speaking, the density of interface states (D_{it}) at the SiO_2/SiC interface is at least two to three orders of magnitude higher ($\sim 10^{12}$ eV^{-1} cm^{-2}) [23–26] compared to the relatively matured Si/SiO_2 interface.

To overcome the abovementioned problems associated with SiO_2 on SiC, potential of various high-k gate dielectrics (e.g., Al_2O_3, HfO_2, AlN, La_2O_3, Y_2O_3, Al_2O_3, and Ta_2O_5) [27–43] is being explored recently for SiC MOS technology. Among them, Al_2O_3 is getting more attention as potential substitute for SiO_2 primarily due to its excellent lattice matched with SiC, its compatible high-k value with 4H-SiC, good thermal stability, reasonably high conduction band off-set between 4H-SiC and Al_2O_3, and relatively large dielectric bandgap. Similarly, HfO_2 is considered another interesting candidate for 4H-SiC MOS devices due to its high-k value. However, its bandgap is relatively small compared to SiO_2 or Al_2O_3 gate dielectric. Utilizing high-k value dielectric but with problem of its low bandgap, this issue is being addressed by inserting a sandwiched layer of SiO_2 between HfO_2 and 4H-SiC [35, 43, 44] or between Al_2O_3 and 4H-SiC [36, 38, 40, 41]. On the other side, a significant reduction in leakage current density has been observed in high-k La_2O_3 structure (i.e., smaller bandgap than that of SiO_2) when a 6 nm thick thermal nitrided SiO_2 has been inserted between La_2O_3 and SiC [45]. Recent device performance of using high-k gate dielectric for 4H-SiC based MOSFETs [38–41] shows overall good progress for future SiC device commercialization.

Similar to conventional low power devices using high-k gate dielectric over Si, one of the major challenges is the significantly lower channel mobility for the SiC-MOSFETs due to high interface state density and interface surface roughness through scattering and trapping effects for the carriers at SiO_2/SiC interface. This surface roughness also poses threat to the gate oxide reliability and further leads to instability of the threshold voltage. While for matured SiO_2/Si interface channel mobility lies close to the universal mobility curve, mobility for high-k gate dielectric interface with Si lies well below the universal curve and the real cause is not well understood yet [25]. Channel mobility values of SiC MOSFETs with thermally grown SiO_2 as gate dielectric

present unacceptably low value of 10 $cm^2/V \cdot s$ due to high density of interface traps [12, 13]. Postoxidation annealing of the gate oxide under NO/N_2O (nitric/nitrous oxide) [14] and $POCl_3$ [15] environment helps to further improve this value to 50 [32] and 89 $cm^2/V \cdot s$ [15], respectively. A record peak channel mobility of 150 $cm^2/V \cdot s$ has also been reported after performing oxidation process in the presence of alumina [11] for SiO_2/SiC interface based MOSFETs. More recently, the improvement of channel mobility over 100 $cm^2/V \cdot s$ using combined "Sb + NO" process which is associated with the dual mechanisms of counterdoping by Sb (i.e., antimony) and interface trap passivation by NO has been reported [46] for 4H-SiC MOSFETs. Ultrahigh channel mobility has been demonstrated for 4H-SiC-MOSFETs with Al_2O_3 gate insulators fabricated at low temperatures by metal-organic chemical-vapor deposition [38]. Relatively high field effect channel mobility of 64 $cm^2/V \cdot s$ is obtained when Al_2O_3 gate insulator is deposited at 190°C. Furthermore, extremely high field effect mobility of 284 $cm^2/V \cdot s$ was obtained for a MOSFET fabricated with an ultrathin thermally grown SiO_x layer inserted between the Al_2O_3 and SiC [38]. Yet, in another investigation, the same group [38, 40] has demonstrated a remarkable increase in the channel mobility by inserting a thin SiO_2 layer (1-2 nm) between Al_2O_3 and SiC [41]. A maximum channel mobility in stacked dielectric of $Al_2O_3/SiO_2/SiC$ based MOSFETs as high as 300 $cm^2/V \cdot s$ was reported using low deposition temperature of the gate-insulator film. Note that three positive outcomes as reported in the scientific literature [35–43] are generally expected using thin SiO_2 layer between high-k material and SiC surface underneath; namely, (i) thin layer of SiO_2 acts as a barrier layer against unwanted chemical reaction with the SiC substrate during high-k growth, (ii) interface state density should be lower than that of growing high-k material directly over SiC surface, and (iii) inserted SiO_2 layer would further reduce the effect of Coulomb scattering from the fixed charges present in the high-k film. All in all, these preliminary findings using high-k stacked dielectric [35–43] present a fair advancement in SiC MOSFET technology development through reduction in the interface state density (D_{it}) compared to pure SiO_2/SiC interface.

A list of various dielectrics [23–26, 44, 45] and their physical properties is illustrated in Table 1. Figure 1 shows the analytical trend of the bandgap energy as a function of the dielectric constant of the material [23–25, 44, 45]. Note that the energy bandgap of dielectric material has a direct correlation with the gate leakage current through the conduction band edge offset values where a wider bandgap energy of dielectric material means a better chance for getting larger conduction (ΔE_C) or valence band (ΔE_V) offsets at the interface of semiconductor and the gate dielectric. Interestingly, lower conduction band offsets are predicted with respect to dielectric material due to smaller bandgap difference at the interface of dielectric and SiC compared to Si material counterpart as shown in experimentally extracted values from the literature [23, 24, 44] in Table 1. It is therefore objective of this work to investigate various dielectrics that can be employed as a potential candidate for 4H-SiC based

TABLE 1: List of physical properties of various dielectric materials reported in the literature.

Material	Dielectric constant (k)	Bandgap E_g (eV)	ΔE_C (eV) with respect to Si	ΔE_C (eV) with respect to 4H-SiC	Structure	Preparation method
SiO$_2$	3.9	8.9	3.2	2.2–2.7	Amorphous	Thermal, PECVD
Si$_3$N$_4$	7.0	5.1	2.0	—	Amorphous	Thermal, LPCVD, MOCVD
SiON	4.0–7.0	5.0–9.0 (O/N ratio)	2.8	—	Amorphous	Thermal, PECVD
Al$_2$O$_3$	9.0	8.7	2.8	1.7	Amorphous	Sputtering, ALCVD
HfO$_2$	25	5.7	1.5–1.7	0.54 and 0.7–1.6	Mono, tetra, cubic	Sputtering, ALCVD
ZrO$_2$	25	7.8	1.4	1.6	Mono, tetra, cubic	ALCVD
Ta$_2$O$_5$	26	4.5	1–1.5	—	Orthorhombic	MOCVD
Y$_2$O$_3$	15.0	5.6	2.3	—	Cubic	—
La$_2$O$_3$	30	4.3	2.3	—	Cubic	—
AlN	9.14	6.2	2.2	1.7	Wurtzite	MOCVD

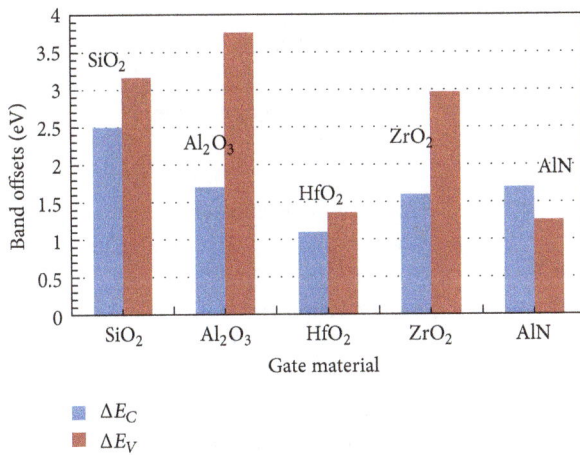

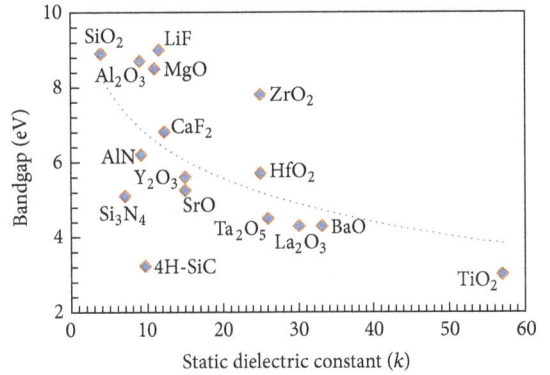

(a)   (b)

FIGURE 1: Conduction band (ΔE_C) and valence band (ΔE_V) offsets of various dielectrics with respect to 4H-SiC material (a) and bandgap energy (E_g) as a function of dielectric constant (k) of various materials reported in the literature (b).

MOSFETs. Influence of dielectric constants, dielectric thicknesses, and interface state densities has been studied using two-dimensional numerical computer simulation.

2. Device Simulation Setup

A schematic cross sectional view of the simulated 4H-SiC based MOSFET device along with the net doping profile is shown in Figure 2. For simplicity, only left half of the device with horizontal dimension of $4\,\mu$m is simulated with a channel length of $0.8\,\mu$m. A drift layer thickness of $25\,\mu$m with a doping concentration 5×10^{14} cm^{-3} is used to get a hypothetical device of blocking voltage of at least 1700 V. Device simulation was executed by considering

bandgap narrowing model [47], Auger recombination model [47], Shockley-Read-Hall (SRH) recombination [47], doping and temperature dependent field mobility models [47], and incomplete ionization model [47]. In addition, all the simulations were carried out using Fermi Dirac statistics. The used physical models (e.g., bandgap, incomplete ionization, mobility model, and carrier lifetime) and their parameters in this work have earlier been applied for 4H-SiC devices [6, 19, 20, 28]. The total number of mesh points was 40,000 while mesh resolution at the surface interfaces and p-n junction areas was 0.2-0.3 nm. The material parameters used in present simulations are listed in Tables 2 and 3 and have earlier been used and verified in previous simulation papers for 4H-SiC BJT [5–7] and 4H-SiC IGBT [19, 20] devices. The carrier lifetime in different regions of the device was simulated

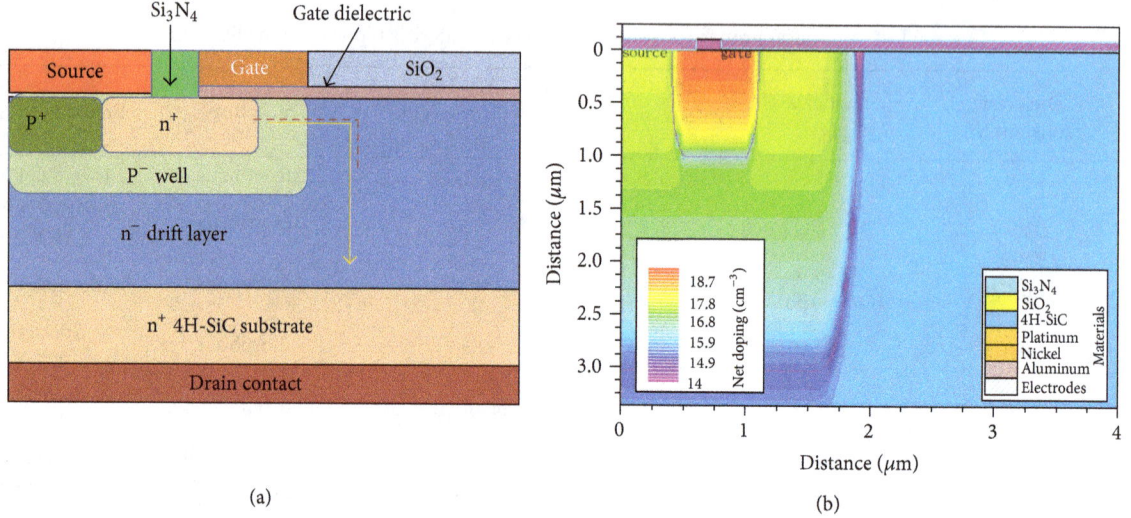

FIGURE 2: A schematic cross section of simulated device layer structure (a) and corresponding net doping profile showing top portion of 4H-SiC based MOSFET (b).

TABLE 2: Parameters used in 4H-SiC based MOSFET device simulation.

Eg_{300} (eV)	3.24	Bandgap at 300 K
Eg_{alpha} (eV/K)	4.15×10^{-4}	Parameter of bandgap model
Eg_{beta} (eV/K)	-131	Parameter of bandgap model
Permittivity	9.66	Permittivity
Affinity (χ)	4.2	Affinity
A_{UGN} (cm^6/s)	5×10^{-32}	Auger recombination parameter for electrons
A_{UGP} (cm^6/s)	2×10^{-32}	Auger recombination parameter for holes
Ea_b (eV)	0.2	Acceptor energy level
Ed_b (eV)	0.1	Donor energy level
Gv_b (eV)	4	Degeneracy factor for valence band
Gc_b (eV)	2	Degeneracy factor for conduction band
LT.TAUN	5	Lifetime model parameter for electrons
LT.TAUP	5	Lifetime model parameter for holes
N_{srhn} (cm^{-3})	3×10^{17}	SRH concentration-dependent lifetime for electrons
N_{srhp} (cm^{-3})	3×10^{17}	SRH concentration-dependent lifetime for holes
A_{RICHN} (A/K^2 cm^2)	110	Effective Richardson constant for electrons
A_{RICHP} (A/K^2 cm^2)	30	Effective Richardson constant for holes

by considering doping and temperature dependent carrier lifetime model [6, 20]:

$$\tau_{n,p} = \frac{\tau_{max,n,p}(T/300)^{1.72}}{1 + \left(N/(3 \times 10^{17})\right)^{0.3}}. \quad (1)$$

A wide spread in the band offset values of various dielectric materials with respect to 4H-SiC has been reported in the literature [21–27, 44, 45]. For few potential high-k dielectric materials (e.g., see Table 1), the band offset values with respect to 4H-SiC material are still unknown. In present simulation, the electron affinity values were adjusted to get the right conduction band offsets of respective dielectric material equivalent to its experimentally extracted value with respect to 4H-SiC. For example, experimentally extracted

values of ΔE_C of 2.5, 1.7, 1.1, 1.6, and 1.7 eV are used in present simulations for SiO$_2$, Al$_2$O$_3$, HfO$_2$, ZrO$_2$, and AlN material, respectively, that correspond to 44, 31, 45, 35, and 57%, respectively, of the bandgap difference of the respective dielectric with respect to 4H-SiC. Since ΔE_C of few dielectrics (e.g., Si$_3$N$_4$, Y$_2$O$_3$, and La$_2$O$_3$) is still unknown, 50% of the bandgap difference has been associated with conduction band, which is almost similar but closer to other known dielectric materials (e.g., SiO$_2$, HfO$_2$).

Note that time dependent dielectric breakdown (TDDB) in power semiconductor devices is considered a potential reliability concern for dielectric layers. Since Si and SiC based MOS devices generally use SiO$_2$ as a gate dielectric, the resultant barrier height and conduction band offset difference between SiC and SiO$_2$ is fairly smaller than that

TABLE 3: Mobility parameters used in 4H-SiC based MOSFET device simulation.

μ_{1n}.caug	40	cm²/V·s
μ_{2n}.caug	950	cm²/V·s
ncrn.caug	2×10^{17}	cm⁻³
δ_n.caug	0.73	Arbitrary
γ_n.caug	−0.76	Arbitrary
α_n.caug	0	Arbitrary
β_n.caug	−2.4	Arbitrary
μ_{1p}.caug	53.3	cm²/V·s
μ_{2p}.caug	105.4	cm²/V·s
N_{critp}.caug	2.2×10^{18}	cm⁻³
δ_p.caug	0.7	Arbitrary
γ_p.caug	0	Arbitrary
α_p.caug	0	Arbitrary
β_p.caug	−2.1	Arbitrary
v_{satn}	2×10^7	cm²/s
v_{satp}	2×10^7	cm²/s
β_n	2	—
β_p	1	—

of Si counterpart [29–32]. This leads to higher tunneling current in 4H-SiC/SiO₂ based MOS system than that in Si/SiO₂ based MOS devices. For thick gate oxides (>5.0 nm), Fowler-Nordheim (FN) tunneling mechanism has been suggested to contribute to dielectric breakdown, particularly at high electric fields. Since FN tunneling has earlier been used to characterize SiC based MOS capacitors [23, 29, 30, 32, 45] with thick dielectric layers (5–50 nm), the present work therefore takes into account FN tunneling model for various dielectric assessment. Direct tunneling through the dielectric has therefore been ruled out in this work since it dominates for very thin dielectrics (<3 nm) where the tunneling current increases exponentially with reduction in the oxide thickness. The Fowler-Nordheim current density equation that represents the tunneling current through the gate dielectric is expressed as

$$J_{\text{FN}} = F_{\text{AN}} \cdot E^2 \cdot \exp\left[-\frac{F_{\text{BN}}}{E}\right],$$

$$J_{\text{FH}} = F_{\text{AH}} \cdot E^2 \cdot \exp\left[-\frac{F_{\text{BH}}}{E}\right], \tag{2}$$

where E specifies the magnitude of the electric field in the gate dielectric. F_{AN} and F_{BN} are 1.8×10^{-7} and 1.92×10^8, respectively, and are adjustable modeling parameters for electrons. Similarly, F_{AH} and F_{BH} are 1.83×10^{-7} and 1.91×10^8, respectively, and are modeling parameters for holes. Principally, $F_{\text{AN}}(F_{\text{AH}})$ and $F_{\text{BN}}(F_{\text{BH}})$ depend on the tunneling barrier height (i.e., $F_{\text{AN}}/F_{\text{AH}} \propto 1/\phi_b$ and $F_{\text{BN}}/F_{\text{BH}} \propto \phi_b^{3/2}$) and effective mass of the tunneling electrons (holes) [31] and where the barrier height is defined as the difference between the electron affinities of the metal/semiconductor and the dielectric.

3. Results and Discussion

Figure 3 illustrates the conduction band energy diagram (a), current voltage characteristics of SiC MOSFET (b), electric field at the interface, and gate current density at various gate biases. While dielectric constant of HfO₂ is larger than that of SiO₂ (see Table 1), a narrow bandgap of HFO₂ results in smaller conduction band offsets with respect to SiC material (Figure 3(a)). With these small conduction band offset values, the probability of carriers tunneling through the dielectric is increased significantly and hence it may limit the purpose of using this high-k material. Tanner et al. [23] have earlier reported a conduction band offset of 0.7–0.9 eV at the HfO₂/4H-SiC interface that results in insufficient barrier height causing unacceptably high leakage current values. A high leakage current density as a result of small conduction band offset and increased density of surface trap at HfO₂/SiC interface has been reported [34] and could have an impact on the electron transport properties of the MOSFET. Note that, for low power electronics, HfO₂ is a promising gate oxide material for Si-MOSFETs due to its high dielectric constant. However, its small bandgap value of 5.7 eV presents prohibitive feelings for 4H-SiC based MOSFET/IGBT devices. On the other side, a significant reduction in the electric field is induced in the dielectric material as a result of its higher dielectric constant values. The low band offsets at the dielectric/SiC interface have recently been addressed by introducing an ultrathin SiO₂ interfacial layer between SiC and high-k HfO₂ dielectrics [35, 44] for SiC based MOS structures. A barrier height of 1.5 eV has been extracted from the Schottky emission characteristics, which is higher than the reported value for HfO₂ on SiC surface without additional interfacial SiO₂. Thus, presence of an interfacial SiO₂ layer increases band offsets to reduce the leakage current characteristics. While gate leakage current has been reduced significantly with sandwiched SiO₂ [44], further optimization of this new layer stack is required to get a clean and abrupt interface morphology with minimization of formation of intermixing layer at the interface. Note that simulation of new dielectric layer stack with additional SiO₂ has been ignored here primarily because band alignment values are not available for the whole layer stack of HfO₂/SiO₂/4H-SiC [35]. This may in fact be the objective of future work where new dielectric combinations are assessed for surface passivation and as gate dielectric material for SiC based MOSFETs/IGBTs. With fixed drain-source bias, gate current density increases with the increase of gate bias. Current voltage characteristics at different gate-source biases of simulated MOSFET show reduction in the drain current in saturation region and increase in the ON-resistance (i.e., R_{ON} is defined as a slope of the I-V characteristics in the linear region of device operation for a given gate bias above threshold: see Figure 3(b)) with temperature, consistent with the experimental findings of 4H-SiC based MOSFETs [12–14], and hence allow paralleling of the MOSFET devices for high power applications. Increase in the ON-resistance with temperature is attributed mainly to decrease in the bulk carrier mobility for thick drift layer MOSFETs.

(a)

(b)

(c)

(d)

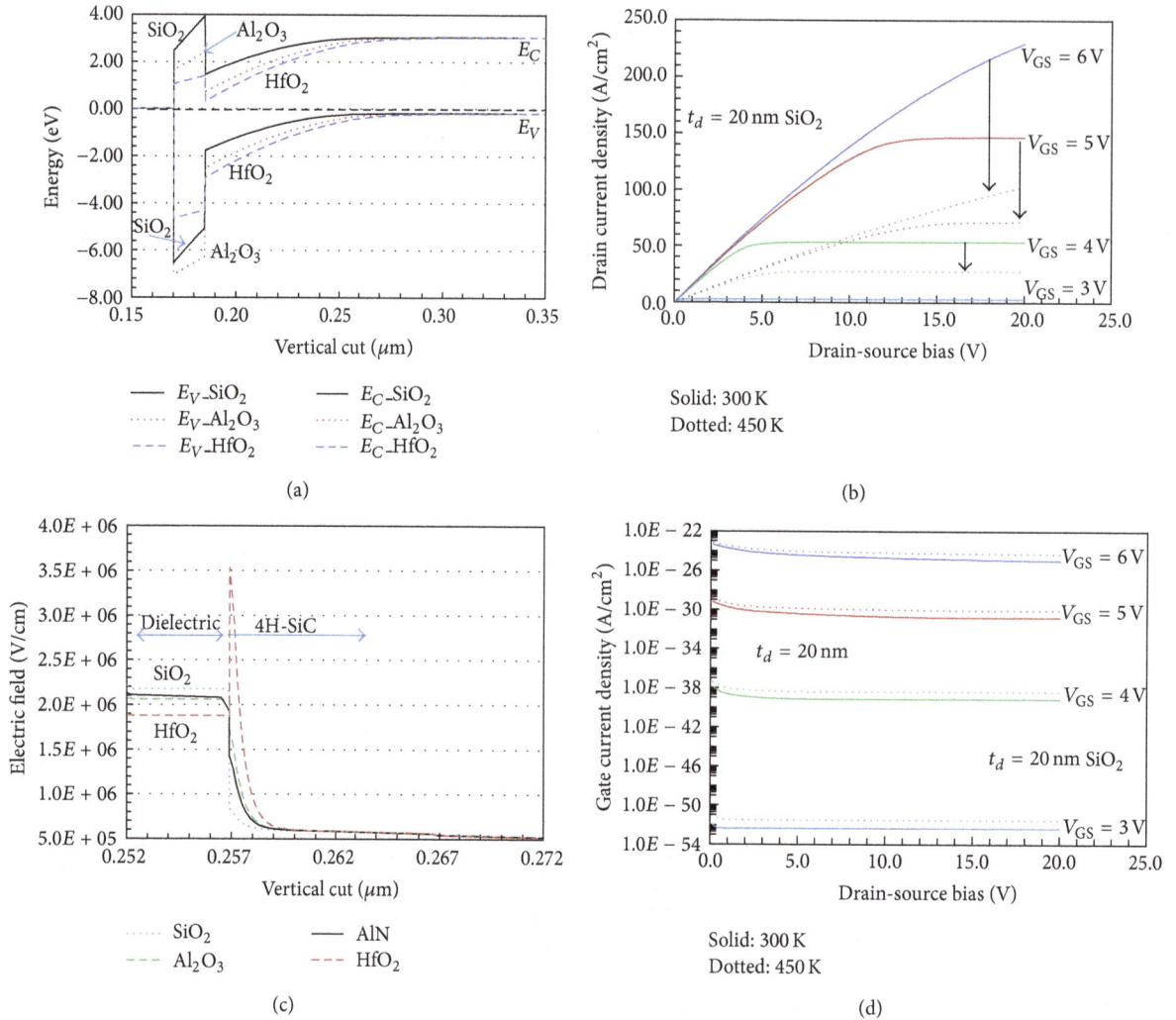

FIGURE 3: Conduction and valence band energy diagram (a), current-voltage characteristics at different temperatures for 20 nm thick SiO_2 (b), electric field at the semiconductor-dielectric interface for various dielectrics (c), and gate current density at different gate-source biases for 20 nm SiO_2 gate dielectric (d).

For fixed dielectric thicknesses of 5 nm each, gate current density is plotted in Figure 4 as a function of gate-source bias at constant drain-source voltage of 10 V for various dielectric materials. With varying gate bias at constant drain-source bias, the MOSFET goes from accumulation (negative bias) to depletion and then to inversion region (positive gate bias) as expected. In the range from 6 to 12 MV/cm, the gate current density decreases with the increase of dielectric constant of the respective material. For example, current density of 3.9×10^{-4} (1.3×10^{-8}), 1.3×10^{-4} (2.5×10^{-9}), 1.2×10^{-5} (6.8×10^{-10}) and 1.1×10^{-5} (1.3×10^{-10}) A/cm^2 for SiO_2, Al_2O_3, AlN and HfO_2, respectively is obtained at 10 (7) MV/cm with equal dielectric thickness. While dielectric constant of HfO_2 is larger than that of other high-k dielectrics (e.g., Al_2O_3, AlN), this advantage is fairly negated due to smaller bandgap and hence smaller conduction band offset of HfO_2 with 4H-SiC. Simulations predict that materials (AlN, Y_2O_3) with moderate/high dielectric constant along with smaller bandgaps tend to suffer from higher leakage current

at negative bias (accumulation) as a result of smaller valence band offsets. This has earlier been experimentally witnessed for TiO_2 ($E_g = 3.5$ eV) material that has shown significantly higher gate leakage current density in accumulation region in comparison to A_2O_3 ($E_g = 8.7$ eV) material for 4H-SiC based MOS capacitors [27].

AlN and Al_2O_3 are two promising compatible (almost similar dielectric constant with 4H-SiC) candidates as gate dielectric with 4H-SiC materials. However, lower bandgap of AlN (6.2 eV) in comparison with Al_2O_3 (8.7 eV) or SiO_2 (8.9 eV) might be disappointing for 4H-SiC devices, but a lattice mismatch to SiC of only 1% along with almost the same thermal expansion coefficient of up to 1000°C and a high dielectric constant are more encouraging parameters. Similar to HfO_2 and Al_2O_3 [35, 36, 44] as discussed earlier, a thin SiO_2 as a buffer layer has been inserted [33] between SiC and AlN as additional barrier layer to prevent electron injection from semiconductor to dielectric, which may further decrease leakage current. Figure 5 illustrates the gate

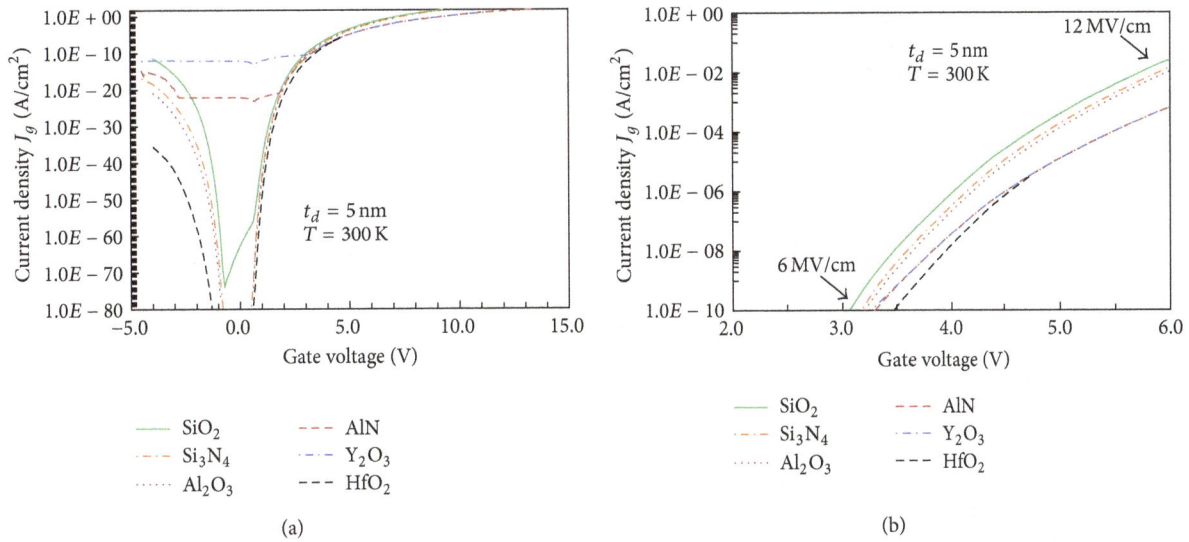

FIGURE 4: Gate current density as a function of gate bias for various gate dielectrics of thickness 5 nm, each at 300 K (a). Zoomed-in view over limited gate bias range is also shown with an electric field span of 6 to 12 MV/cm (b).

current density for SiO_2 and Al_2O_3 dielectrics with various thicknesses at 300 K. Gate current density decreases with increasing dielectric thicknesses as expected. Compared to SiO_2, a lower gate current density is achieved for Al_2O_3 material for equal gate dielectric thickness. Al_2O_3 material belongs to the family of wide bandgap (8.7 eV) and possesses a potential barrier of 2.8 eV and 1.7 eV with Si and 4H-SiC conduction band, respectively. Although conduction band offsets with 4H-SiC material are smaller than that of Si, this value is high enough to effectively suppress the carrier injection at interface. A gate current density of $2.0e-3(1.7e-3)$ and $3.5e-3(3.0e-3)$ is obtained for Al_2O_3 and SiO_2 respectively, at 10 MV/cm field assuming 20 (10) nm thick gate dielectric. Similarly, a gate current density of $3.0e-13(1.7e-13)$ and $5.0e-13(5.6e-13)$ is predicted for Al_2O_3 and SiO_2, respectively, at 5 MV/cm field with 20 (10) nm thick gate dielectric. A gate leakage current density of 10^{-3} A/cm^2 at 8 MV/cm of 4H-SiC MOS capacitor has been obtained for amorphous Al_2O_3 film grown on 4H-SiC surface by atomic layer deposition technique [23]. A barrier height of 1.58 eV has been extracted using Fowler-Nordheim tunneling model for Al_2O_3/4H-SiC interface. The amorphous Al_2O_3 films [23] further show superior leakage current density characteristics compared with many other high-k materials and stacks (i.e., Ta_2Si, SiO_2/TiO_2, Gd_2O_3, SiO_2/HfO_2, AlN, and Si_3N_4) investigated on 4H-SiC surface.

Drain-source current density and threshold voltage (i.e., the value of the gate-to-source voltage V_{GS} needed to create or induce the conducting channel to cause surface inversion) of simulated 4H-SiC MOSFET device are illustrated in Figure 6 for various gate dielectrics. The value of the threshold voltage for MOSFET device is generally dependent on some physical parameters of the device structure such as the gate material, the thickness and type of dielectric layer, substrate doping concentration, oxide-interface fixed charge concentration (or density), and channel length and channel width. For

a given gate dielectric material, threshold voltage of a 4HSiC-MOSFET increases linearly with the dielectric thickness as expected. However, the amount of variation in threshold voltage is suppressed using high-k dielectric material (e.g., Al_2O_3, HfO_2) with fixed other physical parameters of the device structure, an aspect that is favorable from device manufacturability point of view. On the other side, threshold voltage decreases with the increase of dielectric constant of a gate material for a fixed gate dielectric thickness indicating that a trade-off is required with the substrate doping to adjust the threshold voltage to higher value for a SiC based MOSFET power device. For example, a total shift in the threshold voltage of about 2.5 V has been observed with variation in gate dielectric material from SiO_2 to HfO_2 (see Figure 6(a)). This further resulted in the increase in the transconductance of the device with the increase of the dielectric constant from SiO_2 to HfO_2. For example, a maximum device transconductance of 87, 69, 68, 64, and 45 $\mu S/\mu m$ has been observed for HfO_2, AlN, Al_2O_3, Si_3N_4, and SiO_2, respectively. MOSFET device transconductance is simply defined as $(W \cdot \mu_n \cdot C_{ox} \cdot V_{DS})/L$, where W is the gate width, L is the gate length, μ_n is the channel mobility of electrons, and $C_{ox}(\approx \varepsilon_r/t_{ox})$ is the dielectric capacitance of the respective material. Considering μ_n and other parameters to be constant in our simulations, changing dielectric constant only helps to improve the maximum device transconductance as observed in our numerical simulations and also is found out to be consistent with the experimental findings [42]. Higher transconductance might help to improve the switching capability of the power device. Note that threshold voltages of Si_3N_4, Y_2O_3, and AlN are clustered with each other as a result of smaller difference in the bandgap (and hence conduction band discontinuity) of these materials for fixed dielectric thickness. Generally speaking for a given channel length of a MOSFET, as the physical thickness of the gate dielectric material increases, the number of electric

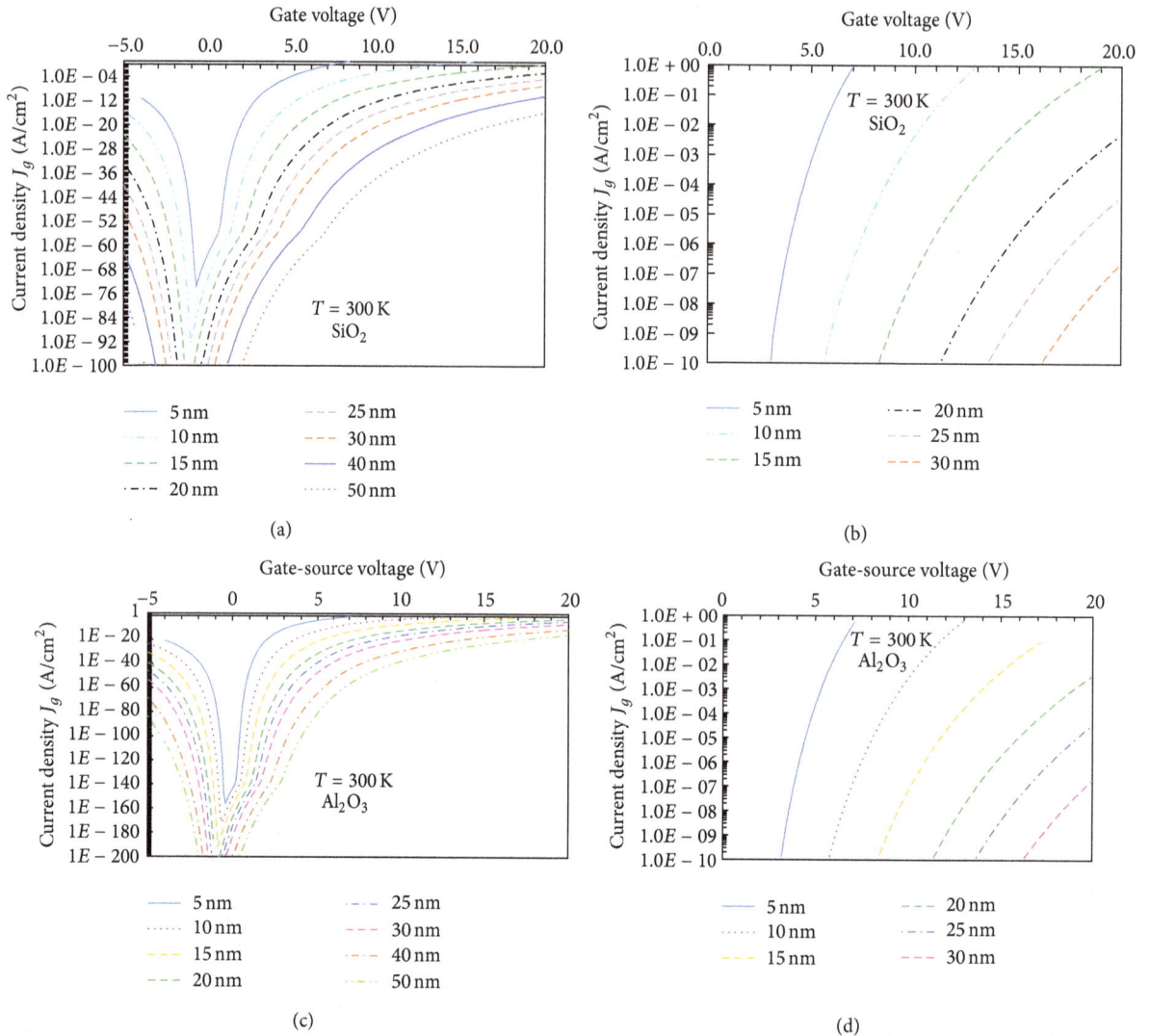

FIGURE 5: Gate current density at 300 K for different gate dielectric thicknesses using SiO_2 as a gate dielectric ((a), (b)) and Al_2O_3 as a gate dielectric ((c), (d)). Zoomed-in view is also shown for SiO_2 (b) and Al_2O_3 (d) at higher gate current densities.

field lines originating from the bottom of the gate electrode and terminating on the source and possible drain regions increases. These field lines form an electric field from source to channel region and thereby decrease the potential barrier height between the channel and the source. A lower potential barrier height here reflects a lower threshold voltage of a device for a constant channel length. Note that a potential distribution along the channel surface is strongly dependent on gate dielectric permittivities that in fact define the amount of variation in the potential barrier height. The high-k dielectrics allow reducing gate leakage current while keeping a very low electrical equivalent oxide thickness, an aspect that is especially critical for Si based low power electronics. For high power electronics using SiC based MOSFETs where the conduction band discontinuity of gate dielectric is relatively small compared to Si counterpart, a thick gate dielectric of the order of 20–30 nm (i.e., maximum of 6.5–10 MV/cm considering 20 V bias at the gate; commercial SiC-MOSFETs

today use 4-5 MV/cm with 20 V gate bias and 40–50 nm SiO_2 gate dielectric) is far sufficient to simultaneously fix the threshold voltage to large positive value with reduced gate leakage current density for high-k gate material.

One reliability concern for 4H-SiC MOSFETs is the quality and reliability of the gate-dielectric interface that has severely been affected by the presence of traps and carrier energy interface states. The origin of these states is primarily linked with the imperfect nature of 4H-SiC dielectric interfaces due to presence of carbon clusters and dangling Si and C bonds. Consequently, the channel electrons scatter with these energy states and get trapped there, hence increasing the channel resistance. The location and density of interface states within the bandgap affect not only the channel electron mobility but also the FN tunneling currents at the SiC dielectric interface. Note that the threshold voltage may also be affected by the so-called fast surface states at the semiconductor-dielectric interface and by fixed charges

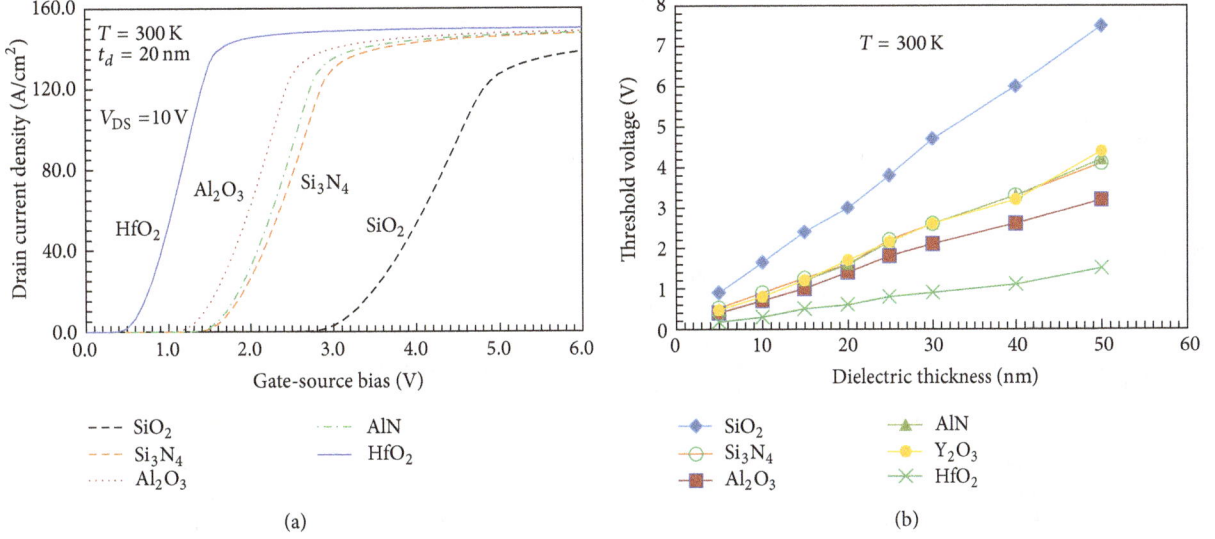

FIGURE 6: Drain current density as a function of gate-source bias for various gate dielectrics using 20 nm thick dielectric material (a) and threshold voltage shift as a function of dielectric thickness for various gate dielectrics (b). Threshold voltages of Si_3N_4, Y_2O_3, and AlN are clustered with each other as a result of smaller difference in the bandgap of these materials. Simulations have been performed for 20 nm thick gate dielectric at 300 K.

in the insulator layer of the gate. As earlier said, the density of charged interface states in 4H-SiC/SiO_2 structures is 2-3 order of magnitude higher than that at Si/SiO_2 MOS interface. While this may not be a significant concern with modern day Si based fabrication technology, depending on the type and growth mechanism of dielectric film on 4H-SiC MOSFET surface, these energy states of the order of 1.0×10^{11}–5.0×10^{12} cm^{-2} are routinely measured in real devices [23–26]. The threshold voltage of nonideal MOS capacitor is simply defined as

$$V_{TH} = \varphi_{ms} + 2\phi_F - \frac{Q_B}{C_{ox}} - \frac{Q_{ox}}{C_{ox}}, \tag{3}$$

where φ_{ms} is the metal-semiconductor work function difference, ϕ_F is the Fermi potential, Q_B is the depletion charge due to ionized impurities, C_{ox} is the oxide capacitance, and Q_{ox} is the total oxide charge which is zero for ideal MOS capacitor. Note that Q_{ox} is the sum of interface trapped charge (Q_{int}: these may be positive or negative and located at the interface), fixed oxide charge (Q_f: these may be positive or negative and located very close to the interface), oxide trapped charge (Q_{ot}: these may be positive or negative due to hole and electron trapped in the bulk of the gate oxide), and mobile ionic charge (Q_m: origin is due to presence of ionic impurities in the oxide film but far from the interface) and its overall impact causes a voltage drop across the oxide. Depending on the growth mechanism and interface material, the polarity of this fixed oxide charge induces a shift in the voltage (i.e., V_{TH} decreases if Q_{ox} is positive or vice versa). Considering arbitrary growth conditions, positive and negative interface state charge density have been introduced here at 4H-SiC/SiO_2 (positive), 4H-SiC/HfO_2 (positive), and

4H-SiC/Al_2O_3 (negative) to study their influence on drain-source current. Earlier experimental findings report that depositing Al_2O_3 or HfO_2 on 4H-SiC [23, 34–36] drastically shifts the flat band voltage to positive voltages, meaning that negative charges are generated either at the interface or near the interface and/or in the bulk of the insulating layer. Thus effective fixed oxide charge of the MOS capacitor is considered negative for pure Al_2O_3 and HfO_2 on 4H-SiC [23, 34–36, 43] surface and positive for another dielectric such as pure SiO_2 on SiC [35, 37]. More interestingly, stacking gate dielectric of HfO_2/SiO_2/SiC [35] and Al_2O_3/SiO_2/SiC [36] scheme also reveals negative effective oxide charges at the respective interfaces for SiC MOS capacitors. Figure 7 illustrates the influence of these interface state densities on drain-source characteristics. Positive interface density induces a negative shift in the threshold (i.e., V_{TH} decreases) while negative interface density increases the threshold voltage of the device. Furthermore, using a high-k dielectric material, threshold voltage variation is shrunk for a given dielectric thickness predicted by the numerical simulation.

4. Conclusions

Influence of various possible dielectric materials on 4H-SiC MOSFETs has been studied in this work. For fixed dielectric thickness, numerical device simulation predicts a smaller shift in the threshold voltage for high-k dielectrics. Similarly, a smaller shift in the threshold voltage is expected for high-k dielectric material with variation in the interface charge densities. Compared to conventional SiO_2 as gate dielectric used today in SiC MOSFETs, high k-gate dielectric reduces significantly the amount of electric field in the gate dielectric with equal gate dielectric thickness and hence

(a)

(b)

(c)

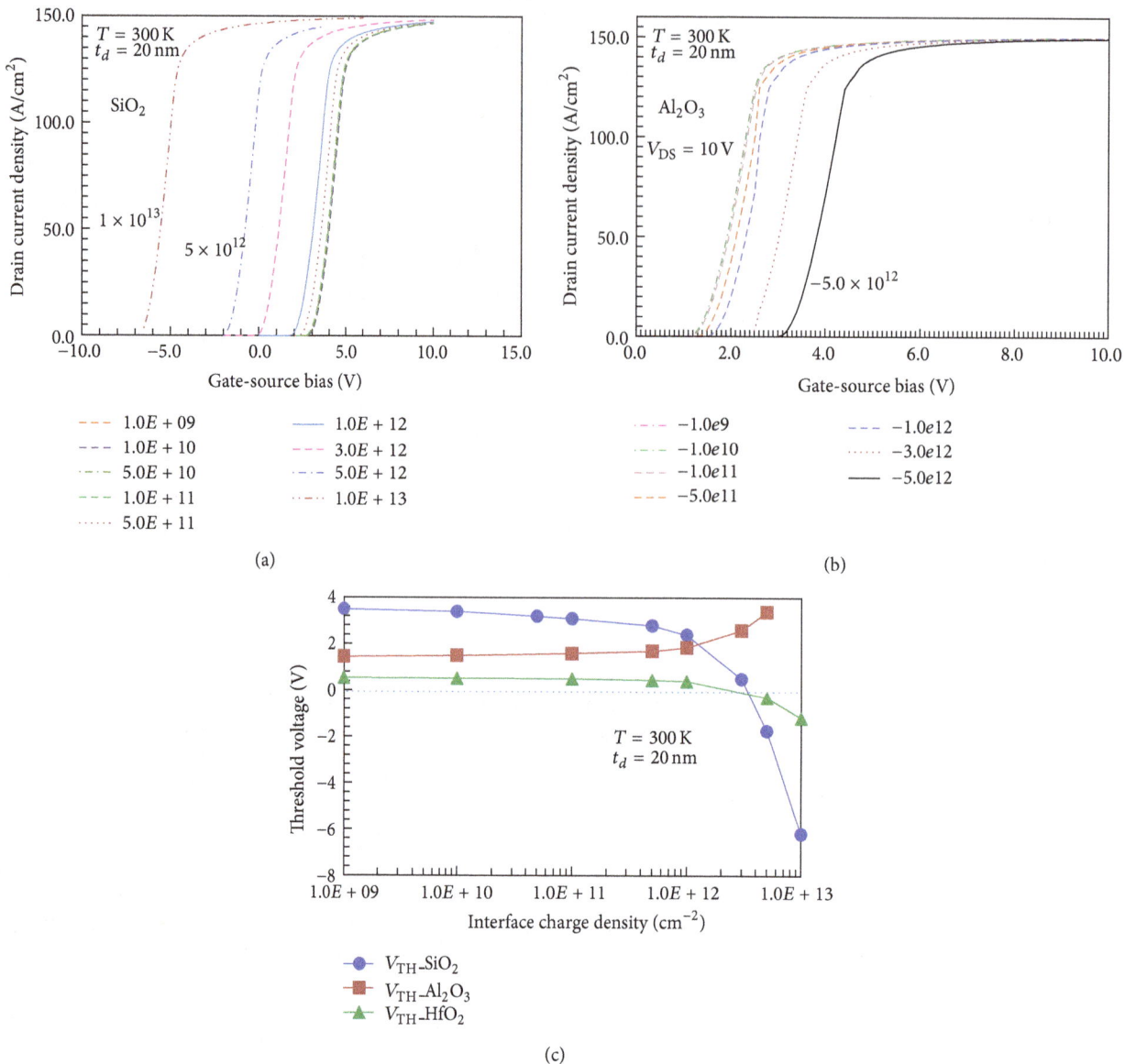

FIGURE 7: Drain current density as a function of gate-source bias for various interface state densities using SiO_2 as a gate dielectric (a) and Al_2O_3 as a gate dielectric (b). Threshold voltage shift is also shown for various gate dielectrics as a function of interface state densities (c). Simulations have been performed for 20 nm thick gate dielectric at 300 K.

the overall gate current density. To realize full potential of high-k dielectric over 4H-SiC MOSFET surface, a clean and abrupt dielectric/semiconductor interface morphology is prerequisite for reliable device operation. The numerical data presented in this work will not only provide a useful guideline to device and circuit designer but also support technology development when different dielectric options are considered.

Conflict of Interests

The author declares that there is no conflict of interests regarding the publication of this paper.

Acknowledgment

The author would like to thank University Graduate Centre (UNIK) at Kjeller, Norway, for providing computer simulation facility.

References

[1] S. Madhusoodhanan, K. Hatua, S. Bhattacharya et al., "Comparison study of 12kV n-type SiC IGBT with 10kV SiC MOSFET and 6.5kV Si IGBT based on 3L-NPC VSC applications," in *Proceedings of the IEEE Energy Conversion Congress and Exposition (ECCE '12)*, pp. 310–317, 2012.

[2] J. A. Cooper, T. Tamaki, G. G. Walden, Y. Sui, S. R. Wang, and X. Wang, "Power MOSFETs, IGBTs, and thyristors in SiC: Optimization, experimental results, and theoretical performance," in *Proceedings of the International Electron Devices Meeting (IEDM '09)*, pp. 149–152, December 2009.

[3] L. Zhu and T. P. Chow, "Advanced high-voltage 4H-SiC Schottky rectifiers," *IEEE Transactions on Electron Devices*, vol. 55, no. 8, pp. 1871–1874, 2008.

[4] M. Nawaz, C. Zaring, J. Bource et al., "Assessment of high and low temperature performance of SiC BJTs," *Materials Science Forum*, vol. 615–617, pp. 825–828, 2009.

[5] M. Nawaz, "On the assessment of few design proposals for 4H-SiC BJTs," *Microelectronics Journal*, vol. 41, no. 12, pp. 801–808, 2010.

[6] B. Buono, R. Ghandi, M. Domeij, B. G. Malm, C.-M. Zetterling, and M. Ostling, "Modeling and characterization of current gain versus temperaturein 4H-SiC power BJTs," *IEEE Transactions on Electron Devices*, vol. 57, no. 3, pp. 704–711, 2010.

[7] M. Usman, M. Nawaz, and A. Hallen, "Position-dependent bulk traps and carrier compensation in 4H-SiC bipolar junction transistors," *IEEE Transactions on Electron Devices*, vol. 60, no. 1, pp. 178–185, 2013.

[8] Y. Li, P. Alexandrov, and J. H. Zhao, "1.88-mΩ-cm^2 1650-V normally on 4H-SiC TI-VJFET," *IEEE Transactions on Electron Devices*, vol. 55, no. 8, pp. 1880–1886, 2008.

[9] A. Ritenour, I. Sankin, N. Merret et al., "High temperature electrical characteristics of 20 A, 800 V enhancement mode SiC VJFETs," in *Proceedings of the IMAPS High Temperature Electronics Conference (HiTEC '08)*, pp. 103–108, 2008.

[10] L. Cheng, A. K. Agarwal, C. Capell et al., "15 kV, large area (1 cm^2), 4H-SiC p-type gate turn-off thyristors," *Materials Science Forum*, vol. 740–742, pp. 978–981, 2013.

[11] H. Ö. Ólafsson, G. Gudjónsson, P.-Å. Nilsson et al., "High field effect mobility in Si face 4H-SiC MOSFET transistors," *Electronics Letters*, vol. 40, no. 8, pp. 508–510, 2004.

[12] J. Wang, T. Zhao, J. Li et al., "Characterization, modeling, and application of 10-kV SiC MOSFET," *IEEE Transactions on Electron Devices*, vol. 55, no. 8, pp. 1798–1806, 2008.

[13] R. S. Howell, S. Buchoff, S. van Campen et al., "A 10-kV large-area 4H-SiC power DMOSFET with stable subthreshold behavior independent of temperature," *IEEE Transactions on Electron Devices*, vol. 55, no. 8, pp. 1807–1815, 2008.

[14] V. Tilak, K. Matocha, G. Dunne, F. Allerstam, and E. Ö. Sveinbjornsson, "Trap and inversion layer mobility characterization using Hall effect in silicon carbide-based MOSFETs with gate oxides grown by sodium enhanced oxidation," *IEEE Transactions on Electron Devices*, vol. 56, no. 2, pp. 162–169, 2009.

[15] D. Okamoto, H. Yano, K. Hirata, T. Hatayama, and T. Fuyuki, "Improved inversion channel mobility in 4H-SiC MOSFETs on Si face utilizing phosphorus-doped gate oxide," *IEEE Electron Device Letters*, vol. 31, no. 7, pp. 710–713, 2010.

[16] Q. Zhang, M. Das, J. Sumakeris, R. Callanan, and A. Agarwal, "12-kV p-channel IGBTs with low on-resistance in 4H-SiC," *IEEE Electron Device Letters*, vol. 29, no. 9, pp. 1027–1029, 2008.

[17] X. Wang and J. A. Cooper, "High-voltage n-channel IGBTs on free-standing 4H-SiC epilayers," *IEEE Transactions on Electron Devices*, vol. 57, no. 2, pp. 511–515, 2010.

[18] S.-H. Ryu, C. Capell, L. Cheng et al., "High performance, ultra high voltage 4H-SiC IGBTs," in *Proceedings of the IEEE Energy Conversion Congress and Exposition (ECCE '12)*, pp. 3603–3608, Raleigh, NC, USA, September 2012.

[19] M. Nawaz and F. Chimento, "On the assessment of temperature dependence of 10–20 kV 4H-SiC IGBTs using TCAD," *Materials Science Forum*, vol. 740–742, pp. 1085–1088, 2013.

[20] M. Usman and M. Nawaz, "Device design assessment of 4H-SiC n-IGBT—a simulation study," *Solid-State Electronics*, vol. 92, pp. 5–11, 2014.

[21] S. K. Haney, V. Misra, D. J. Lichtenwalner, and A. Agarwal, "Investigation of nitrided atomic-layer-deposited oxides in 4H-SiC capacitors and MOSFETs," *Materials Science Forum*, vol. 740–742, pp. 707–710, 2013.

[22] T. Kimura, T. Ishikawa, N. Soejima, K. Nomura, and T. Sugiyama, "Effect of suppressing reoxidation at SiO$_2$/SiC interface during post-oxidation annealing in N$_2$O with Al$_2$O$_3$ capping layer," *Materials Science Forum*, vol. 740–742, pp. 737–740, 2013.

[23] C. M. Tanner, Y.-C. Perng, C. Frewin, S. E. Saddow, and J. P. Chang, "Electrical performance of Al$_2$O$_3$ gate dielectric films deposited by atomic layer deposition on 4H-SiC," *Applied Physics Letters*, vol. 91, no. 20, Article ID 203510, 2007.

[24] M. Wolborski, *Characterization of dielectric layers for passivation of 4H-SiC devices [Ph.D. thesis]*, KTH, 2006.

[25] J. Robertson, "High dielectric constant oxides," *The European Physical Journal: Applied Physics*, vol. 28, no. 3, pp. 265–291, 2004.

[26] L. A. Lipkin and J. W. Palmour, "Insulator investigation on sic for improved reliability," *IEEE Transactions on Electron Devices*, vol. 46, no. 3, pp. 525–532, 1999.

[27] Q. Shui, M. S. Mazzola, X. Gu, C. W. Myles, and M. A. Gundersen, "Investigation of Al$_2$O$_3$ and TiO$_2$ as gate insulators for 4H-SiC pulsed power devices," in *Proceedings of the 26th International Power Modulator Symposium*, pp. 501–504, May 2004.

[28] T. Tamaki, G. G. Walden, Y. Sui, and J. A. Cooper, "Numerical study of the turnoff behavior of high-voltage 4H-SiC IGBTs," *IEEE Transactions on Electron Devices*, vol. 55, no. 8, pp. 1928–1933, 2008.

[29] R. Singh and A. R. Hefner, "Reliability of SiC MOS devices," *Solid-State Electronics*, vol. 48, no. 10-11, pp. 1717–1720, 2004.

[30] A. K. Agarwal, S. Seshadri, and L. B. Rowland, "Temperature dependence of Fowler-Nordheim current in 6H- and 4H-SiC MOS capacitors," *IEEE Electron Device Letters*, vol. 18, no. 12, pp. 592–594, 1997.

[31] L. Yu, *Simulation, modeling and characterization of SiC devices [Ph.D. thesis]*, Graduate School-New Brunswick Rutgers, The State University of New Jersey, 2010.

[32] M. Gurfinkel, J. C. Horst, J. S. Suehle et al., "Time-dependent dielectric breakdown of 4H-SiC/SiO$_2$ MOS capacitors," *IEEE Transactions on Device and Materials Reliability*, vol. 8, no. 4, pp. 635–640, 2008.

[33] O. Biserica, P. Godignon, X. Jordà, J. Montserrat, N. Mestres, and S. Hidalgo, "Study of AlN/SiO$_2$ as dielectric layer for SiC MOS structures," in *Proceedings of the International Semiconductor Conference*, vol. 1, pp. 205–208, October 2000.

[34] A. Taube, S. Gierałtowska, T. Gutt et al., "Electronic properties of thin HfO$_2$ films fabricated by atomic layer deposition on 4H-SiC," *Acta Physica Polonica A*, vol. 119, no. 5, pp. 696–698, 2011.

[35] K. Y. Cheong, J. M. Moon, T.-J. Park et al., "Improved electronic performance of HfO$_2$/SiO$_2$ stacking gate dielectric on 4H SiC," *IEEE Transactions on Electron Devices*, vol. 54, no. 12, pp. 3409–3413, 2007.

[36] K. Y. Cheong, J. H. Moon, D. Eom, H. J. Kim, W. Bahng, and N.-K. Kim, "Electronic properties of atomic-layer-deposited Al_2O_3/thermal-nitrided SiO_2 stacking dielectric on 4H SiC," *Electrochemical and Solid-State Letters*, vol. 10, no. 2, pp. H69–H71, 2007.

[37] H. Yano, F. Katafuchi, T. Kimoto, and H. Matsunami, "Effects of wet oxidation/anneal on interface properties of thermally oxidized SiO_2/SiC MOS system and MOSFET's," *IEEE Transactions on Electron Devices*, vol. 46, no. 3, pp. 504–510, 1999.

[38] S. Hino, T. Hatayama, J. Kato, E. Tokumitsu, N. Miura, and T. Oomori, "High channel mobility 4H-SiC metal-oxide-semiconductor field-effect transistor with low temperature metal-organic chemical-vapor deposition grown Al_2O_3 gate insulator," *Applied Physics Letters*, vol. 92, no. 18, Article ID 183503, 2008.

[39] A. Pérez-Tomás, M. R. Jennings, P. M. Gammon et al., "SiC MOSFETs with thermally oxidized Ta_2Si stacked on SiO_2 as high-k gate insulator," *Microelectronic Engineering*, vol. 85, no. 4, pp. 704–709, 2008.

[40] A. Shima, K. Watanabe, T. Mine, N. Tega, H. Hamamura, and Y. Shimamoto, "Reliable 4H-SiC MOSFET with high threshold voltage by Al_2O_3-inserted gate insulator," in *Proceedings of the European Conference on Silicon Carbide & Related Materials (ECSCRM '14)*, 2014.

[41] T. Hatayama, S. Hino, N. Miura, T. Oomori, and E. Tokumitsu, "Remarkable increase in the channel mobility of SiC-MOSFETs by controlling the interfacial SiO_2 layer between Al_2O_3 and SiC," *IEEE Transactions on Electron Devices*, vol. 55, no. 8, pp. 2041–2045, 2008.

[42] A. Perez Tomas, *Novel materials and processes for gate dielectrics on silicon carbide [Ph.D. thesis]*, Facultat De Ciencies, Department de Fisica, Universitat Autonoma De Barcelona, Barcelona, Spain, 2005.

[43] K. Y. Cheong, J. H. Moon, H. J. Kim, W. Bahng, and N.-K. Kim, "Current conduction mechanisms in atomic-layer-deposited HfO_2/nitrided SiO_2 stacked gate on 4H silicon carbide," *Journal of Applied Physics*, vol. 103, no. 8, Article ID 084113, 2008.

[44] R. Mahapatra, A. K. Chakraborty, A. B. Horsfall, N. G. Wright, G. Beamson, and K. S. Coleman, "Energy-band alignment of HfO_2 SiO2 SiC gate dielectric stack," *Applied Physics Letters*, vol. 92, no. 4, Article ID 042904, 2008.

[45] J. H. Moon, K. Y. Cheong, D. Eom et al., "Electrical properties of atomic-layer-deposited La_2O_3/thermal-nitrided SiO_2 stacking dielectric on 4H-SiC(0001)," *Materials Science Forum*, vol. 556-557, pp. 643–646, 2007.

[46] S. Dhar, A. Modic, G. Liu, A. C. Ahyi, and L. C. Feldman, "Channel mobility improvement in 4H-SiC MOSFETs using a combination of surface counter-doping with Antimony and NO annealing," in *Proceedings of the European Conference on Silicon Carbide & Related Materials (ECSCRM '14)*, 2014.

[47] Silvaco data system Inc, *Atlas User Manual, ver 5.15.31.C*, 2009.

2

A New CMOS Controllable Impedance Multiplier with Large Multiplication Factor

Munir A. Al-Absi

EE Department, King Fahd University of Petroleum & Minerals, Dhahran, Saudi Arabia

Correspondence should be addressed to Munir A. Al-Absi; mkulaib@kfupm.edu.sa

Academic Editor: S. M. Rezaul Hasan

This paper presents a new compact controllable impedance multiplier using CMOS technology. The design is based on the use of the translinear principle using MOSFETs in subthreshold region. The value of the impedance will be controlled using the bias currents only. The impedance can be scaled up and down as required. The functionality of the proposed design was confirmed by simulation using BSIM3V3 MOS model in Tanner Tspice 0.18 μm TSMC CMOS process technology. Simulation results indicate that the proposed design is functioning properly with a tunable multiplication factor from 0.1- to 100-fold. Applications of the proposed multiplier in the design of low pass and high pass filters are also included.

1. Introduction

A capacitance multiplier circuit is a useful building block in many very large-scale integration (VLSI) analog circuits, especially for active RC filter and oscillator designs and for cancellation of parasitic elements. Signal processing for biomedical applications is one of the areas where very low frequency filters are used [1–11]. In such a filter, a large time constant is required, which means large values capacitors and/or resistors are required. However, in integrated circuit design, implementing such a large time constant will not be acceptable due to a required large area on the chip and large power consumption. A more viable solution is to use small physical capacitor or resistor and it is scaled up using a simple circuit. There are many impedance-scaling circuits published in the open literature [2–9]. In [2, 3], an operational transconductance amplifier (OTA) based tunable C-multiplier is developed. The design is for capacitor scaling-up only and it uses three OTAs in which the multiplication factor is tuned using the OTAs' bias currents. An impedance scaler is presented in [4, 5] using MOSFETs. This would require a small area on the chip. However, the scaling factor is controlled by the aspect ratios of the transistors used. This means that, once fabricated, the scaling factor cannot be controlled. The design reported in [6] used three current-controlled current amplifiers in addition to an external

resistor. A universal immittance function simulator using a current conveyor is reported in [7]. In this design three CCIIs are used. Moreover, external resistors are used to control the multiplication factor. In [8] current conveyor based R- and-C multiplier circuits are developed. The values of R and C are controlled by two other resistors. In [9] an enhanced grounded capacitor multiplier is presented. The design is based on using the differential amplifier with exponential current scaling. In [10, 11] current conveyors and dual-x current conveyors are used.

In this paper, a new impedance scaler is proposed. The design can scale up and down the capacitance and the resistance.

2. Proposed Impedance Multiplier

The block diagram of the proposed design is shown in Figure 1. It consists of a current amplifier, a voltage buffer, and the impedance to be scaled Z. With reference to Figure 1, the equivalent impedance seen by the voltage source V_x is given by

$$Z_{eq} = \frac{V_x}{I_x} = \frac{V_x}{i_o}. \tag{1}$$

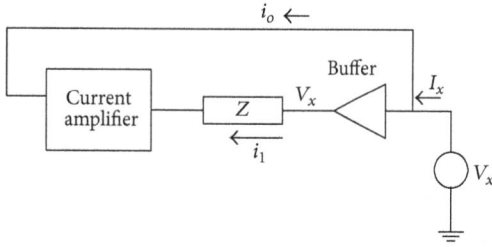

FIGURE 1: Block diagram of the proposed design.

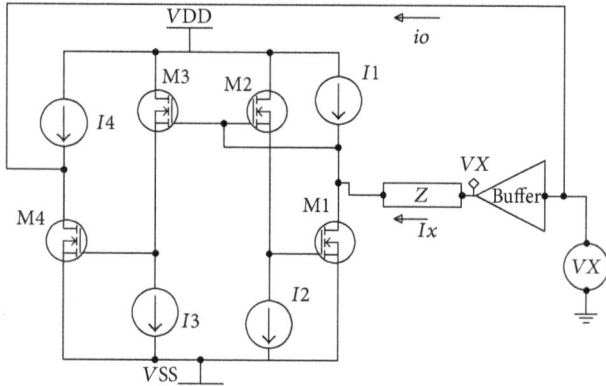

FIGURE 2: Circuit diagram of the proposed design.

The amplifier output is given by

$$i_0 = G \times i_1, \tag{2}$$

where G is the gain of the amplifier. If the input impedance of the current amplifier is small compared with Z, then the current i_1 passing through the impedance Z can be approximated by

$$i_1 = \frac{V_x}{Z}. \tag{3}$$

Combining (1), (2), and (3), the equivalent impedance is given by

$$Z_{eq} = \frac{Z}{G}. \tag{4}$$

The circuit diagram of the proposed design is shown in Figure 2. Four MOSFETs M1–M4 form a translinear loop with regulated cascade input to lower the input impedance in series with Z. The MOSFETs are biased in the subthreshold region and this will provide high output impedance and hence enhance the lower corner frequency. All biased currents are designed using simple current mirrors. The buffer used is a two-MOSFET buffer and is shown in Figure 3.

With reference to Figure 2, applying KVL to the translinear loop yields

$$V_{GS1} + V_{GS2} = V_{GS3} + V_{GS4}. \tag{5}$$

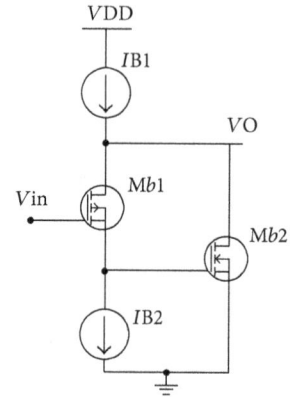

FIGURE 3: Circuit diagram for the buffer.

The drain current of an NMOS operating subthreshold is given by

$$I_D = I_{D0}e^{((V_{GS}-V_{Th})/nV_T)}, \tag{6}$$

where I_{D0} is the saturation current, n is the slop factor, and V_T is the thermal voltage.

For the MOSFET to operate in subthreshold mode, the following condition must be satisfied:

$$\frac{I_{D0}}{I_D} \ll 1, \tag{7}$$

$$V_{DS} > 4V_T.$$

From (6), the gate-to-source voltage is given by

$$V_{GS} = nV_T \ln\left(\frac{I_D}{I_{D0}}\right) - V_{Th}. \tag{8}$$

Combining (5) and (8), it is easy to write

$$I_{D1}I_{D2} = I_{D3}I_{D4}. \tag{9}$$

The equivalent impedance Z_{eq} seen at terminal x can be obtained if an AC voltage source is applied and the AC currents i_1 and i_0 are included in the analysis. Thus, (9) can be rewritten as

$$(i_{o+}I_4) * I_3 = I_2 * (I_1 + i_1) \tag{10}$$

or

$$(i_{o+}I_4) = G * (I_1 + i_1), \tag{11}$$

where $G = I_2/I_3$.

If $I_4 = G * I_1$, then $i_o = G * i_1$.

With reference to Figure 2, the impedance at node x is given by

$$Z_{eq} = \frac{v_x}{i_x} = \frac{v_x}{i_o} = \frac{v_x}{Gi_1} = \frac{v_x}{i_1}\frac{1}{G}. \tag{12}$$

Since Z is much greater than the impedance at the drain of M1, then $v_x/i_1 \approx Z$, and (12) can be written as

$$Z_{eq} = Z\frac{1}{G}. \tag{13}$$

TABLE 1: Performance comparison.

	Technology (μm)	Multiplication factor	Power consumption	Area (mm^2)	Frequency range	Experimental/simulation
Ref [3]	0.35	10	NA	—	400 Hz–70 KHz	Simulation
Ref [4]		10	10.8 mW	0.0297	NA	Simulation
Ref [9]	0.50	28	1.32 mW	0.07	NA	Simulation & experimental
Ref [11]	0.35	50	0.2 mW	NA	NA	Simulation
This work	0.18	Controllable 0.1–200	20 μW	0.0030	10 Hz–7 KHz	Simulation

It is evident from (13) that the circuit implements a tunable impedance scaler, which is tuned using the control parameter $G = I_2/I_3$. If Z is replaced by a capacitor, then

$$Z_{\text{eq}} = \frac{1}{sC}\frac{1}{G}. \tag{14}$$

It is clear from (14) that a capacitance multiplier is achieved.

If Z is replaced by a resistor, then

$$Z_{\text{eq}} = R\frac{1}{G}. \tag{15}$$

Equation (15) implements a tunable resistor that can be tuned by the control variable G. The resistor can be scaled up or down as required.

3. Simulation Results

The proposed circuit was simulated using Tanner Tspice in 0.18 μm TSMC CMOS technology and BSIM3v3 MOSFET model. To prove the concept, the circuit is configured as low pass filter with $R = 10$ MΩ and the capacitance that can be scaled up and down is 5 pF. The transistors aspect ratio is 7/2 and the bias currents for the buffer are set to $I_{B1} = 1 \mu$A and $I_{B2} = 0.2 \mu$A. The circuit is operated from ± 0.75 V. The bias currents $I_1 = I_3 = 5$ nA, the current $I_2 = I_4 = G * I_1$, and G are swept from 0.1 to 100. Plots of the simulated result of the proposed design and the theory are shown in Figure 4.

It is evident from the plots that the proposed c-multiplier is working well. The 5 pF capacitor is scaled up to 500 pF and down to 0.5 pF.

As it appears from Figure 4 there is a deviation between ideal case and the proposed design. This deviation is due to the approximation made in (12) where we assume the input impedance at the node of M1 is much smaller than Z, in addition to the output impedance of the buffer circuit.

The proposed design can be used in the frequency range from 10 Hz to 7 KHz as shown in Figure 5.

The proposed circuit was simulated for transient analysis. An input signal of 100 mv amplitude and 5 KHz frequency was applied to the input of an ideal and simulated circuit. The output voltage for the ideal and simulated design is shown in Figure 6. It is clear that the proposed circuit is functioning properly.

The performance of the proposed design is compared with previously published works and is summarized in Table 1. It can be seen from the table the proposed design is superior to all in terms of controllability, area on chip, and frequency range, where the lower limit is 10 Hz, which make it attractive in very low frequency applications such as very low frequency filters.

4. Nonideal Analysis

The error shown in Figure 4 was investigated through the small signal analysis. The small signal equivalent circuit for Figure 1 is shown in Figure 7. The parasitic capacitance is not included because this design is suitable for low frequency applications.

Using routine analysis, the equivalent impedance seen at the node V_X is given by

$$Z_{\text{eq}} = \frac{V_X}{i_X} = \frac{V_{d4}}{i_X}$$

$$= \frac{Z}{Zg_{ds4} + Zg_{m4}g_{m3}\left(2g_{m2b} + g_{m1b}g_{m2b}/g_{ds2b}\right)/\left(g_{ds3} + g_{m3}\right)/\left(Zg_{ds1} + 1 + Zg_{m1}g_{m2}/\left(g_{ds2} + g_{m2}\right)\right)\left(g_{ds1b} + 2g_{m2b} - g_{m1b} + g_{m1b}g_{m2b}/g_{ds2b}\right)}. \tag{16}$$

Comparing (16) with (13), the control variable G is given by

$$G = Zg_{ds4} + \frac{Zg_{m4}g_{m3}\left(2g_{m2b} - g_{m1b}g_{m2b}/g_{ds2b}\right)/\left(g_{ds3} + g_{m3}\right)}{\left(Zg_{ds1} + 1 + Zg_{m1}g_{m2}/\left(g_{ds2} + g_{m2}\right)\right)\left(g_{ds1b} + 2g_{m2b} + g_{m1b} - g_{m1b}g_{m2b}/g_{ds2b}\right)}. \tag{17}$$

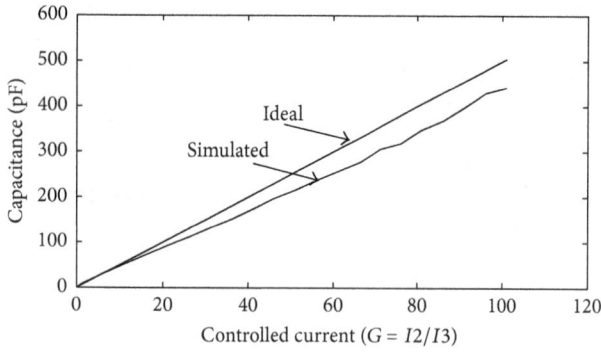

FIGURE 4: Plot of simulated capacitance and ideal case.

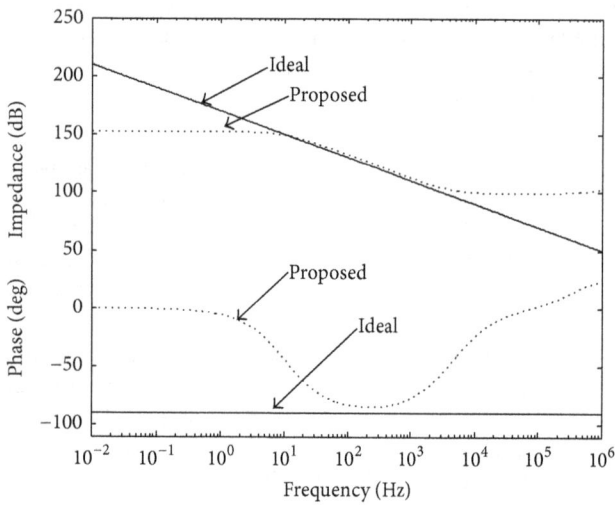

FIGURE 5: Compared impedance and phase for ideal and the proposed design.

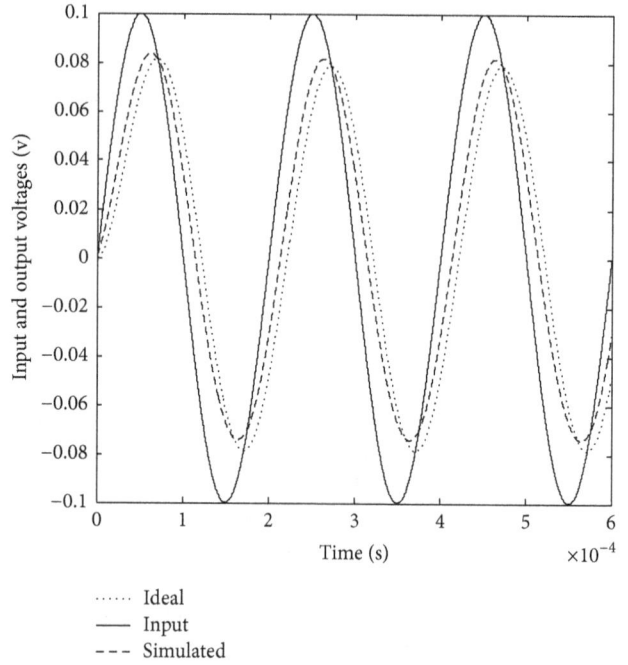

FIGURE 6: Simulation for the transient response.

Equation (16) was simulated using MATLAB and the simulation results coincide with Tanner simulation. This confirms the correctness of the analysis.

To make it easy for designers to make use of (16), simplification was carried out as follows:

The transconductance and the output admittance of the MOSFET operating in subthreshold are given by the following: $g_m = I_D/nV_T$, $g_{ds} = \lambda I_D$; then using routine analysis, (17) can be reduced to

$$Z_{eq} = \frac{Z}{G} + \frac{nV_T}{GI_1}. \tag{18}$$

It is clear from (18) that the second term is the source of the error and it will be a function of the bias current I_1. The error can be minimized if the bias current I_1 is increased.

4.1. Stability Analysis. The proposed circuit was designed for low frequency applications using MOSFETs operating in the subthreshold mode. Therefore, the parasitic capacitances will not affect the stability of the circuit. Using (16) and replacing Z with the capacitance, there is only one pole:

$$w_p = \frac{g_{m1} + 1/r_{o1}}{g_{m4}r_{o4}} \approx \frac{g_{m1}}{g_{m4}r_{o4}}. \tag{19}$$

But $g_{m4} = Gg_{m1}$.

Equation (19) can be written as

$$w_p = \frac{1}{G \times r_{o4}}. \tag{20}$$

It is clear from (19), the pole depends on multiplication factor G and r_{o4}.

5. Applications

The proposed design was used in the design of an RC low pass filter with cutoff frequency of 31.8 Hz. The parameters used in the proposed design are $C = 5$ pF, $R = 10$ Meg resistor, bias current $I_1 = I_3 = 2.5$ nA, and the multiplication factor $G = 100$. It is evident from the simulation results shown in Figure 8 that the filter designed using the proposed c-multiplier is in a close agreement in the frequency response with passive RC low pass filter. It is also obvious that the proposed design will work properly in the low frequency applications such as biomedical circuits and systems.

The proposed design was used as a resistance multiplier in the design of RC high pass filter with controllable cutoff frequency. The capacitance used was 5 pf, the resistor to be scaled was 10 Meg, and the bias current is $I_1 = 100$ nA. The control parameters are $G = 0.1$, 0.5, and 0.9. The simulation results shown in Figure 9 indicate that the proposed design is in close agreement with passive RC high pass filter in both the gain and the phase shift.

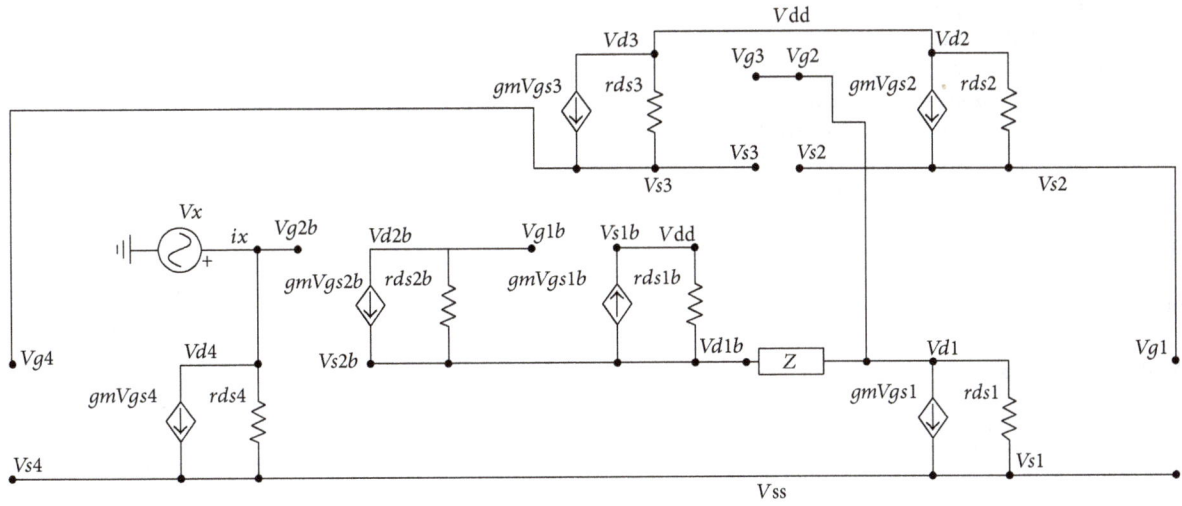

FIGURE 7: Small signal equivalent circuit of the proposed C-multiplier.

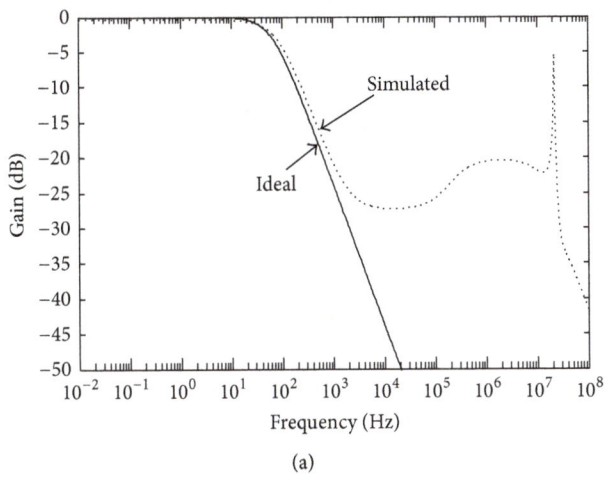

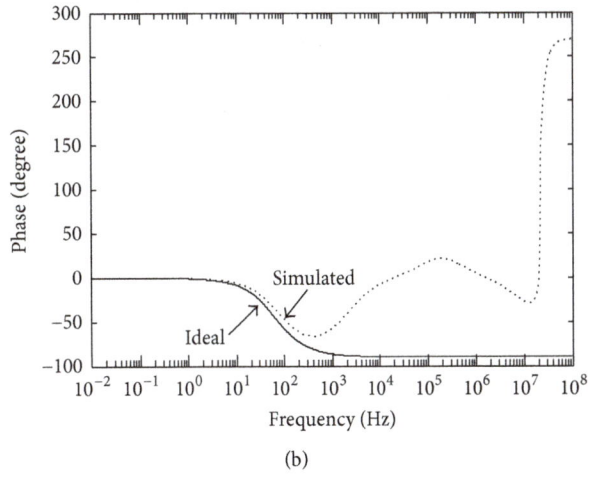

(a)

(b)

FIGURE 8: Frequency response for the low pass filter: (a) gain and (b) phase shift.

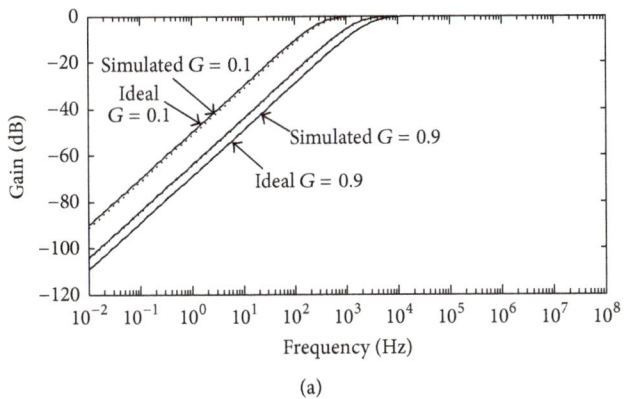

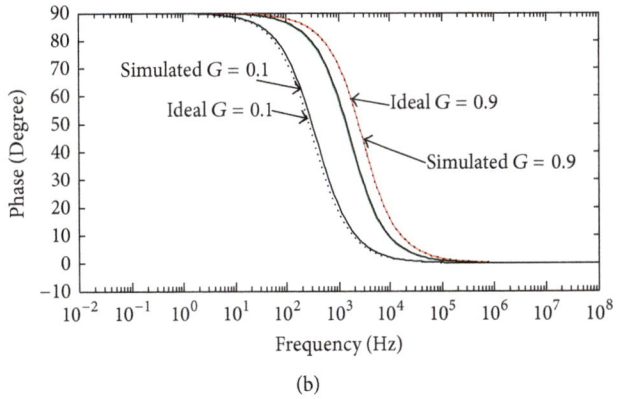

(a)

(b)

FIGURE 9: Frequency response for high pass filter: (a) gain and (b) phase shift.

6. Conclusion

A new simple and compact impedance multiplier was developed. The design is free of passive elements. The multiplication factor is controllable in the range from 0.1 to 100, which is large compared with previously reported designs. The proposed circuit can be used to scale either the capacitance or the resistance. We believe the developed design will be an excellent building block in integrated circuit design for applications where large time constant is required.

Conflicts of Interest

The author declares that he has no conflicts of interest.

Acknowledgments

This work is a partial result of the research work funded by KFUPM, Project no. IN 131066.

References

[1] L. J. Stotts, "Introduction to Implantable Biomedical IC Design," *IEEE Circuits and Devices Magazine*, vol. 5, no. 1, pp. 12–18, 1989.

[2] M. T. Ahmed, I. A. Khan, and N. Minhaj, "Novel electronically tunable C-multipliers," *Electronics Letters*, vol. 31, no. 1, pp. 9–11, 1995.

[3] J. Silva-Martinez and A. Vazquez-Gonzalez, "Impedance scalers for IC active filters," in *Proceedings of the IEEE International Symposium on Circuits and Systems (ISCAS '98)*, pp. 151–154, Monterey, Calif, USA, June 1998.

[4] M. T. Abuelma'Atti and N. A. Tasadduq, "Electronically tunable capacitance multiplier and frequency-dependent negative-resistance simulator using the current-controlled current conveyor," *Microelectronics Journal*, vol. 30, no. 9, pp. 869–873, 1999.

[5] S. Solís-Bustos, J. Silva-Martínez, F. Maloberti, and E. Sánchez-Sinencio, "A 60-dB dynamic-range CMOS sixth-order 2.4-Hz low-pass filter for medical applications," *IEEE Transactions on Circuits and Systems II: Analog and Digital Signal Processing*, vol. 47, no. 12, pp. 1391–1398, 2000.

[6] O. Çiçekolu, A. Toker, and H. Kuntman, "Universal immittance function simulators using current conveyors," *Computers and Electrical Engineering*, vol. 27, no. 3, pp. 227–238, 2001.

[7] A. A. Khan, S. Bimal, K. K. Dey, and S. S. Roy, "Current conveyor based R- and C-multiplier circuits," *AEU-Archiv fur Elektronik und Ubertragungstechnik*, vol. 56, no. 5, pp. 312–316, 2002.

[8] T. Kulej, "Regulated capacitance multiplier in CMOS technology," in *Proceedings of the 16th International Conference on Mixed Design of Integrated Circuits and Systems (MIXDES '09)*, pp. 316–319, Lodz, Poland, June 2009.

[9] I. Padilla-Cantoya and P. M. Furth, "Enhanced grounded capacitor multiplier and its floating implementation for analog filters," *IEEE Transactions on Circuits and Systems II: Express Briefs*, vol. 62, no. 10, pp. 962–966, 2015.

[10] M. Siripruchyanan and W. Jaikla, "Floating capacitance multiplier using DVCC and CCCIIS," in *Proceedings of the International Symposium on Communications and Information Technologies (ISCIT '07)*, pp. 218–221, October 2007.

[11] I. Myderrizi and A. Zeki, "Electronically tunable DXCCII-based grounded capacitance multiplier," *AEU—International Journal of Electronics and Communications*, vol. 68, no. 9, pp. 899–906, 2014.

A New Fractal Multiband Antenna for Wireless Power Transmission Applications

Taoufik Benyetho (ID),[1] **Jamal Zbitou,**[1] **Larbi El Abdellaoui** (ID),[1] **Hamid Bennis** (ID),[2] **and Abdelwahed Tribak** (ID)[3]

[1]*LMEET, FST of Settat, Hassan 1st University, Settat, Morocco*
[2]*TIM Research Team, EST of Meknes, Moulay Ismail University, Meknes, Morocco*
[3]*Microwave Team, INPT, Rabat, Morocco*

Correspondence should be addressed to Taoufik Benyetho; t.benyetho@gmail.com

Academic Editor: Gerard Ghibaudo

The Microwave Power Transmission (MPT) is the possibility of feeding a system without contact by using microwave energy. The challenge of such system is to increase the efficiency of transmitted energy from the emitter to the load. This can be achieved by rectifying the microwave energy using a rectenna system composed of an antenna of a significant gain associated with a rectifier with a good input impedance matching. In this paper, a new multiband antenna using the microstrip technology and fractal geometry is developed. The fractal antenna is validated into simulation and measurement in the ISM (industrial, scientific, and medical) band at 2.45 GHz and 5.8 GHz and it presents a wide aperture angle with an acceptable gain for both bands. The final antenna is printed over an FR4 substrate with a dimension of $60 \times 30 \text{ mm}^2$. These characteristics make the antenna suitable for a multiband rectenna circuit use.

1. Introduction

The wireless power transmission concept was introduced in the last decade of the 19th century by Nicola Tesla's experiment in which he tried to light bulbs wirelessly by transmitting energy from distant oscillators operated to 100 MV at 150 KHz, but he could not implement his system for commercial use due to its very low efficiency [1]. After this contribution, researchers in Japan [2] and the United States [3] continued to improve the efficiency of wireless power transmission in 1920s and 1930s.

1950 has known the true start of wireless power transmission thanks to the development of high power microwave tubes by Raytheon company [4]. In 1958, a 15 KW microwave tube had measured 81% DC to RF conversion efficiency [5, 6]. The first rectenna, conceived also by Raytheon company, came in the early 1960s. This Rectifying Antenna was composed of a half-wave dipole antenna associated with a balanced bridge or semiconductor diode placed over a reflecting plane. A resistive load was then connected to the output of the rectenna. The 2.45 GHz as transmitting frequency was privileged due to its advanced and efficient technology base, its minimal attenuation through the atmosphere even in bad weather, and its location at the center of an industrial, scientific, and medical (ISM) band.

The rectenna's greatest conversion efficiency ever recorded was in 1977 by Brown in Raytheon company [7]. The rectenna operated at 2.45 GHz and reached 90.6% conversion efficiency with an input power level of 8 W. This rectenna element used GaAs-Pt Schottky barrier diode and aluminum bars to construct the dipole and transmission line. In 1982, 85% conversion efficiency was achieved, by using a rectenna printed thin film operated at 2.45 GHz [8].

Other frequencies were used to design rectenna circuit like 35 GHz frequency [9, 10] which can decrease the transmitting and rectenna aperture areas and then increase the transmission range, but the problem is the high cost and inefficiency of components which generate high power at 35 GHz.

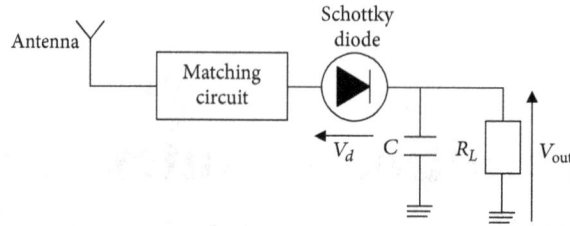

FIGURE 1: Block diagram of a rectenna circuit [11].

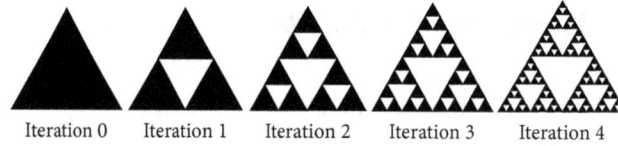

Iteration 0 Iteration 1 Iteration 2 Iteration 3 Iteration 4

FIGURE 2: Sierpiński triangle at iterations from 0 to 4.

Figure 1 shows basic elements of a rectenna system. A rectenna circuit is composed of an antenna which collects the RF energy and a Schottky diode which rectifies this energy. The matching circuit between the antenna and the diode is a low pass filter designed to pass the fundamental frequency and reject the higher order harmonics generated from the nonlinear Schottky diode. The capacitor between the Schottky diode and the load behaves as a DC pass filter which protects the load from the HF harmonics.

In order to have an efficient rectenna circuit, the antenna must present good performance. A good return loss to avoid reflected energy, a big gain, and wide aperture angles to maximize RF harvested energy. The rectenna conversion efficiency η is defined by (1) in [12]:

$$\eta = \frac{P_{DC}}{P_{DC} + P_{LOSS}}, \tag{1}$$

where P_{DC} is the output power and P_{LOSS} is the loss power. Because the rectenna output is a DC power, the output power could be defined by

$$P_{DC} = \frac{V_{out}^2}{R_L}, \tag{2}$$

where V_{out} is the output voltage of the rectifier and R_L is the load.

When only conduction losses of the diode are considered and all the other losses are neglected, the conversion efficiency can be determined by [12]

$$\eta = \frac{1}{1 + V_d/2V_{out}}. \tag{3}$$

V_d is the voltage drop across the conducting diode.

Since its invention, the rectenna was used for various applications like RFID [16], SHARP (Stationary High Altitude Relay Platform) the microwave powered aircraft [17], and Solar Power Satellite (SPS). This last application concept is based on the construction of solar stations with great photovoltaic panels in space which will produce electricity that would be sent directly to Earth by microwaves to replace towers and power lines [18]. Recently, researchers introduced power through Wi-Fi, which permits charging batteries by Wi-Fi routers transmission [19].

2. Fractal Antenna Theory

Fractal antenna is an antenna based on fractal geometry. This term was first used by the French mathematician Mandelbrot in 1975 to describe a fractal shape that can be subdivided in many parts; each one of them is a reduced-size copy of the whole. Fractal term is derived from Latin word "Fractus" meaning "broken" [20]. Since the achievement of the first fractal antenna in 1995 [21], much progress proved that fractal geometry helps reduce antenna size and gives it a multiband behavior [21–25].

The fractal dimension D defined by (4) calculates the irregularity degree and fragmentation of a natural object or a geometric assembly [20]:

$$D = \frac{\ln\,(\text{Number of self similar pieces})}{\ln\,(1/\text{magnification factor})}. \tag{4}$$

The "Number of self similar pieces" represents the number of copies identical to the original shape when applying a fractal aspect from one step (or iteration) to another. The "magnification factor" signifies the scaling value between an iteration and the next one when applying a fractal technique.

The fractal antenna developed in this paper is based on Sierpiński triangle fractal geometry introduced by the Polish mathematician Sierpiński in 1916 [26]. The Sierpiński triangle is obtained from an equilateral triangle by a repeated process or iteration. The first iteration consists of subdividing the equilateral triangle into four smaller congruent equilateral triangles and removing the central one. The result as illustrated in Figure 2 is 3 equilateral triangles with half the size of the original one.

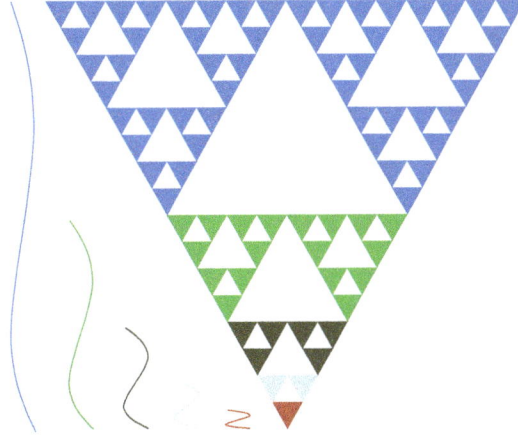

FIGURE 3: Sierpiński triangle multiband behavior.

To reach the second iteration, the same process is applied to the 3 equilateral triangles obtained in the first iteration; the result is 9 triangles and so on.

The number of copies from an iteration to the next one is multiplied by 3 and the size of the triangles is divided by 2. In consequence, the Sierpiński triangle fractal dimension is

$$D = \frac{\ln(3)}{\ln(1/(1/2))} = 1.58. \tag{5}$$

The Sierpiński triangle is widely used in antenna design due to its multiband behavior. The resonant frequency of an antenna is related to its length. When applying Sierpiński fractal concept to an equilateral triangle, different equilateral triangles with diverse sizes are created that lead to the multiband aspect of Sierpiński triangle antenna as explained in Figure 3.

3. Antenna Design and Measurement

3.1. Antenna Design. As explained above, Sierpiński triangle antenna is known for its multiband behavior. Some researches present accurate equations that predict the resonance frequencies of a standard Sierpiński antenna [27]. The problem is that the frequencies are related to the chosen dimensions that are defined in advance, so in order to design a Sierpiński antenna which resonates in frequencies different from those defined by equations, some modification of the standard design must be applied.

As illustrated in Figure 4, three antennas based on Sierpiński triangle at iterations 1, 2, and 3 are designed in order to study the fractal impact over this structure. The designed antennas differ from the standard Sierpiński triangle by the ground which is also a Sierpiński triangle, symmetric to the radiating patch. The substrate is an FR4 with relative permittivity equal to 4.4, 1.6 mm for thickness, and 0.025 for loss tangent. The simulation and optimization of the antenna were done by using CST Microwave Studio software.

Table 1 summarizes the antennas parameters.

Figure 5 shows that the simulated antennas return losses at the three iterations.

TABLE 1: Antennas' dimensions in mm.

Parameter	Length
a	65
b	30
c	26
d	25.8
e	1.3
f	15
g	4.7
i	3
j	3

The three iterations present almost the same behavior in the simulated frequency range. The −10 dB simulated return loss bandwidths cover ISM 2.4 GHz and 5.8 GHz bands. From iteration 1 to iteration 3 the resonance frequencies decrease slightly, while the input impedance matching decreases at the lower band and increases at the higher band. It is deduced that applying a fractal aspect over the new Sierpiński triangle designed structure has a very low effect on the return loss results. For more detailed study the X-Z plane and the Y-Z plane radiation pattern as well as the current distribution of the three designed antennas are compared.

Figure 6 shows the antenna radiation pattern comparison of the three iterations at 2.45 GHz and 5.8 GHz.

We notice that at 2.4 GHz the three iterations present exactly the same behavior: an omnidirectional propagation in the YZ plane (a) and a bidirectional radiation through the Z-axis ($0°$ and $180°$) in the XZ plane (b).

At 5.8 GHz, the radiation pattern difference between the three iterations is very small. In the YZ plane, the propagation is almost omnidirectional at iterations 1 and 2 with some attenuation through the Y-axis and perfectly omnidirectional at iteration 3 but with smaller gain. In the XZ plane, even if the radiation is no longer bidirectional, the maximum of propagation is still through the Z-axis ($0°$ and $180°$).

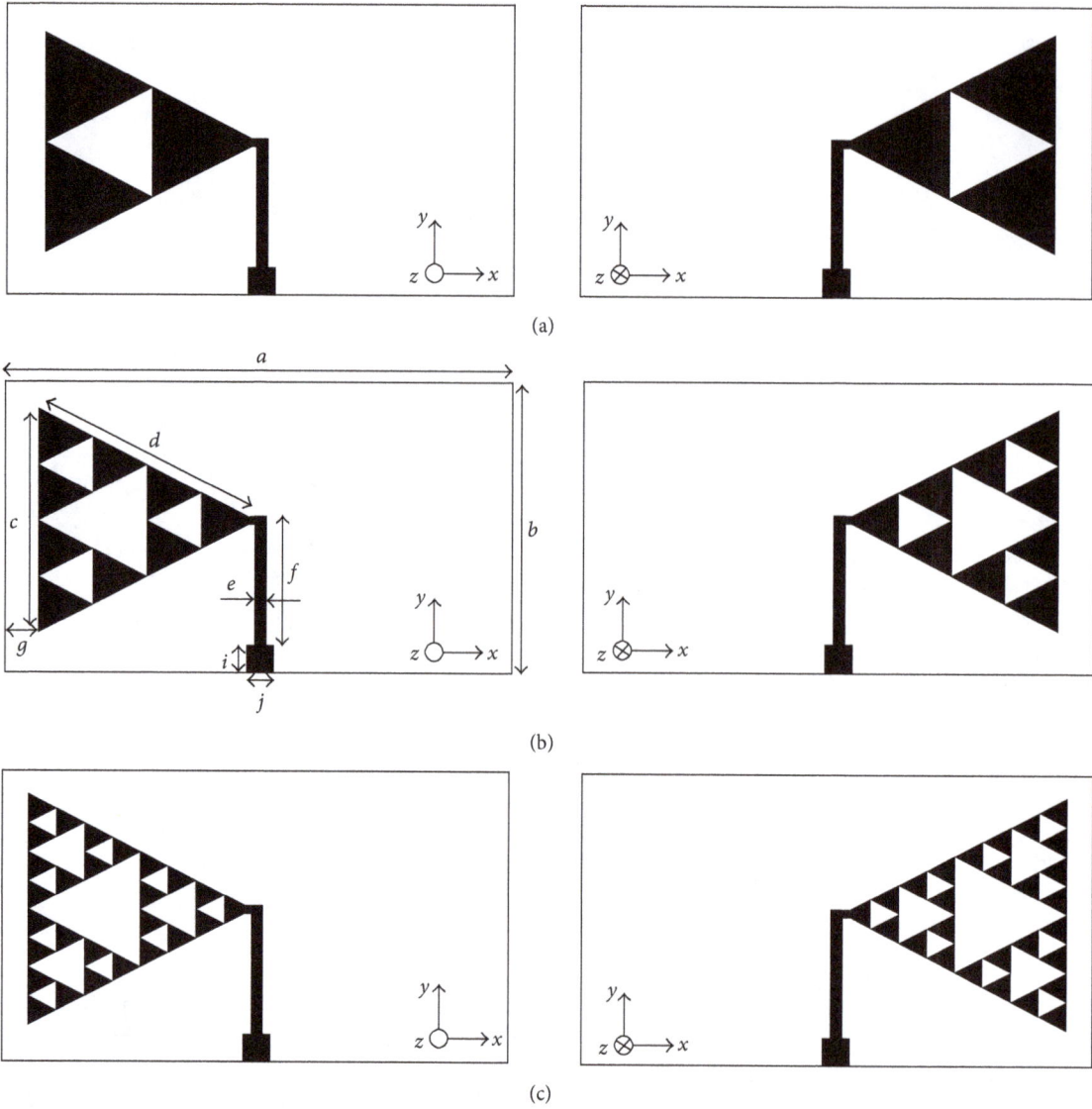

FIGURE 4: The patch (left) and the ground (right) of the designed antennas: (a) iteration 1, (b) iteration 2, and (c) iteration 3.

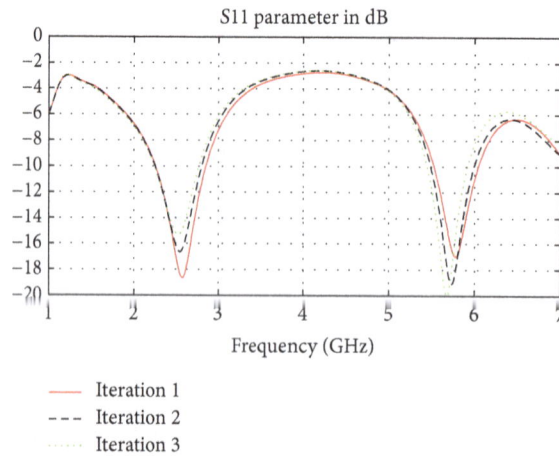

FIGURE 5: The designed antennas return loss at iterations 1, 2, and 3.

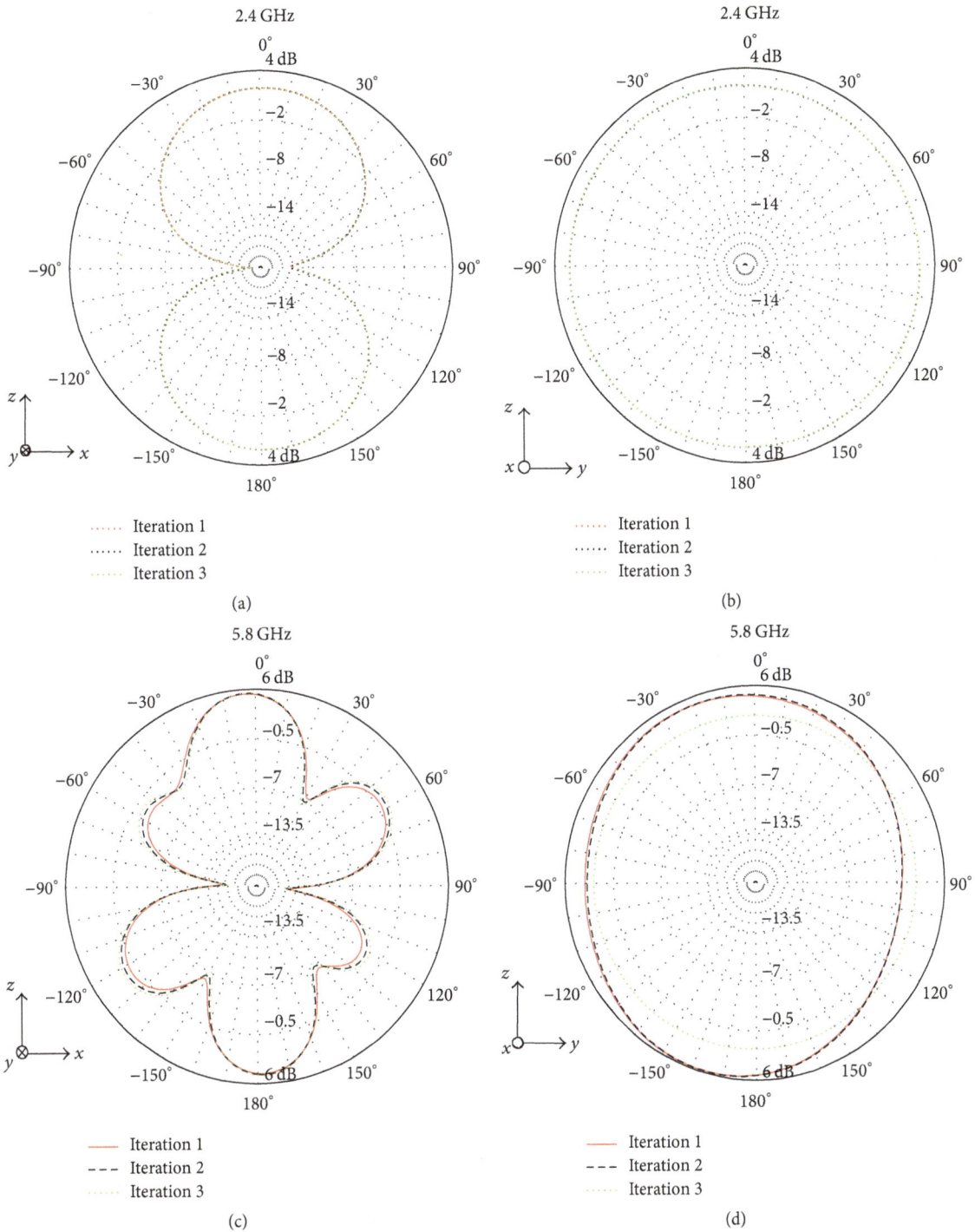

FIGURE 6: Antenna radiation pattern comparison of the three iterations at 2.4 GHz (XZ plane (a) and YZ plane (b)) and 5.8 GHz (XZ plane (c) and YZ plane (d)).

The simulation shows that the antenna radiation pattern characteristics stay almost stable at both resonance frequencies for the three iterations.

Figure 7 compares the antenna radiation pattern at 2.4 GHz and 5.8 GHz in the XZ plane (a) and YZ plane (b). In XZ plane, the low gain (2.1 dBi) at the 2.4 GHz is compensated by an aperture angle of 77° compared to 33° and

a gain of 5.5 dBi at 5.8 GHz. In the YZ plane, the aperture angles at both resonance frequencies are very large. The gain is constant at 2.4 GHz (2.2 dBi) and attains 5.8 dBi at 5.8 GHz with little attenuation through the Y-axis.

Figure 8 illustrates the current distribution of the three iterations designed antennas for both bands 2.4 GHz and 5.8 GHz.

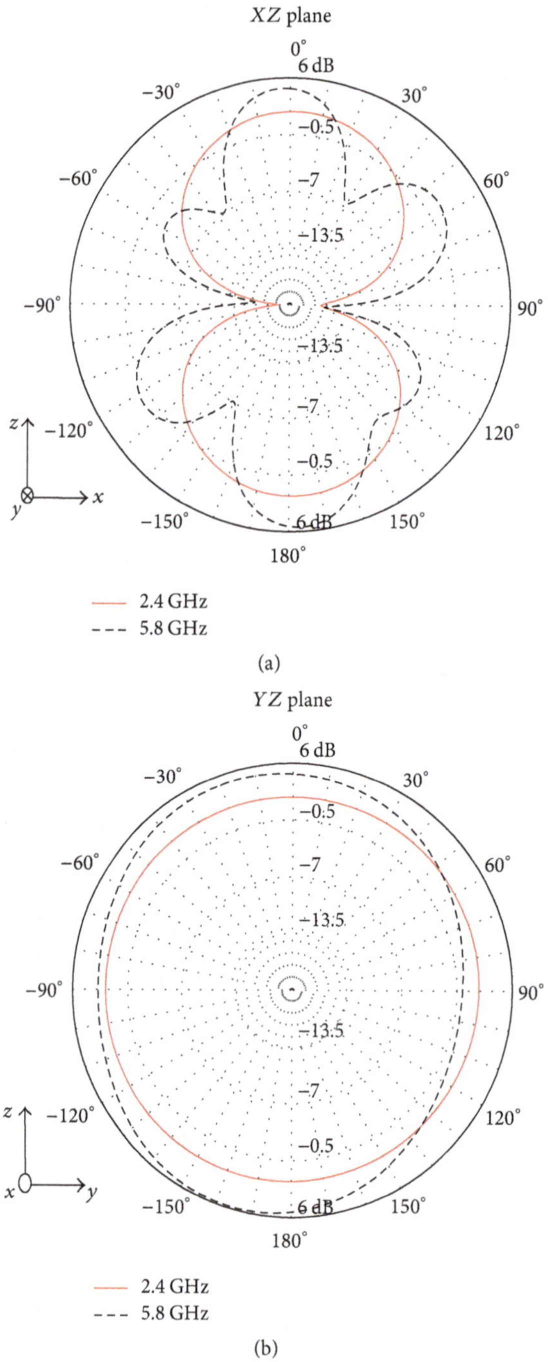

XZ plane

———— 2.4 GHz
- - - 5.8 GHz

(a)

YZ plane

———— 2.4 GHz
- - - 5.8 GHz

(b)

FIGURE 7: X-Z plane and Y-Z plane designed antennas radiation pattern at 2.45 GHz (a) and 5.8 GHz (b).

The current distribution is the same at the three iterations. At 2.4 GHz the current is distributed over all the structure while at 5.8 GHz it is more distributed in the half of the structure near to the feeding line.

3.2. *Results and Discussion.* After a study of the designed Sierpiński triangle antennas at the three first iterations, we deduced that the antennas characteristics do not change too much. We then chose to realize the structure at the second iteration as illustrated in Figure 9.

Figure 10 shows a comparison between simulated and measured return loss.

The simulated and measured return losses show good agreement. The slight difference is generally related to connector use which is not considered during simulation. Table 2 gives a numerical comparison of the validated bands.

The −10 dB measured return loss bandwidths (19.5% and 11.7% at the low and high resonance frequencies, resp.) cover ISM 2.4 GHz (2.4–2.5 GHz) and 5.8 GHz (5.725–5.875 GHz) bands.

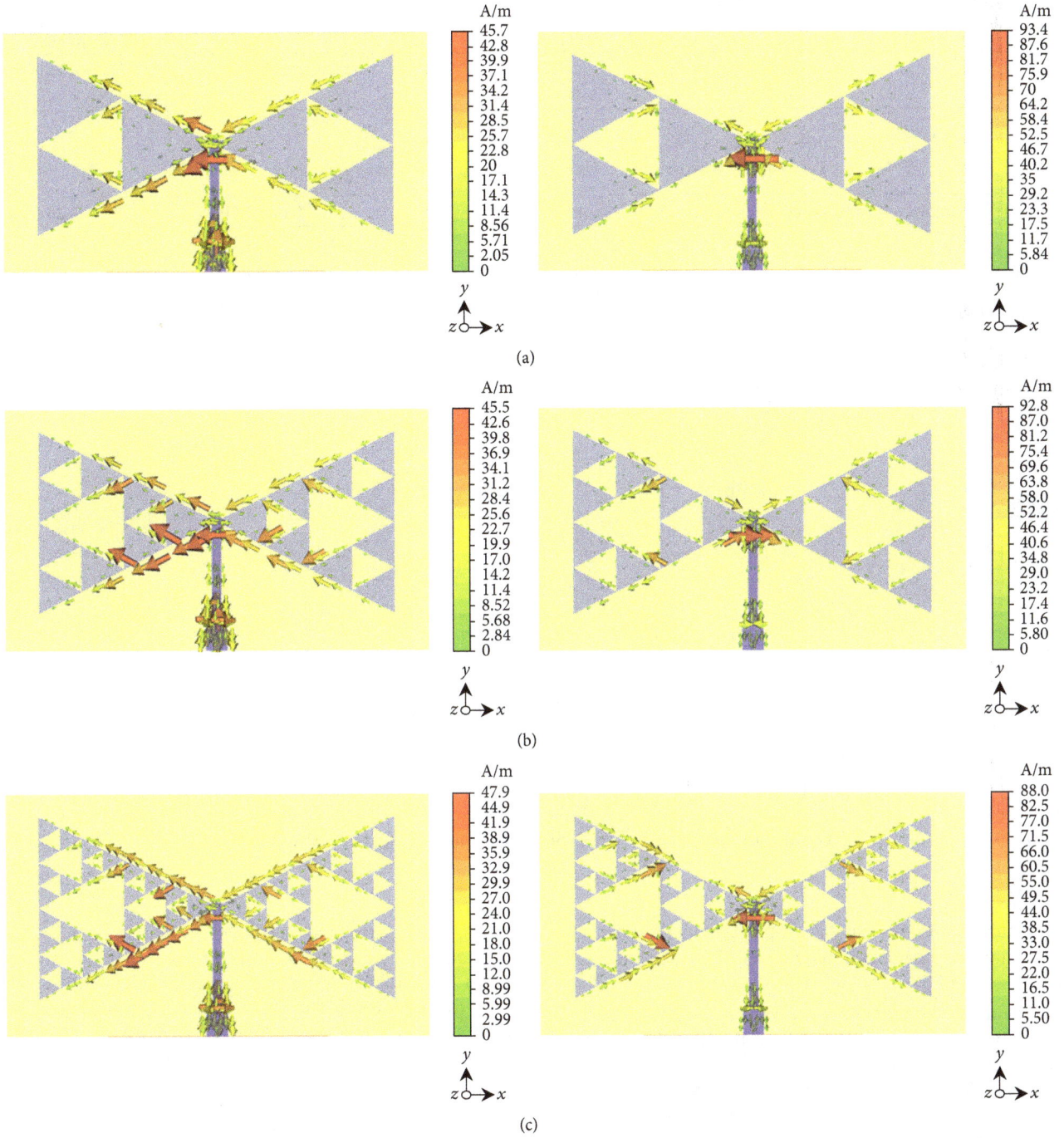

FIGURE 8: Designed antennas current distribution at 2.45 GHz (left) and 5.8 GHz (right) in iterations 1 (a), 2 (b), and 3 (c).

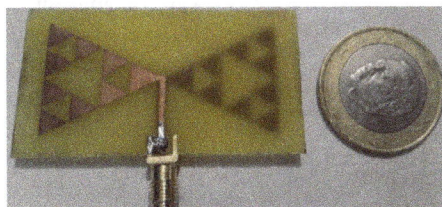

FIGURE 9: Realized antenna picture.

TABLE 2: Comparison between simulated and measured antenna return losses.

Bands	Simulation	Measurement
Band 1	[2.25–2.8] GHz	[2.35–2.86] GHz
Band 2	[5.5–6] GHz	[5.48–6.16] GHz

TABLE 3: Performance comparison between the proposed antenna and other compact antennas.

Published literature versus proposed antenna	Antenna size (mm^2)	Bands (GHz)	Gain (dBi)
This work	60×30	2.4	2.2
		5.8	5.8
Reference [13]	45×30	5.2	4.4
		5.8	6.6
Reference [14]	45×45	1.8	3.9
Reference [15]	135×93	2.4	10

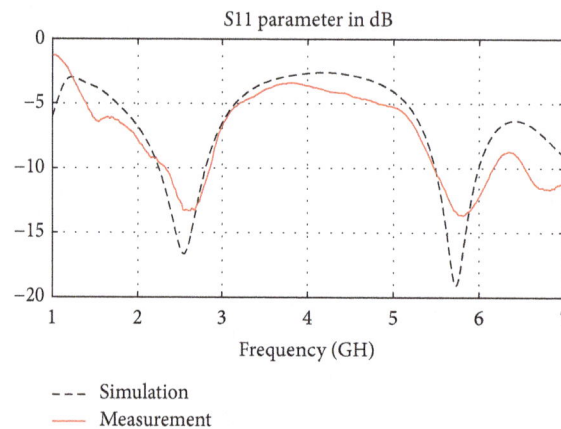

FIGURE 10: Comparison between simulated and measured antenna return loss.

Figure 11 presents the XZ (E-plane) and YZ (H-plane) planes radiation pattern at frequencies 2.5 GHz (a and c) and 5.8 GHz (b and d).

The measured radiation at 2.5 GHz is distributed over all directions but presents a maximum through the Z-axis at 180° while it is concentrated in the upper part of the XZ plane at 5.8 GHz. In the YZ plane, the antenna radiates in all directions except around 180° through the Z-axis at both 2.5 GHz and 5.8 GHz.

We notice that there is considerable radiation attenuation around 180° relative to the antenna back in the YZ plane and at 5.8 GHz in the XZ plane. After analysis we deduced that this attenuation is related to mechanical support of measurement that reflects the energy received from the horn antenna when the receiving antenna is turned back.

The achieved antenna presents good characteristics that are suitable for wireless power transmission (WPT) applications. The size is small (60×30 mm^2). The dual ISM bands covered by the antenna in this work are commonly used for WPT. The radiation pattern is almost omnidirectional in both bands, which permits harvesting a maximum of energy. The antenna gain could be improved by several techniques.

In [28], authors designed an antenna array (4 elements) with multiple superstrates to improve the Sierpiński triangle antenna gain. The antenna operates at 860 MHz for RFID application. This method could be applied to our structure in order to enhance the gain and then the rectenna efficiency.

Table 3 compares the achieved antenna to other planar antennas used for rectenna application.

4. Conclusion

A new planar multiband fractal antenna based on Sierpiński triangle is presented. In the first Sierpiński triangle, three iterations are designed and studied. The second iteration structure was printed over an FR4 substrate of 60×30 mm^2 as dimension, a relative permittivity equal to 4.4, 1.6 mm of thickness, and 0.025 of loss tangent. The measurements present good performance at ISM 2.4 GHz and 5.8 GHz. The structure is simple to fabricate, low cost, and easy to associate with integrated circuits. These characteristics are suitable for wireless power transmission applications.

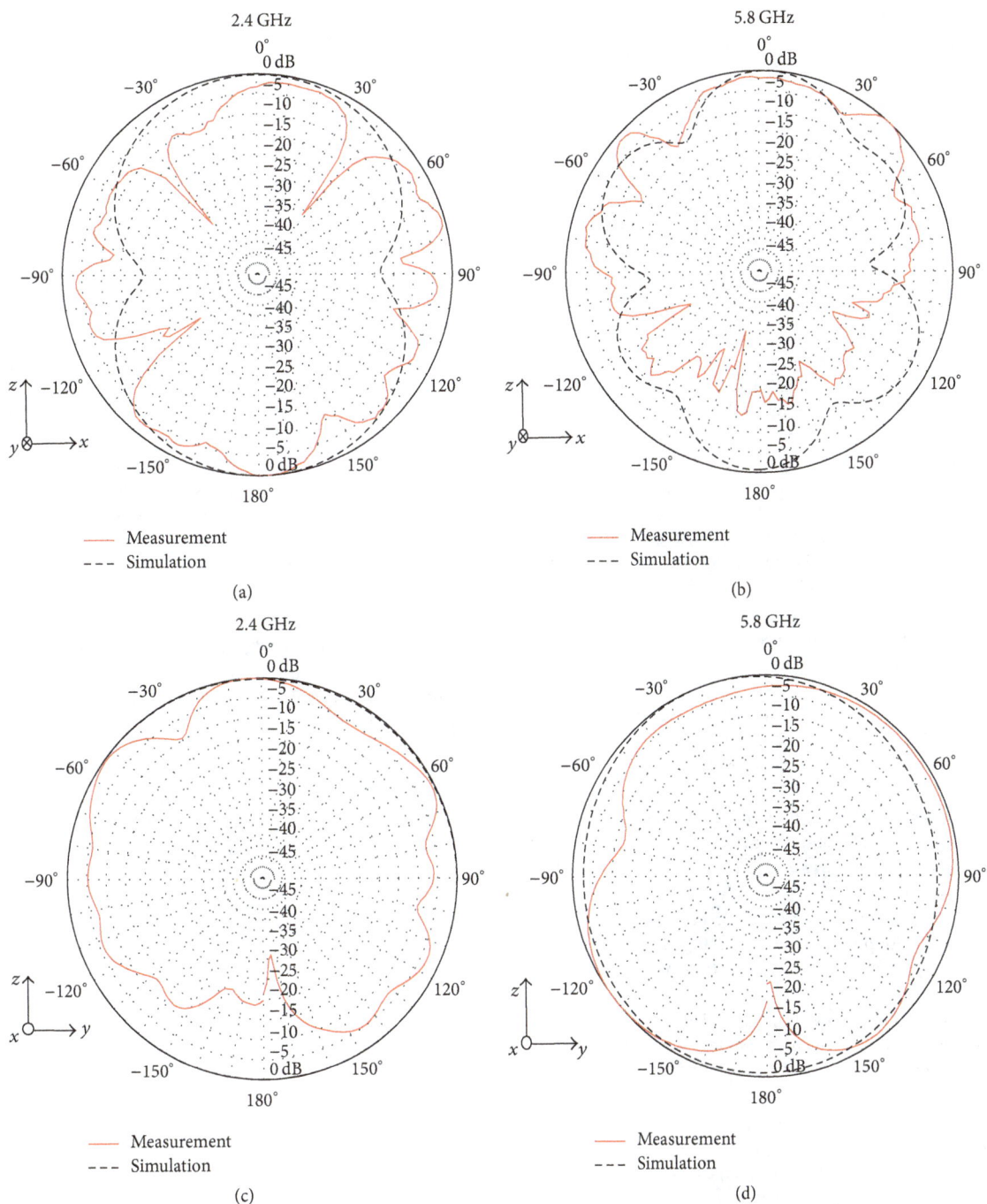

FIGURE 11: Measured radiation pattern of the realized antenna at 2.5 GHz (*XZ* plane (a) and *YZ* plane (c)) and 5.8 GHz (*XZ* plane (b) and *YZ* plane (d)).

Conflicts of Interest

The authors declare that they have no conflicts of interest.

Acknowledgments

The authors have to thank Mr. Mohamed Latrach, Professor in ESEO, Engineering Institute in Angers, France, for allowing them to use all the equipment and electromagnetic solvers available in his laboratory.

References

[1] M. Cheney, *Tesla Man Out of Time*, Prentice-Hall, Englewood Cliffs, NJ, USA, 1981.

[2] H. Yagi and S. Uda, "On the feasibility of power transmission by electric waves," in *Proceedings of the third Pan-Pacific Science Congress*, vol. 2, pp. 1305–1313, Tokyo, Japan, 1926.

[3] U.S department of commerce, "Electric light without current," *The Literary Digest*, vol. 112, no. 3, p. 30, 1932.

[4] W. C. Brown, "The history of power transmission by radio waves," *IEEE Transactions on Microwave Theory and Techniques*, vol. 32, no. 9, pp. 1230–1242, 1984.

[5] W. C. Brown, "The amplitron: a super power microwave generator," *Electron Progress*, vol. 5, no. 1, pp. 1–5, 1960.

[6] W. C. Brown, "The history of the crossed-field amplifier," *IEEE MTT-S Newsletters*, no. 141, pp. 29–40, 1995.

[7] W. C. Brown, "Electronic and mechanical improvement of the receiving terminal of a free-space microwave power transmission system," NASA Report CR-135194 4964, Raytheon Company, Wayland, MA, USA, August 1977.

[8] W. Brown and J. Triner, "Experimental thin-film, etched-circuit rectenna," in *Proceedings of the MTT-S International Microwave Symposium Digest*, pp. 185–187, Dallas, TX, USA, June 1982.

[9] P. Koert, J. Cha, and M. Machina, "35 and 94 GHz rectifying antenna systems," in *Proceedings of the 2nd International Symposium on SPS 91—Power from Space*, Gif-sur-Yvette, France, August 27–30, 1991.

[10] S. S. Bharj, R. Camisa, S. Grober, F. Wozniak, and E. Pendleton, "High efficiency C-band 1000 element rectenna array for microwave powered applications," in *Proceedings of the 1992 IEEE MTT-S International Microwave Symposium Digest Part 2 (of 3)*, pp. 301–303, June 1992.

[11] J. Zbitou, M. Latrach, and S. Toutain, "Hybrid rectenna and monolithic integrated zero-bias microwave rectifier," *IEEE Transactions on Microwave Theory and Techniques*, vol. 54, no. 1, pp. 147–152, 2006.

[12] T.-W. Yoo and K. Chang, "Theoretical and experimental development of 10 and 35 GHz rectennas," *IEEE Transactions on Microwave Theory and Techniques*, vol. 40, no. 6, pp. 1259–1266, 1992.

[13] P. Lu, X. S. Yang, and J. L. Li, "A compact frequency reconfigurable rectenna for 5.2- and 5.8-ghz wireless power transmission," *IEEE Transactions on Power Electronics*, vol. 30, no. 11, pp. 6006–6010, 2015.

[14] M. Zeng, A. S. Andrenko, X. Liu, Z. Li, and H.-Z. Tan, "A compact fractal loop rectenna for RF energy harvesting," *IEEE Antennas and Wireless Propagation Letters*, vol. 16, pp. 2424–2427, 2017.

[15] M.-J. Nie, X.-X. Yang, G.-N. Tan, and B. Han, "A compact 2.45-GHz broadband rectenna using grounded coplanar waveguide," *IEEE Antennas and Wireless Propagation Letters*, vol. 14, pp. 986–989, 2015.

[16] S. V. Georgakopoulos and O. Jonah, "Optimized wireless power transfer to RFID sensors via magnetic resonance," in *Proceedings of the 2011 IEEE International Symposium on Antennas and Propagation and USNC/URSI National Radio Science Meeting (APSURSI '11)*, pp. 1421–1424, Spokane, Wash, USA, July 2011.

[17] J. J. Schlesak, A. Alden, and T. Ohno, "Microwave powered high altitude platform," in *Proceedings of the 1988 IEEE MTT-S International Microwave Symposium Digest: Microwaves—Past, Present and Future*, pp. 283–286, 1988.

[18] S. Sasaki, K. Tanaka, and K. Maki, "Microwave power transmission technologies for solar power satellites," *Proceedings of the IEEE*, vol. 101, no. 6, pp. 1438–1447, 2013.

[19] V. Talla, B. Kellogg, B. Ransford, S. Naderiparizi, S. Gollakota, and J. R. Smith, "Powering the next billion devices with Wi-Fi," in *Proceedings of the the 11th ACM Conference*, pp. 1–13, Heidelberg, Germany, May 2015.

[20] B. B. Mandelbrot, *Fractals: Form, Chance and Dimension, les Objets Fractals : Forme Hasard et Dimension*, Nouvelle Bibliothèque Scientifiques, 1975.

[21] E. L. Barreto and L. M. Mendonça, "A new triple band microstrip fractal antenna for C-band and S-band applications," *Journal of Microwaves, Optoelectronics and Electromagnetic Applications*, vol. 15, no. 3, pp. 210–224, 2016.

[22] P. N. Rao and N. V. S. N. Sarma, "Koch fractal boundary single feed circularly polarized microstrip antenna," *Journal of Microwaves, Optoelectronics and Electromagnetic Applications*, vol. 6, no. 2, pp. 406–413, 2007.

[23] A. N. Jabbar, "Studying the effect of building block shape on sierpinski tetrahedron fractal antenna behavior using FDTD-equivalent electric circuits," *Journal of Microwaves, Optoelectronics and Electromagnetic Applications*, vol. 11, no. 1, pp. 162–173, 2012.

[24] N. Kushwaha and R. Kumar, "Study of different shape electromagnetic band gap (EBG) structures for single and dual band applications," *Journal of Microwaves, Optoelectronics and Electromagnetic Applications*, vol. 13, no. 1, pp. 16–30, 2014.

[25] T. Benyetho, J. Zbitou, L. El Abdellaoui, H. Bennis, A. Tribak, and M. Latrach, "A novel design of elliptic fractal multiband planar antenna for wireless applications," in *Proceedings of the 1st URSI Atlantic Radio Science Conference (URSI AT-RASC '15)*, Gran Canaria, Spain, May 2015.

[26] W. Sierpiński, "On curves which contain the image of any given curve," *Matematicheskii Sbornik*, vol. 30, pp. 267–287, 1916.

[27] R. K. Mishra, R. Ghatak, and D. R. Poddar, "Design formula for sierpinski gasket pre-fractal planar-monopole antennas," *Journals & Magazines, IEEE Antennas and Propagation*, vol. 50, no. 3, pp. 104–107, 2008.

[28] B. R. Franciscatto, T.-P. Vuong, and G. Fontgalland, "High gain Sierpinski Gasket fractal shape antenna designed for RFID," in *Proceedings of the SBMO/IEEE MTT-S International Microwave & Optoelectronics Conference (IMOC '11)*, pp. 239–243, IEEE, Natal, Brazil, October-November 2011.

4

DRV Evaluation of 6T SRAM Cell using Efficient Optimization Techniques

Vinod Kumar Joshi ⓘ **and Chetana Nayak**

Department of Electronics and Communication Engineering, Manipal Institute of Technology, Manipal Academy of Higher Education, Manipal 576104, India

Correspondence should be addressed to Vinod Kumar Joshi; vinodkumar.joshi@manipal.edu

Academic Editor: Gerard Ghibaudo

An optimization based method which uses bisection search algorithm has been proposed to evaluate the accurate value of Data Retention Voltage (DRV) of a 6T Static Random Access Memory (SRAM) cell using 45 nm technology in the presence of process parameter variations. Further, we incorporate an Artificial Neural Network (ANN) block in our proposed methodology to optimize the simulation run time. The highest values obtained from these two methods are declared as the DRV. We noted an increase in DRV with temperature (T) and process variations (PVs). The main advantage of the proposed technique is to reduce the DRV evaluation time and for our case, we observe improvement in evaluation time of DRV by ≈ 46, ≈ 27, and ≈ 8 times at 25°C for 3 σ, 4 σ, and 5 σ variations, respectively, using ANN block to without using ANN block.

1. Introduction

Memory structures are now present not just as stand-alone memory chips but also an integral part of complex VLSI systems [1]. SRAM plays a major role in random access memory design, but its leakage currents reduction has become a major concern in past decade. Various architectures of SRAM cell have been also proposed in this regard [2, 3]. The most straightforward and easier approach for reducing the leakage power is to reduce the supply voltage (V_{dd}) of the SRAM cell. Moreover, reducing it below a certain limit may result in the detrition of the stored data due to T and PVs. In SRAM cell, the critical V_{dd} above which a data-bit is retained reliably is called the DRV of the cell. Figure 1 shows the reduction of leakage current with the V_{dd}. Hence, operating the SRAM cell with the voltage higher than its DRV helps in reducing leakage current in standby mode [4]. However, some of the circuit mismatches result in the variation of the V_t of transistors, which causes shifts in the DRV value. Hence, accurate estimation of DRV is a major challenge in low power SRAM design [5, 6].

Qin et al. developed an analytical model for DRV to get a substantial reduction in leakage current by suppressing the

V_{dd} to DRV [7]. The most straightforward method being used to obtain the DRV is by running Monte-Carlo (MC) simulations until a desired failure probability level is reached [8]. However, this method has many disadvantages.

Since obtaining the failure point is a rare event which makes MC simulation time-consuming [9]. Another issue is the time to find a large number of samples to get the accurate value of the tail of the DRV distribution as shown in Figure 2(a). Importance sampling (mixture importance and sequential importance) methods are developed to improve the speed of simulation. These methods have been proved to be more effective than MC samples in obtaining the failure point [10–12]. Wang et al. proposed two methods to evaluate DRV [13]. In the first method, they propose a statistical model for DRV evaluation which uses the relationship between DRV and SNM. Mean and variance of the SNM distribution have been obtained using MC simulations, and DRV is evaluated as the value of the V_{dd} at which SNM reaches zero. In the second method, a generic tail model from recursive statistical blockage has been proposed. Postfabrication methods are also developed which uses canary replica cells [14, 15] and built-in self-test [16, 17] to obtain DRV. However, the optimization based method proposed by G. Huang et al. [18] has been

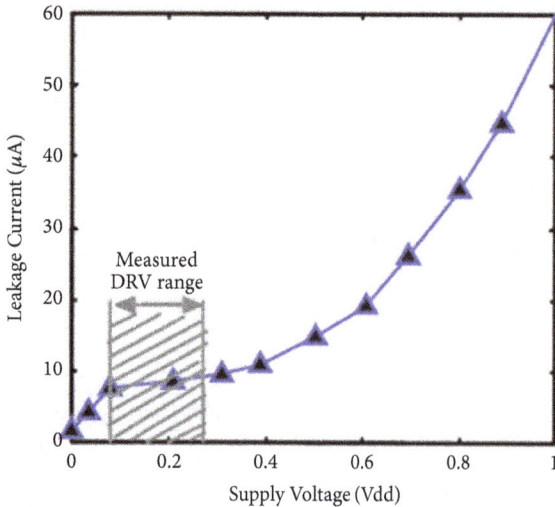

FIGURE 1: Graph showing leakage current versus V_{dd} [1, 4].

claimed to be the fastest evaluation method to obtain DRV. He formulated DRV as a time domain worst performance bound problem and then multistart point (MSP) optimization strategy is developed to evaluate the failure bound.

We use MATLAB tool (version 2015b) to evaluate DRV using optimization based method. A MATLAB code is written for the node voltage equations (Q, Q_B) of MOSFET operating in the subthreshold region [19, 20]. Further, we use bisection search algorithm [21] to search the optimum V_{dd}, and SNM is evaluated using rotation algorithm [22]. PVs are incorporated by generating 5000 quasirandom samples for the Gaussian distribution of V_t. To reduce the time taken for evaluation, an ANN block is incorporated which predicts the value of SNM for a particular sample point. A set of DRV values are evaluated, and the corresponding histogram is plotted. The highest value of DRV obtained or the tail point of the histogram is considered as the DRV. The procedure used by us has not been claimed so far as per our knowledge. A basic 6T SRAM cell consists of two cross-coupled inverters and two access transistors (M5 and M6) are shown in Figure 2(b). M1 and M3 PMOS transistors are pull-up transistors while M2 and M4 NMOS transistors are known as pull-down transistors. During read or write operation word line (WL) is raised high (transistors M5 and M6 become on) while in hold mode (or retention mode) WL is made low (transistors M5 and M6 turn off) and SRAM store the data present in Q and QB nodes. The ability of the SRAM cell to hold the data in retention mode is determined by the SNM of the SRAM cell. The value of SNM is determined from the butterfly curve of the cell in hold mode. Butterfly curve is a plot of voltages (Q versus QB and QB versus Q), where Q and QB are the node voltages of SRAM cell as shown in Figure 3(a). SNM is evaluated as the length of the diagonal of the maximum square that can be incorporated in the butterfly curve. Figure 3(a) shows the butterfly curve plotted at V_{dd} = 1V and Figure 3(b) shows the butterfly curve drawn by varying V_{dd}. It can be observed from Figure 3(b) that the butterfly curve shrinks as the V_{dd} is reduced and the SNM of the cell reduces to zero at V_{dd} = 0.048 V.

2. Proposed Method

The block diagram used for DRV evaluation has been shown in Figure 4 with different colors. The evaluation procedure has four major blocks.

(1) Bisection search algorithm is used for optimizing the value of V_{dd} (blue color in Figure 4).

(2) Quasi MC sample generation block is used to incorporate process parameter variation or variation of the threshold voltage (V_t) (red color in Figure 4).

(3) Seevinck's rotation algorithm is used for SNM evaluation (green color in Figure 4).

(4) ANN block is used to optimize simulation time (yellow color in Figure 4).

2.1. Bisection Search Algorithm [21]. This algorithm helps to evaluate the accurate value of the DRV by searching an optimum solution of V_{dd} at which the SRAM cell fails. First, we define a rough range of V_{dd} from 0 to 1 V based on the initial guess of the DRV. Suppose, if the range is defined as $V_{dd}1$ and $V_{dd}2$, the average between these two points is evaluated as $V_{dd}m$ and this value is used in the analysis phase to evaluate the SNM of the SRAM cell. If the SNM point is evaluated as zero under PVs, it means that the failure has occurred, which implies that the DRV is situated above $V_{dd}m$ and the point $V_{dd}1$ is replaced with $V_{dd}m$. On the other hand, if the failure has not occurred, the DRV is located below $V_{dd}m$ and $V_{dd}2$ is replaced with $V_{dd}m$. The process is repeated as the $V_{dd}1$ and $V_{dd}2$ values get updated. It is continued until the difference $\Delta = V_{dd}2 - V_{dd}1$ evaluates to be less than a defined tolerance (*Tol* = 0.001). Once this condition is met the process ends and the final value of $V_{dd}2$ (or $V_{dd}1$) is declared as the DRV. If the DRV is not located within the defined range, the process repeats for a new range. Table 1 represents the MOSFET constants assumed during the evaluation of SNM which is taken from the 45 nm Predictive Technology Model (PTM) [24].

2.2. Quasi MC Sample Generation. The value of DRV largely depends on V_t of the transistors, *T*, and channel length (*L*). Variation of these parameters affects the value of SNM and hence the DRV. We do the DRV evaluation only by varying V_t of transistors M1, M2, M3, and M4 as shown in Figure 2(b). These values are defined in a Gaussian range with particular mean and variance, and their samples are combined with the Quasi MC samples to obtain the seed points. We take 5000 Quasi MC samples for evaluation to get the better accuracy. The SNM is evaluated for each of these seed points generated by Sobol sequence, and failure analysis is done accordingly. The variance of the Gaussian distribution is calculated by Pelgrom model [25, 26],

$$\sigma = \frac{1.8\,\text{mV} * \mu\text{m}}{\sqrt{W.L}} \quad (1)$$

where *W* is the width of MOSFET and *L* can be used from Table 1. Since during hold mode only transistors M1, M2, M3, and M4 are active, we have employed the variation only for these transistors, which is calculated in Table 2.

TABLE 1: MOSFET parameters used in MATLAB tool for SNM evaluation.

Parameter (unit)	NMOS value	PMOS value
μ_0, Mobility (A/V^2)	$\mu_{0n} = 0.04398$	$\mu_{0p} = 0.0044$
t_{ox}, Thickness of oxide (nm)	$t_{oxn} = 1.75$	$t_{oxp} = 1.85$
ε_r, Relative permittivity	3.9	3.9
n, Subthreshold slope factor	$n_n = 1.042$	$n_p = 1.042$
W, Channel width (μm)	$W_n = 0.12$	$W_p = 0.14$
L, Channel length (nm)	45	45
Mean of V_t variation (V)	$V_{tn} = 0.466$	$V_{tp} = -0.4118$V

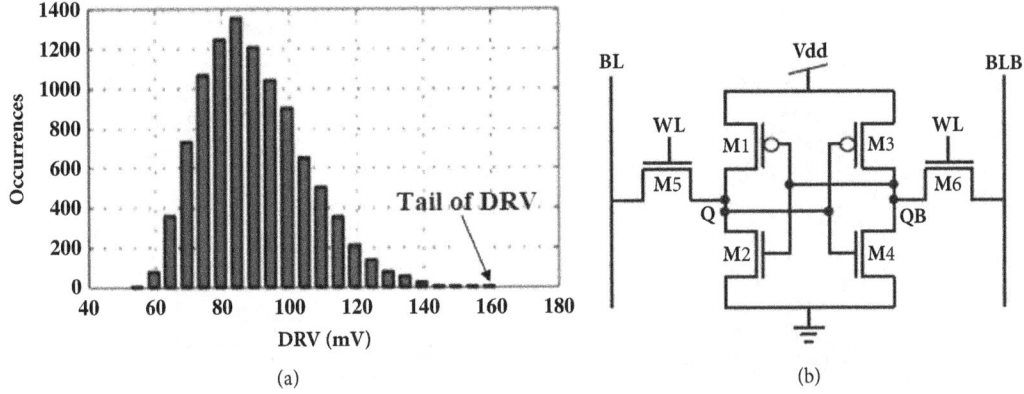

FIGURE 2: (a) DRV distribution obtained by statistical methods using 90 nm technology node for 10K-b SRAM [13]. (b) Schematic of conventional 6T SRAM cell [4].

TABLE 2: The variance of V_t variation (σ) is calculated using (1).

Transistors	Variance of V_t variation (σ)
NMOS	0.02449 V
PMOS	0.0226 V

2.3. SNM Evaluation. We use theoretical equations developed for node voltages (Q, QB) to calculate the SNM using butterfly curve. These equations have been derived by Calhoun et al. which evaluate the node voltages by considering the characteristics of MOSFETS operating in the subthreshold region. The equation for QB is given by [19]

$$QB = V_{th} \frac{n_n n_p}{n_n + n_p} \left(\ln \left(\frac{I_{sp}}{I_{sn}} \right) \right.$$

$$\left. + \ln \left(\frac{1 - \exp \left((-V_{dd} + Q)/V_{th} \right)}{1 - \exp \left(-Q/V_{th} \right)} \right) \right) + \frac{n_n \cdot V_{dd}}{n_n + n_p} \qquad (2)$$

$$+ \frac{n_n \cdot n_p}{n_n + n_p} \left(\frac{V_{tn}}{n_n} - \frac{V_{tp}}{n_p} \right)$$

where I_{sn} and I_{sp} are given by [27]

$$I_{sn} = \mu_{0n} \cdot C_{oxn} \cdot V_{th}^2 \cdot \left(\frac{W_n}{L} \right) \cdot (n_n - 1), \qquad (3)$$

$$I_{sp} = \mu_{0p} \cdot C_{oxp} \cdot V_{th}^2 \cdot \left(\frac{W_p}{L} \right) \cdot (n_p - 1), \qquad (4)$$

where $V_{th} = kT/q$, thermal voltage, n_n, n_p are subthreshold slope factor for NMOS and PMOS transistors, respectively, I_{sn}, I_{sp} are drain current (when $V_{GS} = V_t$) for NMOS and PMOS transistors, respectively, V_{dd} is supply voltage, V_{tn}, V_{tp} are threshold voltage of NMOS and PMOS transistors, respectively, μ_{0n}, μ_{0p} are mobility of NMOS and PMOS transistors, respectively, C_{oxn}, C_{oxp} are oxide capacitance of NMOS and PMOS transistors, respectively, W_n, W_p are width of polysilicon for NMOS and PMOS transistors, respectively, and L is length of polysilicon.

The subthreshold slope factor n is evaluated using (5) by evaluating subthreshold slope (S) [27],

$$S = n \cdot V_{th} \cdot \ln (10) \qquad (5)$$

The value of S is found to be 60 mV/decade at room $T = 25°$C. Its value for typical bulk CMOS can range from 70 to 120 mV/decade [28]. The value of n is evaluated as 1.042 using (5) at room T. The voltage at which node value Q equals QB is known as tripping voltage (V_m). It is the point where curves (Q versus QB and QB versus Q) intersect, as shown in Figure 3(a). Here, we assume the identical cross-coupled inverters to evaluate V_m. The relation of V_m for an inverter is given by (6) (by ignoring the Drain Induced Barrier Lowering (DIBL) effects) as follows [29]:

(a)

(b)

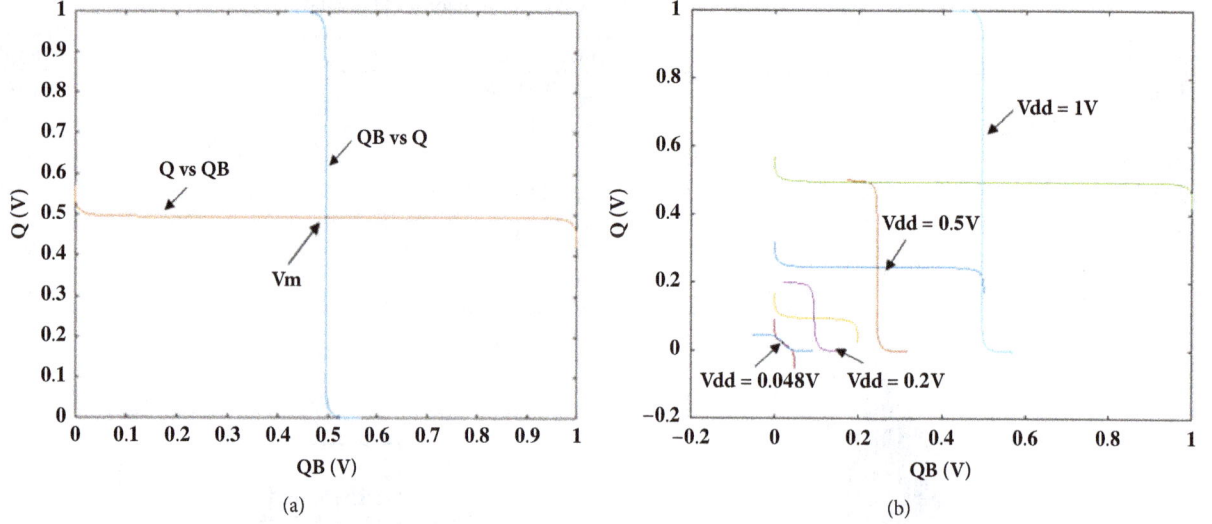

FIGURE 3: Butterfly curve to obtain the SNM of 6T SRAM cell in hold mode (a) at V_{dd} = 1V (b) by varying V_{dd}.

$$V_m = \frac{V_{dd}.n_n}{(n_n + n_p)} + \frac{n_p.V_{tn} - n_nV_{tp}}{(n_n + n_p)} + \frac{n_n.n_p.V_{th}.\ln\left(\left(\left(W_p/L\right).I_{sp}\right)/\left(\left(W_n/L\right).I_{sn}\right)\right)}{(n_n + n_p)}$$
$$+ \frac{n_n n_p.V_{th}.\ln\left(\left(1 - \exp\left(\left(-V_{dd} + V_m\right)/V_{th}\right)\right)/\left(1 - \exp\left(-V_m/V_{th}\right)\right)\right)}{(n_n + n_p)}$$

(6)

All the notations used for (6) are same as mentioned for (2), (3), (4), and (5).

We use the graphical technique proposed by Seevinck [22] to calculate the SNM value as shown in Figure 5. The steps involved in this techniques are as follows:

(1) Obtain Q and QB samples using (2). Figure 6 shows the butterfly curve which is plotted using Q and QB samples for V_{dd} = 0.5V.

(2) Combine Q and QB set into a matrix, X.

(3) Multiply X with rotation matrix, i.e., [U V′] = Rot∗[Q QB], where ∗ indicates matrix multiplication and Rot is the rotational matrix,

$$\text{Rot} = \begin{bmatrix} \cos\left(\frac{\pi}{4}\right) & -\sin\left(\frac{\pi}{4}\right) \\ \sin\left(\frac{\pi}{4}\right) & \cos\left(\frac{\pi}{4}\right) \end{bmatrix}$$ (7)

New axis for the rotated curve is (U, V′). V1 is the matrix corresponding to V′.

(4) Evaluate V2 as V2 = -V1 + (2.√2.V_m), where V_m is obtained using (6). (U, V1) is the rotated version of (Q, QB) and (U, V2) is the rotated version of (QB, Q). Figure 7(a) shows the rotated version of Figure 6.

(5) Take the difference between V2 and V1 samples, i.e., Z = V2 - V1.

(6) Plot (U, Z) as shown in Figure 7(b), and obtain $S1$ = -min (Z), $S2$ = max (Z), and $S3$ = min ($S1$, $S2$).

(7) Finally, SNM is evaluated as SNM = S3/√2.

For a particular value of V_{dd} and each and every sample of Quasi MC seed, the SNM is evaluated.

2.4. ANN Block. Artificial neural networks (ANNs) are a family of learning models that are used to estimate functions that depend on a large number of inputs. ANN is generally presented as system of interconnected "neurons" which exchange messages between each other. This functions approximately as a brain. It consists of three layers input variables, hidden nodes, and outputs. Inputs can be of any number and are provided in the initial learning phase. Outputs are the ones which are obtained after the analysis. During the process, many intermediate hidden nodes are created that are essential for optimizing. However, the user does not have any control over them. The optimization process includes two important stages [23]:

(i) Training or learning phase. This phase uses all the different input signals to predict the possible outcomes of outputs using cautious learning of previous experiments. These can be accomplished either by conducting a large number of experiments or by

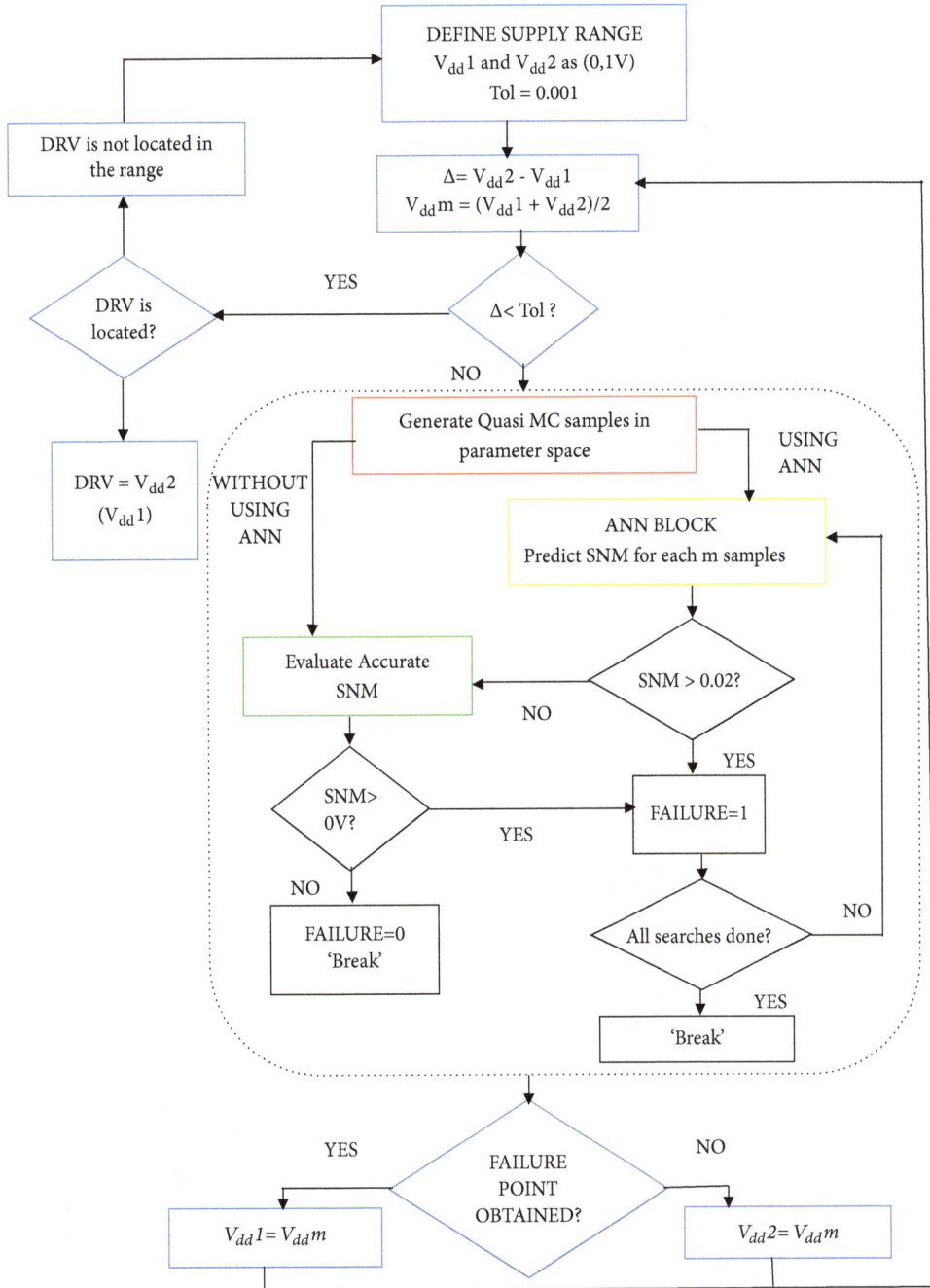

FIGURE 4: General block diagram for DRV evaluation using optimization based method.

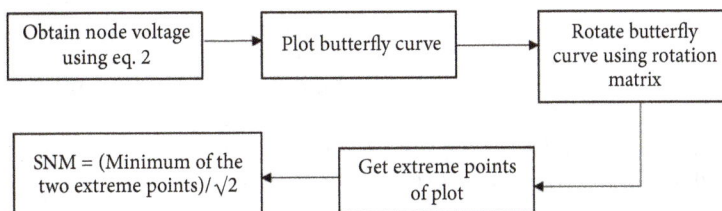

FIGURE 5: General block diagram to evaluate the SNM using optimization method.

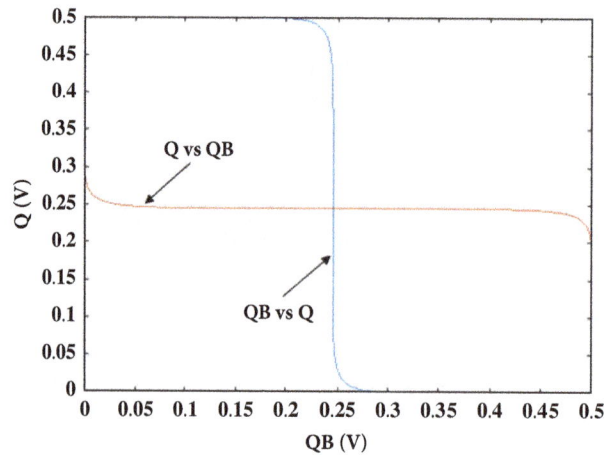

FIGURE 6: Butterfly curve for $V_{dd} = 0.5$V using optimization method.

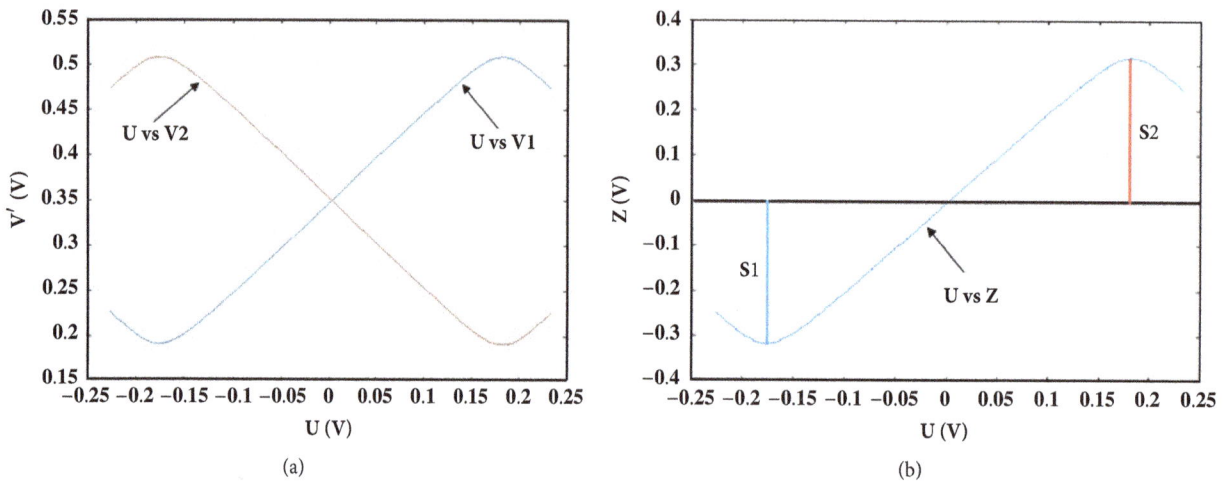

(a)

(b)

FIGURE 7: (a) Rotated butterfly curve at $V_{dd} = 0.5$V. (b) Plot of $Z = V2\text{-}V1$ with respect to U.

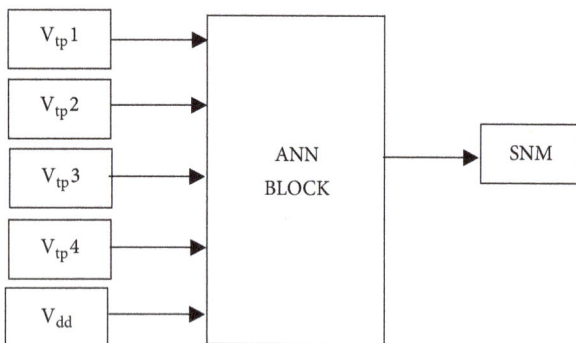

FIGURE 8: General block diagram of ANN block.

using values of the previously conducted experiments. These values will be stored and is used for the upcoming analysis phase.

(ii) Analysis phase. After learning all the calculation procedure from the previous phase, the network is now ready for successfully providing outputs for any inputs provided.

For a 5000-sample space as mentioned in Section 2.2, the process of SNM evaluation is time-consuming. Hence, an ANN block which has been trained to evaluate the SNM is used. This block evaluates the SNM for all the samples and then separates the samples having low SNM (SNM < 0.02V). Only these samples are now sent to the actual analysis block where the accurate value of SNM is evaluated using rotation algorithm.

If the SNM = 0, the sample is declared as the failure sample. Input data set consists of V_t variations of M1, M2, M3, and M4 transistors and the V_{dd}. SNM is the output vector as shown in Figure 8. Fifty data sets are generated using SNM evaluation algorithm, and the network is trained using Radial Basis Function (RBF) network, which is explained in next subsection.

2.4.1. RBF Network [23]. An RBF network uses nonlinear functions to map inputs to the outputs into a high dimensional feature space. A general RBF network consists of three

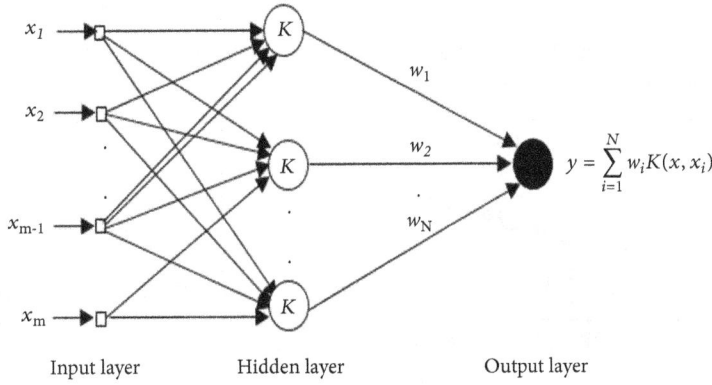

FIGURE 9: Schematic of RBF [23].

layers as shown in Figure 9. The input layer with inputs x_i ($i = 1, 2,...m$) where m is a number of input parameters. The hidden layer is generated by one-to-one correspondence between the training input data x_i and the kernel function $K(x, x_i)$ for $i = 1, 2, ...N$, where N is the number of training samples [23]. In the third layer, the output is evaluated as the linear weighted sum of the kernel functions generated in the hidden layer. The following equations are used by the network:

(i) To evaluate kernel function $K(x_i, x_j)$,

$$K\left(x_i, x_j\right) = e^{|x_i - x_j|/2\sigma'^2} \qquad (8)$$

where x_i, x_j represent input vectors with $i, j = 1, 2, ...m$. Here, σ' denotes the Gaussian bandwidth.

(i) Weight vector $\boldsymbol{w}$ is calculated by

$$(\boldsymbol{K} + \lambda \boldsymbol{I})\,\boldsymbol{w} = \boldsymbol{y_d} \qquad (9)$$

Here $\boldsymbol{K}$ is the kernel matrix, $\boldsymbol{I}$ is the identity matrix of order N, λ is called the regularization parameter, and $\boldsymbol{y_d}$ is the desired response vector.

(ii) To evaluate output of the network y,

$$y = \sum_{i=1}^{N} w_i K\left(x, x_i\right) \qquad (10)$$

w_i is the ith ($i = 1, 2,.... N$) element of the weight vector w and $K(x, x_i)$ is the kernel function.

Kernel used for the control technique is an Exponential Radial Basis Function (ERBF). By considering this ANN block, a considerable reduction in the evaluation time is observed. Four different ANN blocks are generated to evaluate the DRV for $T = 15°C, 25°C, 50°C$, and $100°C$, respectively.

3. Result and Future Work

In this section, we present the results of an optimization based method which evaluates the DRV of a 6T SRAM cell incorporating the process parameter variation by considering the variation of V_t of four transistors.

DRV varies within a range and changes with each run of the experiment. It depends on the samples generated by

Quasi MC simulation. To obtain the actual DRV, we conduct the experiment for 25 runs and the highest value obtained is considered as the DRV. After getting the DRV value from 25 experiments the corresponding histogram is plotted for two cases, (i) by considering the ANN block and (ii) by ignoring the ANN block. Table 3 indicates the V_t variation range for PMOS and NMOS transistors for 3σ, 4σ, and 5σ variation.

Table 4 represents the DRV obtained at 3σ, 4σ, and 5σ variation for T = 15°C, 25°C, 50°C, and 100°C, respectively, using the parameter specifications shown in Table 1 and the methodology followed in Section 2. From Table 4 we can observe that DRV increases with T slightly, while it increases significantly with the variation of V_t. To compare the time taken for DRV evaluation using with and without ANN block we run the MATLAB code at 25°C for 3 σ, 4 σ, and 5 σ variations and note the corresponding time taken for the highest value of DRV for 25 runs as shown in Table 5. Figure 10 represents the corresponding bar chart. The histogram to obtain the DRV at T = 15°C, 25°C, 50°C, and 100°C for 3 σ, 4 σ, and 5 σ variation follows the distribution shown in Figure 2(a) and has been presented in Appendix.

However, the time taken for evaluation depends on the version of MATLAB tool, the machine on which the program is executing and how fast the failure sample is obtained out of the 5000 Quasi MC samples generated. From Table 5, we can observe that ANN block helps in reducing the time taken for DRV evaluation. Since the evaluation, time varies randomly for each run so the comparison of evaluation time cannot be generalized. The method can be extended to evaluate DRV for a memory chip with complex circuit structure. The modification can be made in the algorithm, to obtain the more accurate DRV results with better simulation time. Instead of obtaining the node voltage values using theoretical equations, practical SPICE-level simulation can be used to evaluate the SNM for a given V_{dd} and V_t. Optimization algorithms can be implemented for the node voltages generated by the circuit. The procedure can be extended for other technology nodes by considering other process parameter variations like T and geometry variations in W and L for other cell topologies.

Appendix

See Figure 11.

TABLE 3: V_t variation range used for PMOS and NMOS.

Variation	Range for PMOS	Range for NMOS
3 σ	-0.4796V to -0.344V	0.39253V to 0.53947V
4 σ	-0.5022V to -0.3214V	0.36804V to 0.56396V
5 σ	-0.5248V to -0.2988V	0.34355V to 0.58845V

TABLE 4: DRV values with 3 σ, 4 σ, and 5 σ variation of V_t for T = 15˚C, 25˚C, 50˚C, and 100˚C.

Temperature (˚C)	DRV Without ANN (V)			DRV With ANN (V)		
	3 σ	4 σ	5 σ	3 σ	4 σ	5 σ
15	0.325	0.416	0.4951	0.325	0.4141	0.4961
25	0.331	0.4189	0.5	0.333	0.4189	0.5
50	0.352	0.4229	0.5069	0.352	0.4229	0.5078
100	0.36	0.4443	0.5146	0.36	0.4443	0.5107

TABLE 5: Time elapsed for DRV evaluation at 25˚C for 3 σ, 4 σ, and 5 σ variations.

Temperature (25˚C)	DRV Without ANN (V)			DRV With ANN (V)		
	3 σ	4 σ	5 σ	3 σ	4 σ	5 σ
DRV	0.331	0.4189	0.5	0.333	0.4189	0.5
Time (s)	459	649	783	10	24	103

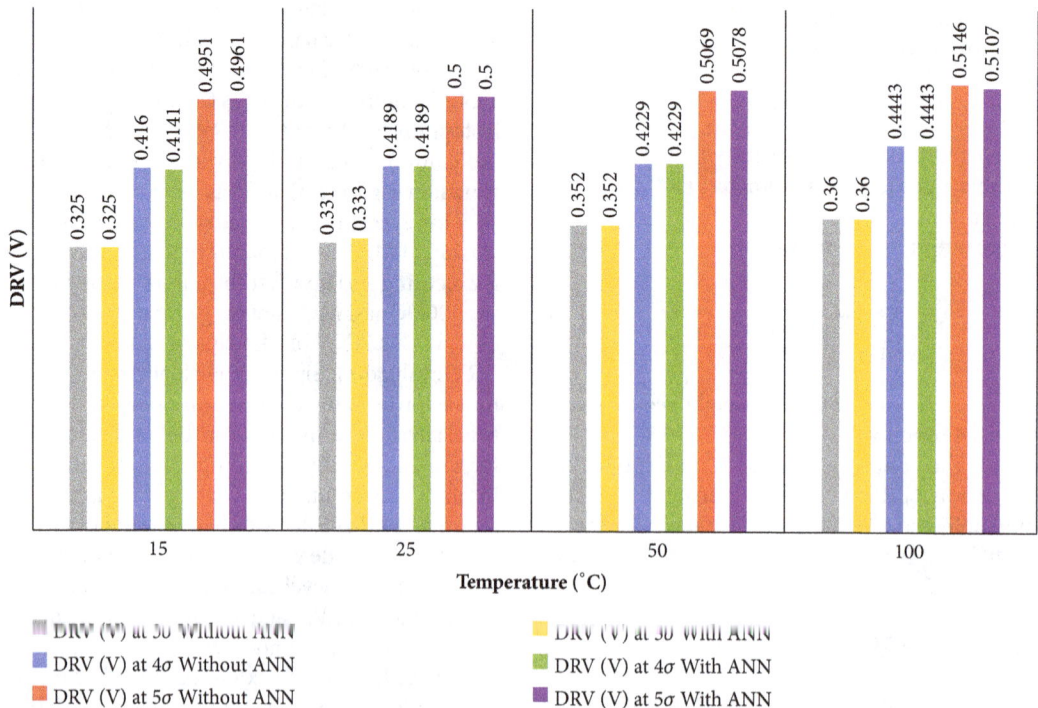

FIGURE 10: Plot of DRV for 3σ, 4σ, and 5σ variation at T = 15˚C, 25˚C, 50˚C, and 100˚C.

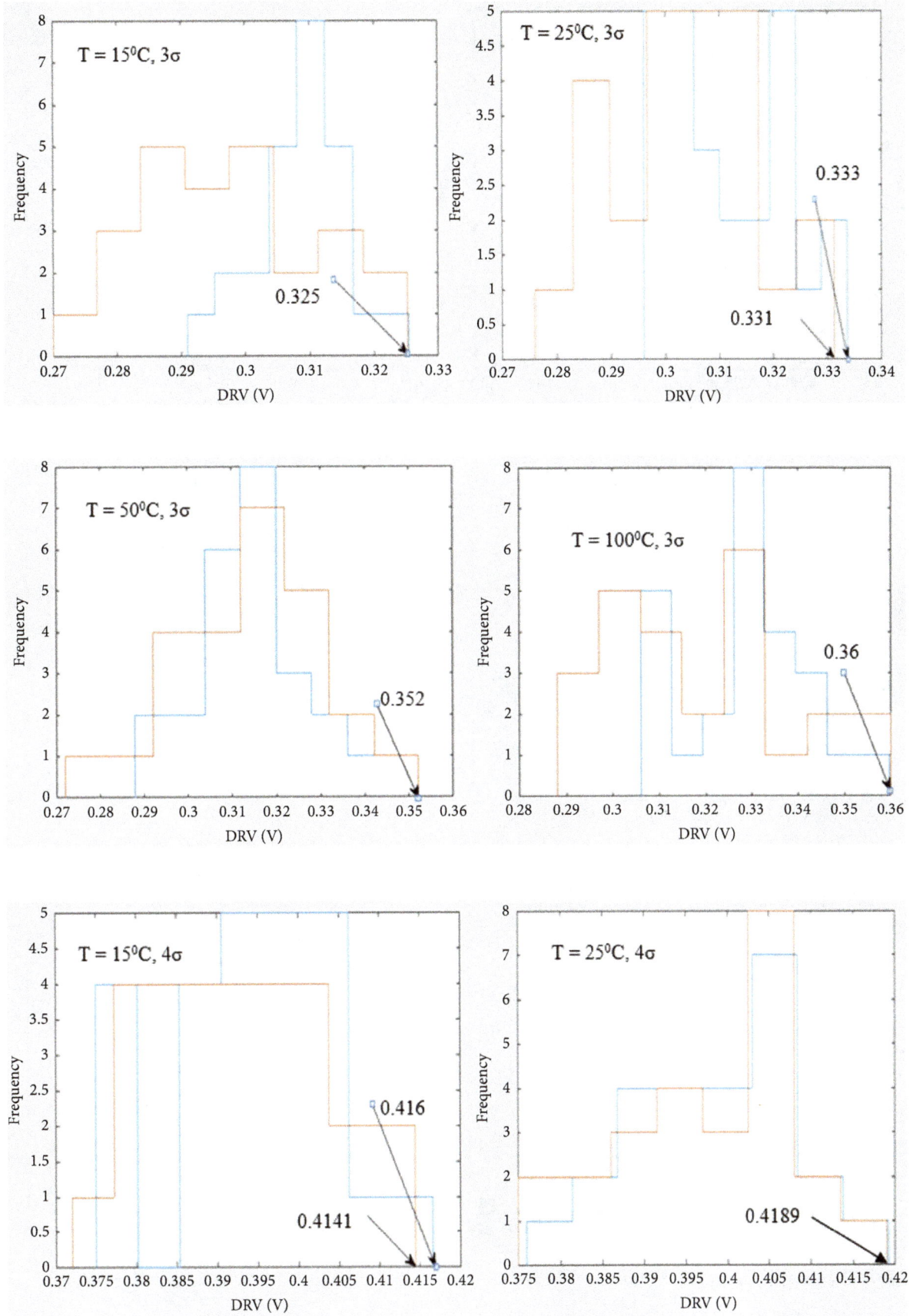

Figure 11: Continued.

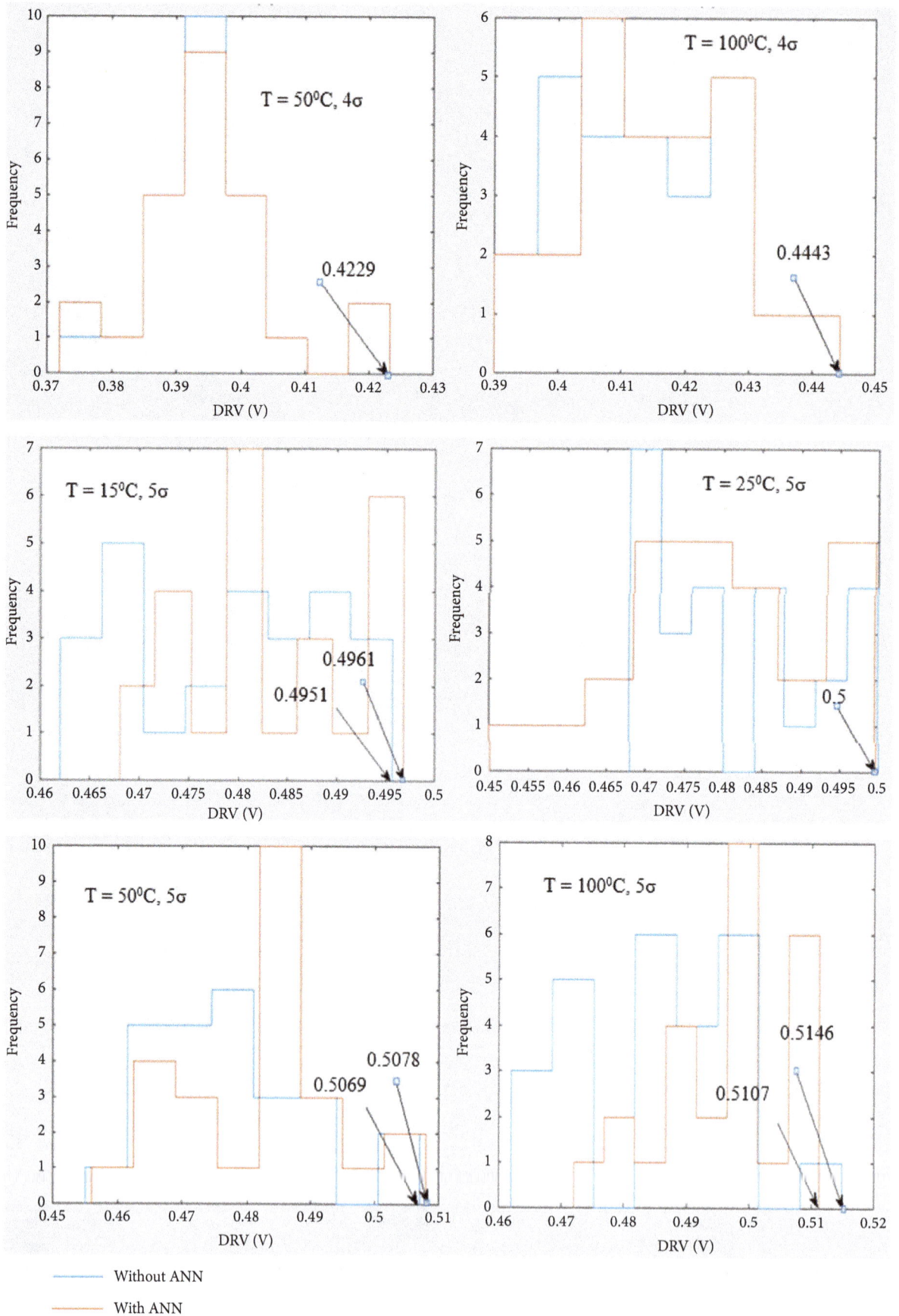

FIGURE 11: DRV for 3 σ, 4 σ, and 5 σ variations at T = 15˚C, 25˚C, 50˚C, and 100˚C with and without ANN block.

Conflicts of Interest

The authors declare that they have no conflicts of interest.

Acknowledgments

The authors kindly acknowledge Department of E & C, Manipal Institute of Technology, Manipal Academy of Higher Education, Manipal, to provide the MATLAB tool facility for simulation. They also acknowledge the PTM website to provide the PTM model file of 45 nm technology.

References

[1] K. Roy and S. C. Prasad, *Low-Power CMOS VLSI Circuit Design*, Wiley, 2009.

[2] D. Nayak, D. P. Acharya, and K. Mahapatra, "A read disturbance free differential read SRAM cell for low power and reliable cache in embedded processor," *AEÜ - International Journal of Electronics and Communications*, vol. 74, pp. 192–197, 2017.

[3] S. Ahmad, N. Alam, and M. Hasan, "Pseudo differential multi-cell upset immune robust SRAM cell for ultra-low power applications," *AEÜ - International Journal of Electronics and Communications*, vol. 83, pp. 366–375, 2018.

[4] H. E. Weste Neil and D. M. Harris, *CMOS VLSI Design-A Circuits and Systems Perspective*, Addison-Wesley, 4th edition, 2010.

[5] Ruchi and S. Dasgupta, "Sensitivity analysis of DRV for various configurations of SRAM," in *Proceedings of the 2015 19th International Symposium on VLSI Design and Test (VDAT)*, pp. 1–5, Ahmedabad, India, June 2015.

[6] Ruchi and S. Dasgupta, "6T SRAM cell analysis for DRV and read stability," *Journal of Semiconductors*, vol. 38, no. 2, pp. 1–7, 2017.

[7] H. Qin, Y. Cao, D. Markovic, A. Vladimirescu, and J. Rabaey, "SRAM leakage suppression by minimizing standby supply voltage," in *Proceedings of the 5th International Symposium on Quality Electronic Design*, pp. 55–60, San Jose, Calif, USA, 2004.

[8] N. Edri, S. Fraiman, A. Teman, and A. Fish, "Data retention voltage detection for minimizing the standby power of SRAM arrays," in *Proceedings of the 2012 IEEE 27th Convention of Electrical and Electronics Engineers in Israel, IEEEI '12*, pp. 1–5, Eilat, Israel, November 2012.

[9] A. Nourivand, A. J. Al-Khalili, and Y. Savaria, "Postsilicon tuning of standby supply voltage in srams to reduce yield losses due to parametric data-retention failures," *IEEE Transactions on Very Large Scale Integration (VLSI) Systems*, vol. 20, no. 1, pp. 29–41, 2012.

[10] L. Dolecek, M. Qazi, D. Shah, and A. Chandrakasan, "Breaking the simulation barrier: SRAM evaluation through norm minimization," in *Proceedings of the 2008 International Conference on Computer-Aided Design, ICCAD '08*, pp. 322–329, USA, November 2008.

[11] R. Kanj, R. Joshi, and S. Nassif, "Mixture Importance Sampling and its Application to the Analysis of SRAM Designs in the Presence of Rare Failure Events," in *Proceedings of the Design Automation Conference*, pp. 69–72, San Francisco, Calif, USA, July 2006.

[12] K. Katayama, S. Hagiwara, H. Tsutsui, H. Ochi, and T. Sato, "Sequential importance sampling for low-probability and high-dimensional SRAM yield analysis," in *Proceedings of the 2010 IEEE/ACM International Conference on Computer-Aided Design (ICCAD)*, pp. 703–708, San Jose, Calif, USA, November 2010.

[13] J. Wang, A. Singhee, R. A. Rutenbar, and B. H. Calhoun, "Two fast methods for estimating the minimum standby supply voltage for large SRAMs," *IEEE Transactions on Computer-Aided Design of Integrated Circuits and Systems*, vol. 29, no. 12, pp. 1908–1920, 2010.

[14] J. Wang and B. H. Calhoun, "Canary replica feedback for near-DRV standby VDD scaling in a 90nm SRAM," in *Proceedings of the 2007 IEEE Custom Integrated Circuits Conference, CICC*, pp. 29–32, San Jose, Calif, USA, September 2007.

[15] J. Wang and B. H. Calhoun, "Techniques to extend canary-based standby V_{DD} scaling for SRAMs to 45 nm and beyond," *IEEE Journal of Solid-State Circuits*, vol. 43, no. 11, pp. 2514–2523, 2008.

[16] F. B. Yahya, M. Mansour, A. Kayssi, and H. Hajj, "Using BIST circuitry to measure DRV of large SRAM arrays," in *Proceedings of the 2010 International Conference on Energy Aware Computing (ICEAC)*, pp. 1–4, Cairo, Egypt, December 2010.

[17] J. Wang, A. Hoefler, and B. H. Calhoun, "An enhanced canary-based system with BIST for SRAM standby power reduction," *IEEE Transactions on Very Large Scale Integration (VLSI) Systems*, vol. 19, no. 5, pp. 909–914, 2011.

[18] G. Huang, L. Qian, S. Saibua, D. Zhou, and X. Zeng, "An efficient optimization based method to evaluate the DRV of SRAM cells," *IEEE Transactions on Circuits and Systems I: Regular Papers*, vol. 60, no. 6, pp. 1511–1520, 2013.

[19] B. H. Calhoun and A. Chandrakasan, "Analyzing static noise margin for sub-threshold SRAM in 65 nm CMOS," in *Proceedings of the 31st European Solid-State Circuits Conference (ESSCIRC '05)*, pp. 363–366, September 2005.

[20] A. Makosiej, O. Thomas, A. Vladimirescu et al., "Stability and Yield-Oriented Ultra-Low-Power Embedded 6T SRAM Cell Design Optimization," in *Proceedings of the Design Automation Test in Europe Conference & Exhibition*, pp. 93–98, Dresden, Germany, March 2012.

[21] M. Bartholomew-Biggs, *Nonlinear Optimization with Engineering Applications*, Springer, 2008.

[22] E. Seevinck, F. J. List, and J. Lohstroh, "Static-noise margin analysis of MOS SRAM cells," *IEEE Journal of Solid-State Circuits*, vol. 22, no. 5, pp. 748–754, 1987.

[23] S. Haykin, *Neural Networks, A Comprehensive Foundation*, Prentice hall, 2nd edition, 2004.

[24] "PTM model for 45nm technology," http://ptm.asu.edu/modelcard/2006/45nm_bulk.pm, assessed February 2016.

[25] M. Qazi, M. Tikekar, and L. Dolecek, "Loop Flattening Spherical Sampling: Highly efficient model reduction techniques for SRAM yield Analysis," in *Proceedings of the Automation & Test in Europe Conference & Exhibition*, pp. 801–806, Dresden, Germany, March 2010.

[26] K. J. Kuhn, "Reducing variation in advanced logic technologies: approaches to process and design for manufacturability of nanoscale CMOS," in *Proceedings of the IEEE International Electron Devices Meeting (IEDM '07)*, pp. 471–474, Washington, DC, USA, December 2007.

[27] A. Wang, B. Calhon, and A. P. Chandrashekharan, *Sub-Thre-shold Design for Ultra-low Power Systems*, Springer, 2006.

[28] K. Roy, S. Mukhopadhyay, and H. Mahmoodi-Meimand, "Leak-age current mechanisms and leakage reduction techniques in deep-submicrometer CMOS circuits," *Proceedings of the IEEE*, vol. 91, no. 2, pp. 305–327, 2003.

[29] J. F. Ryan, J. Wang, and B. H. Calhoun, "Analyzing and modeling process balance for sub-threshold circuit design," in *Proceedings of the 17th Great Lakes Symposium on VLSI, GLSVLSI'07*, pp. 275–280, Italy, March 2007.

5

Design Impedance Mismatch Physical Unclonable Functions for IoT Security

Xiaomin Zheng, Yuejun Zhang, Jiaweng Zhang, and Wenqi Hu

Institute of Circuits and Systems, Ningbo University, No. 818 Fenghua Road, Ningbo 315211, China

Correspondence should be addressed to Yuejun Zhang; zhangyuejun@nbu.edu.cn

Academic Editor: Sourabh Khandelwal

We propose a new design, Physical Unclonable Function (PUF) scheme, for the Internet of Things (IoT), which has been suffering from multiple-level security threats. As more and more objects interconnect on IoT networks, the identity of each thing is very important. To authenticate each object, we design an impedance mismatch PUF, which exploits random physical factors of the transmission line to generate a security unique private key. The characteristic impedance of the transmission line and signal transmission theory of the printed circuit board (PCB) are also analyzed in detail. To improve the reliability, current feedback amplifier (CFA) method is applied on the PUF. Finally, the proposed scheme is implemented and tested. The measure results show that impedance mismatch PUF provides better unpredictability and randomness.

1. Introduction

The Internet of Things is a dynamic living entity, which enables things to exchange information and communication through the networking of physical terminal devices, humans, intelligent buildings, and others [1, 2]. The IoT improves efficiency, accuracy, and economic benefit but is also potentially a huge security risk. Some reports predict that it will spend $547 million on IoT security in 2018 and will involve more than 25% of identified attacks on enterprises by 2020 [3]. Some possible IoT threats are outlined in Figure 1.

According to [4], IoT weakness is so ubiquitous that industrial espionages find it easy to get a good target for attacking. And also, privacy is the other important area of concern. The cybercriminals may recover the personal information, which is potentially residing on IoT networks. In addition, as more and more objects interconnect to today's IoT networks, the physical security of each device is greatly reduced. Attackers could add all kinds of risk scenarios to control systems or change functionality, such as reading, intercepting, or changing the data [5]. To address these problems, there are some methods to increase security for IoT network with the help of security tools, such as identification (ID) authentication, data encryption/decryption, and code obfuscation.

To build a secure and safe IoT, it is very important that the identity of each thing is authenticated. That is a massive challenge, but fortunately there is a way to take full advantage of each thing's unique identifier through Physically Unclonable Function (PUF) technology [6, 7]. PUF generates a unique identifier by exploiting random physical factors introduced in the semiconductor manufacturing process. PUF circuit has the properties of uniqueness, randomness, and unclonability [8–12]. The above features make the PUF circuit an effective defense against intrusion attacks, including a variety of attack patterns. Printed circuit board (PCB) is one of the important hardware carriers of IoT. In the supply chain of PCB, malicious users may make counterfeit PCB come from a variety of sources, such as direct cloning, overproduction, and recycling. In fact, the quality of imitation PCB is poor, such as reliability and performance problems. With the proliferation of fake PCBs and the increase in accidental reports, the problems of the board-level feature recognition technique are becoming more and more important. In this work, we propose an impedance mismatch PUF, which has been exploited to generate a security unique private key to authenticate each thing in an IoT network. According to the characteristic of transmission line and signal transmission theory, the impedance mismatch will cause the transmission signal to

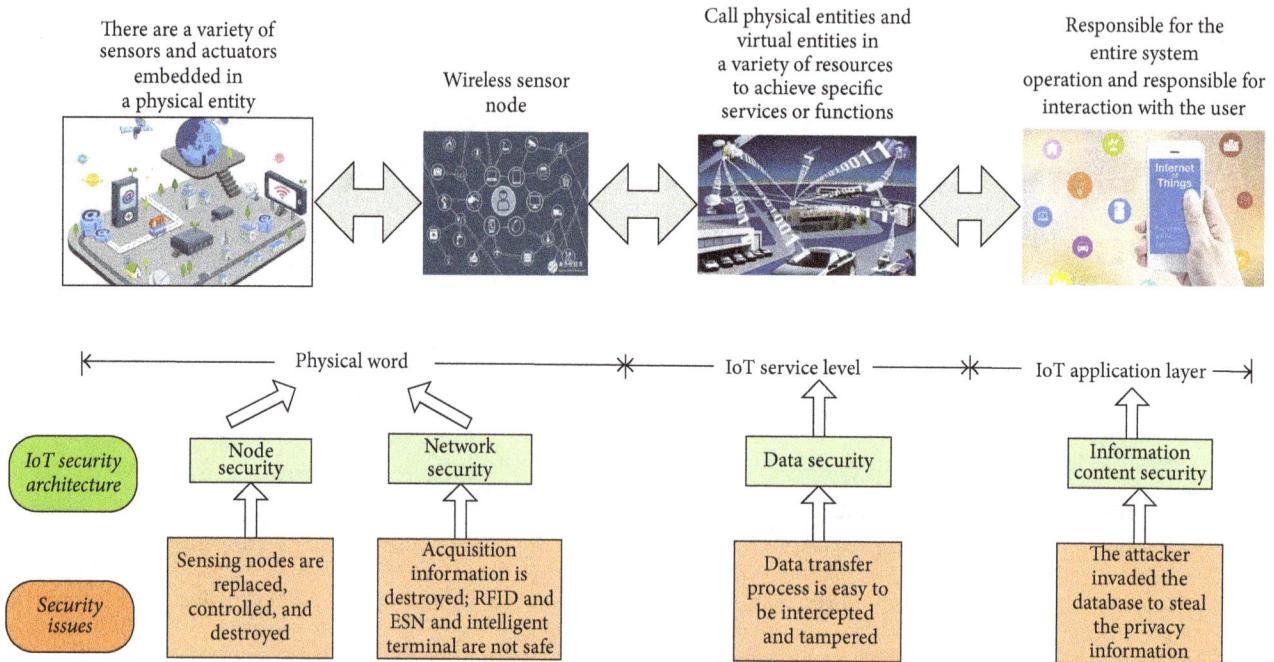

FIGURE 1: The IoT threats model.

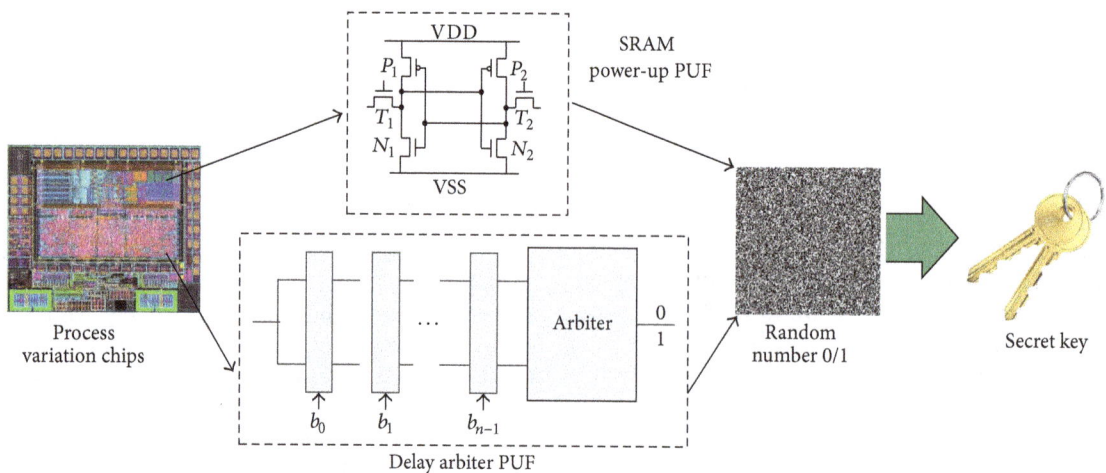

FIGURE 2: Physical Unclonable Functions circuit.

reflect, especially in high-frequency scenarios. The proposed PUF circuit will improve the board-level security of IoT.

This paper is organized as follows: The existing Physically Unclonable Functions circuits are summarized in Section 2. The impedance mismatch effect of transmission line is detailed in Section 3. The designed method of PUF for IoT security is proposed in Section 4. Some experimental results are analyzed in Section 5. This work is concluded in Section 6.

2. Physically Unclonable Functions

SRAM PUF [8, 9] and Arbiter PUF [11–13, 15, 16] are two kinds of typical PUF circuit. SRAM-PUF circuits are produced through the manufacturing process, which introduces a biased digital signal in an integrated circuit. As shown in Figure 2, the SRAM-PUF cell consists of cross coupled inverters and T1 and T2 transmission transistors. SRAM-PUF circuit cells generate a logic level, which is determined by random process deviation threshold V_{th} of the cross coupled inverters. The function relation of SRAM-PUF circuit is easy to affect through the power supply voltage, temperature, aging, and other factors [10]. The output value has stability problems.

The arbiter PUF circuit [11–13, 15, 16] is composed of a delay unit and an arbiter circuit, as shown in Figure 2. The delay unit is composed of two delay paths and switch components. When the left input of the circuit experiences a low level to high level signal rise, the input signal will be conveyed along two paths, each after a data selector signal making two kinds of path selection, as dictated by the control signal b_i.

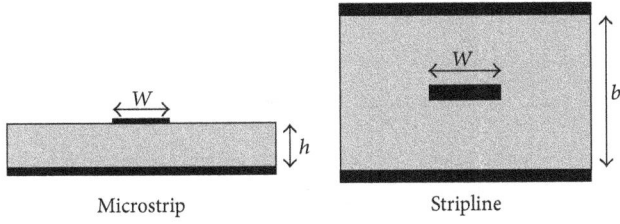

FIGURE 3: Microstrip and Stripline.

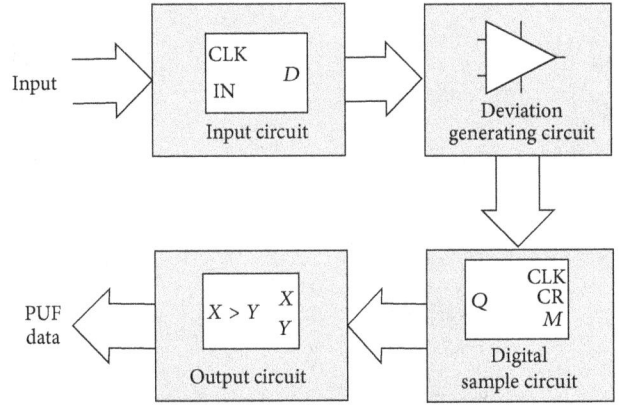

FIGURE 4: IM-PUF circuit model.

If there are m data selectors, this is a signal with 2^m different transmission modes. If the differences in signal transmission to the arbiter through the two delay paths have a time difference, the upper end of the output signal end of a data selector first will output a signal arbiter "1." Otherwise, the output signal of the arbiter is "0." Therefore, the output signal of the arbitrator is determined by the priority arrived signal. The arbiter PUF circuit recognizes model attacks [17, 18].

3. Impedance Mismatch Effect

The definition of characteristic impedance is the ratio of voltage amplitudes and current value on the transmission line. The most important physical factors of characteristic impedance are geometry and materials of the transmission line. It is not dependent on length of transmission line. Under the condition of matching with the load impedance, the signal on transmission line transmits long distant without reflection [19]. If the impedance of transmission line mismatches with the load impedance, it will transmit loss and produce reflection. Impedance mismatch phenomenon means that the impedance of transmission line is different from the characteristic impedance, and transmission signal will be reflected to the opposite direction [20]. If the impedance of transmission line matches with the load impedance, the voltage signal generates positive reflection, and current signal generates negative reflection [21]. On the other hand, when the load impedance is smaller than the characteristic impedance, the voltage signal generates negative reflection, and the current signal generates positive reflection.

There are two types of transmission lines on the PCB board, Microstrip and Stripline (as shown in Figure 3). The impedance calculation formula of Microstrip is shown as follows [22]:

$$Z = \frac{87.0}{\sqrt{\varepsilon_r + 1.41}} \ln\left(\frac{5.98h}{0.8w + t}\right). \tag{1}$$

Among them, Z is the characteristic impedance, ε_r is the relative permittivity, h is the medium wire thickness (mil), w is the wire width (mil), and t is the thickness of the wire (1 oz = 1.5 mil). In (1), relative permittivity ε_r is between 1 and 15; ratio of w/h is between 1 and 15; the width of the ground wire is more than 7 times the width of the signal line. The impedance calculation formula of Stripline is shown as follows [23]:

$$Z_0 = \frac{87}{\sqrt{\varepsilon_r}} \ln \frac{4h}{067\pi (0.8w + t)}. \tag{2}$$

In (2), $w \approx h < 0.35$; relative permittivity ε_r is between 1 and 15; the width of the ground wire is more than 7 times the width of the signal line. From formulas (1) and (2), it is known that the width, thickness, and dielectric constant determine the impedance. Reference [24] shows that the length of the wire, the thickness of the pad, the path of the ground wire, and other nearby wires will also affect the characteristic impedance of the transmission line, especially in high-speed data transmission.

The cut-off frequency calculation formula is shown as follows [25]:

$$f_{\text{cut-off}} = \frac{1}{2\pi RC}, \tag{3}$$

where R and C represent the PCB's equivalent resistance and capacitance, respectively. According to (3), the cut-off frequency does not relate to input signal and power supply. On transmission line of PCB, reflection caused by impedance mismatch may happen. The more the transmission signal reflects, the weaker the output signal is [26]. In experimental testing, the cut-off frequency is described as specific operational frequency that causes the output signal amplitude to reduce 0.707-fold [25]. So, the characteristic impedance Z and Z_0 influence the cut-off frequency of PCB.

4. Proposed Impedance Mismatch PUF Circuit

Comparing the deviation signals present in the same structure, a PUF circuit generates random output response. In PCB circuit, there are random physical factors that affect output signal amplitude, frequency, and bandwidth [27]. The random physical factors can be divided into two categories. The first category is in the integrated circuit, which is produced by the chip fabrication process, such as the ratio of channel width to length, and the threshold voltage. The second is the PCB layout of the processing device, such as the length and the width of wires, capacitors, resistors, and other factors [28]. Thus, the intrinsic characteristics of the PCB may establish a unique and robust fingerprint in these scenarios.

4.1. Impedance Mismatch PUF Model. According to the impedance mismatch theory and the PUF design method, we presented an impedance mismatch PUF (IM-PUF) model, as shown in Figure 4. The model is composed of input

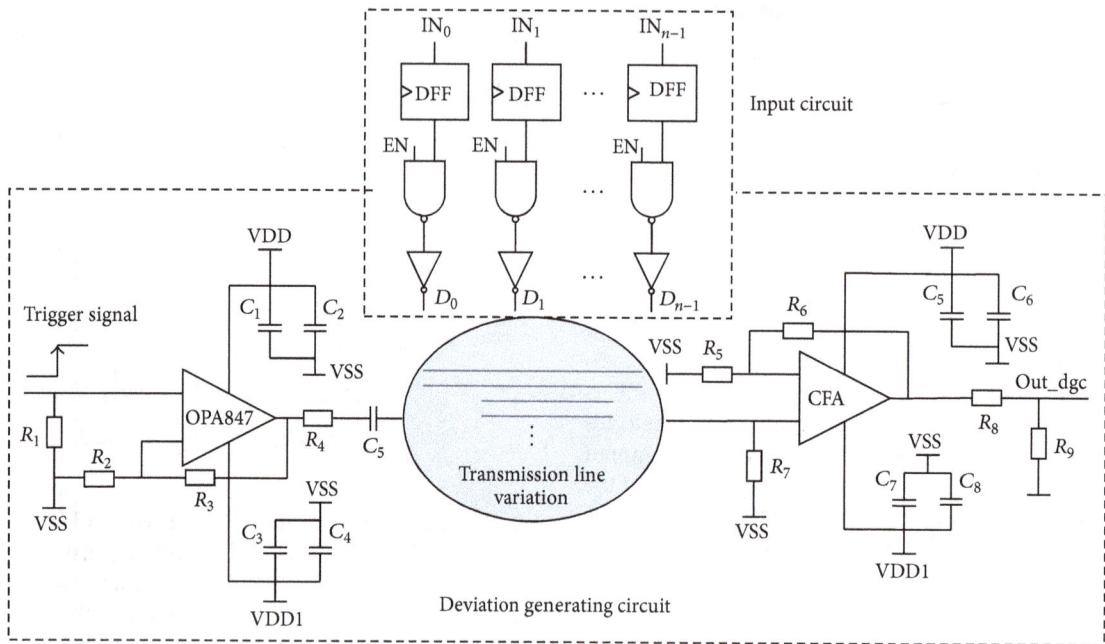

FIGURE 5: Input circuit and deviation generating circuit.

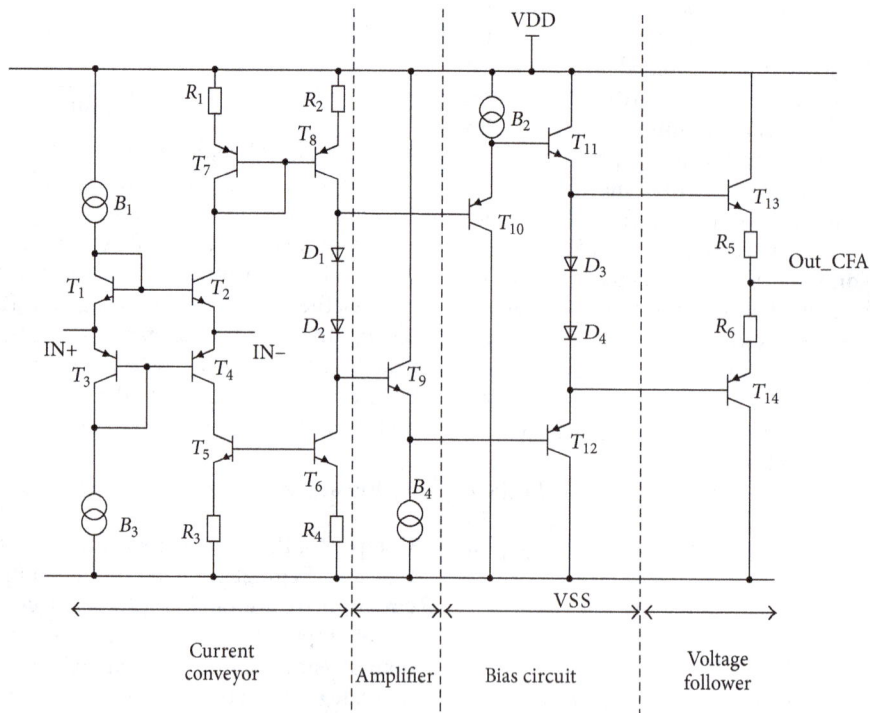

FIGURE 6: CFA circuit structure.

circuit, deviation generating circuit, digital sample circuit, and output circuit. The deviation generating circuit is the core of the PUF circuit, which generates the upper cut-off frequency with the deviation. The upper cut-off frequency is the clock frequency of the digital sample circuit. During the T time, the circuit compares the two signals and generates a "0" or "1" as the response of PUF circuit.

4.2. Input Circuit and Deviation Generating Circuit. The input circuit and deviation generating circuit are shown in Figure 5. The input circuit is composed of D flip-flops (DFF), NAND gates, and inverters, while the deviation generation circuit is composed of an operational amplifier (OPA847), transmission lines, and current feedback amplifier (CFA). The diagram of a current feedback amplifier is shown in Figure 6

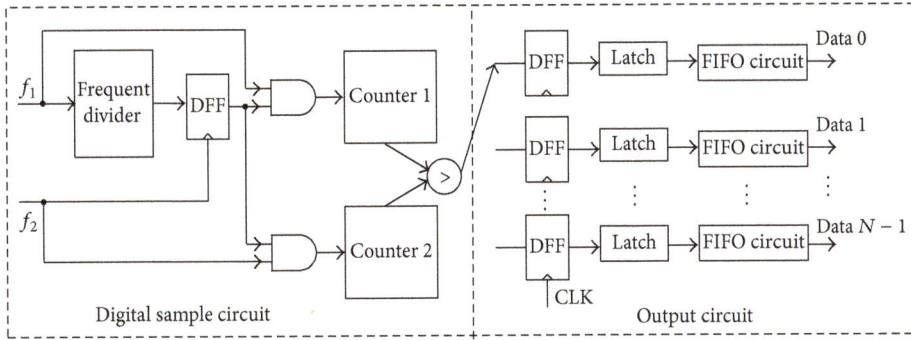

FIGURE 7: Digital sample circuit and output circuit.

[29]. In Figure 6, the class current conveyors are $T_1 \sim T_8$, the amplifier is T_9, the bias circuit is T_{11} and T_{12}, and the voltage follower is T_{13} and T_{14}. In Figure 6, R_3 is a 750-ohm OPA847 feedback resistor, C_1, C_2, C_3, and C_4 are the decoupling capacitors of OPA847. R6 is a 560-ohm CFA feedback resistance; C_5, C_6, C_7, and C_8 are the decoupling capacitors of CFA. R_3 and R_6 are used to set the amplification of the output signal. R_1, R_4, R_7, and R_8 are used in the impedance matching. The amplitude and frequency of the output signal are determined by the process parameters. In this experiment, the deviation of the characteristic impedance of the transmission line makes the upper cut-off frequency of the output signal changed. The amplifier works as follows equation $\beta = \lambda \times (S + \Delta S)$. It means that the deviation signal ΔS is amplified λ times.

4.3. Digital Sample Circuit and Output Circuit.

As shown in Figure 7, the digital sample circuit consists of a frequency divider, a D flip-flop, two AND gates, two counters, and a comparator. Two input signal frequencies f_1 and f_2 serve as the upper cut-off of the output of two transmission lines, respectively. The frequency f_1, though frequency divider, generates a gate control signal TC. During the clock pulse width (named T), f_1 and f_2 behave as two counter clock frequencies. Counter 1 counts the number of N_1, and Counter 2 counts the number of N_2. Comparing N_1 and N_2, if $N_1 > N_2$, the output is "0"; otherwise the output is "1." The output circuit comprises M output units. Each output unit comprises a latch and First Input First Output (FIFO) circuit, as shown in Figure 7.

5. Experimental Results and Analysis

We used many pieces of IM-PUF PCB as designed to measure the upper cut-off frequency in different situations. Figure 8 shows the experiment setup for the IM-PUF measurements. The test platform mainly includes tested PCB board, two MOTECH LPS-305 DC Power Supplies (5 V), SP1461 Type II 300 M Signal Generator, Tektronix MDO3022 200 MHz Oscilloscope, and some wires.

The flow of the experimental measurement is summarized as follows. Four steps are needed in total.

Step 1. Under the peak-to-peak value of 20 mV of the sine wave, measure the voltage amplification values on the original PCB with different frequencies.

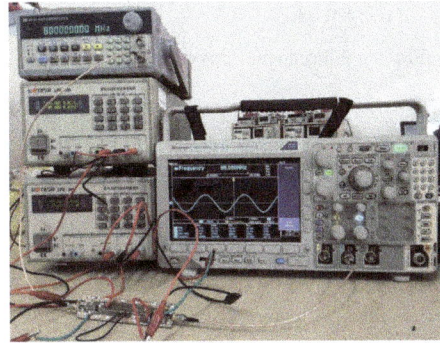

FIGURE 8: The experiment setup for measuring IM-PUF.

Step 2. If the frequency is less than 60 MHz, the output of RMS voltage is about 130 mV.

Step 3. As input frequency increases, voltage values begin to decay.

Step 4. Determine the upper cut-off frequency voltage value as 0.707 times the middle frequency, namely, 91.91 mV; the upper cut-off frequency is 85.6 MHz.

After that, change the length and width of the PUF circuit transmission line, with 7 cm thin wire, 14 cm thin wire, 7 cm thick wire, and 14 cm thick wire in the original circuit on a transmission line. The deviation of 5% more or less than the component's performance shall be allowed, and the fluctuation range of 1 V more or less than power supply shall be allowed. Measure the upper cut-off frequency at 85.5 MHz, 80.4 MHz, 86.5 MHz, and 80.9 MHz. Experimental measurement data is shown in Figure 9. The frequency curve of PUF obviously changed after changing its transmission line. After changing length and width of the transmission line, its frequency changed accordingly. Given the 6 V and 5 V power supply, the cut-off frequencies have nearly equal values with values lower than 100 MHz, as shown in Figure 10. The frequency of the output signal serves as a trigger for a counter. Then, comparing the outputs of counters, the IM-PUF produces a value of 0 or 1.

In statistics, autocorrelation is defined as the correlation among values of random process [30]. In this work, the hypothesis behind calculation of the autocorrelation is that

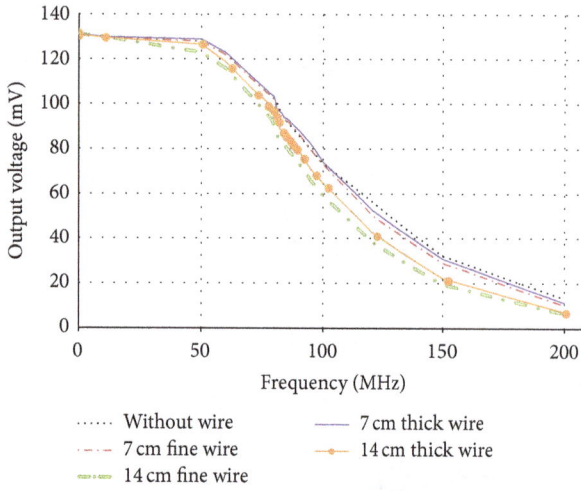

FIGURE 9: Frequency curve of PUF circuit.

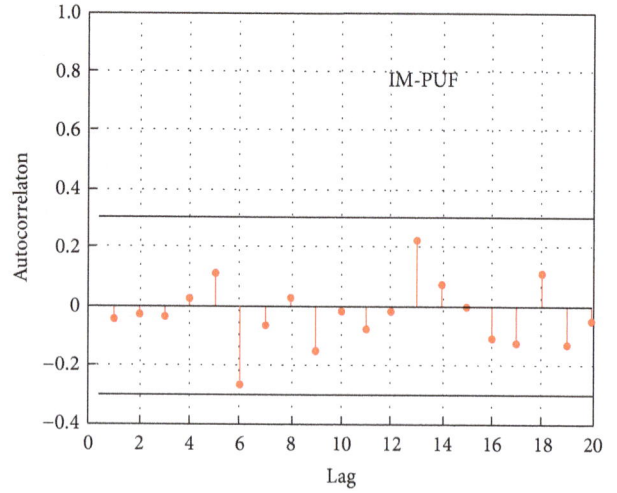

FIGURE 10: IM-PUF with different power supply.

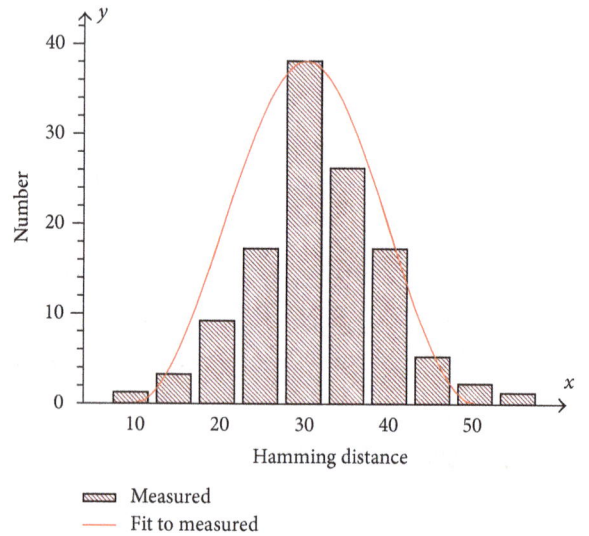

FIGURE 11: Autocorrelation of the IM-PUF output.

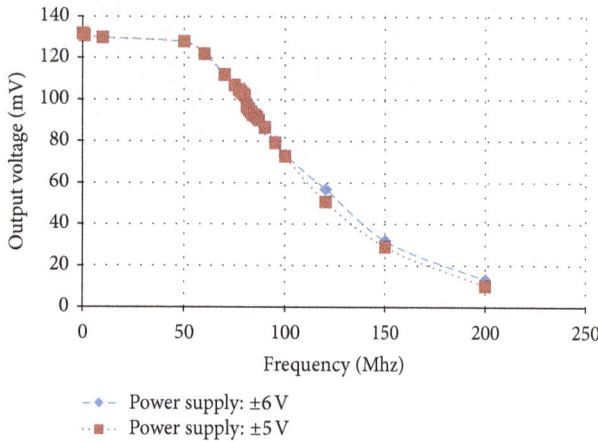

FIGURE 12: Measured hamming distance for IM-PUF.

IM-PUF is a random process. Because IM-PUF is designed according to random variation during PCB manufacturing process, the proposed hypothesis is ok. In other words, the autocorrelation can be used to characterize the performance of antianalysis attack. In the experiment, the sample data of 1# PCB is set as a reference. Figure 11 shows the autocorrelation rates of the IM-PUF circuit. As can be seen, the autocorrelation of the proposed PUF circuit fluctuates between −0.3 and 0.3. The low autocorrelation rates mean that the PUF is resistant to correlation analysis.

The output data of IM-PUF is measured with 60 samples PCB. Recording this data, we use a hamming distance of the IM-PUF output to demonstrate the randomness characteristic. Figure 12 shows the hamming distance of the IM-PUF circuit. As can be seen, the distribution of hamming distance is consistent with standard normal distribution. The Normalized Standard Deviation (σ) of IM-PUF is 0.0611, while the Normalized Standard Deviation of [31, 32] is 0.0818 and 0.0627, respectively. This means the proposed PUF circuit has better randomness characteristic.

TABLE 1: The comparison with other works.

Paper	PUFs type	Frequency (Hz)	Variation source
TNANO, 2015 [8]	MRAM-PUF	—	Material-level
VLSI, 2005 [11]	Arbiter-based PUFs	100 M	Circuit-level
CHES, 2010 [12]	Glitch PUFs	50 M	Circuit-level
IFS, 2011 [13]	Time Bounded	20 M	Circuit-level
Scientific reports, 2015 [14]	mrS-PUFs	25 M	Material-level
This work	IM-PUFs	100 M	Board-level

Key characteristics of implemented PUFs are summarized in Table 1. Our design is the first reported in board-level

PUFs that can read out an ID at each PCB. Because the variation of transmission line will lead to impedance mismatch and signal reflection, the high-frequency signal processing is very hard in board-level. The 100 M frequency of IM-PUF is closed to the best circuit about arbiter-based PUFs. The total number of possible IM-PUF data depends on the number of the transmission lines (2^N). There are so much possible transmission lines in PCB that it is feasible for an adversary to guess the output. And also, under a fixed input, the output data varies across different PCBs, because the IM-PUF responses are designed to be sensitive to circuit delays which are determined by process variation in wires. Since process variation is beyond the manufacturers' control, no one can physically clone the IM-PUF. So the impedance mismatch Physical Unclonable Functions eliminate the problems of a board-level physical feature recognition technique.

6. Conclusion

We proposed a new kind of PUF circuit design based on PCB. Imposing the impedance matching characteristic of high-frequency PCB circuit, changing length and width of the transmission line, causes the output of the upper cut-off frequency to be different. With this frequency as a counter clock frequency, the output produced by the deviation frequency circuit is different, and with the same time T, the count value is also different. Due to the difference in count value, the comparison circuit would output a binary response signal. The function of this PUF on a PCB is unpredictable, so the security of the IoT will be improved.

Competing Interests

The authors declare that they have no competing interests.

Acknowledgments

This work was supported by the National Natural Science Foundation of China (nos. 61404076, 61474068, and 61274132); the Zhejiang Provincial Natural Science Foundation of China (no. LQ14F040001); The S&T Plan of Zhejiang Provincial Science and Technology Department (no. 2015C31010); China Spark Program (no. 2015GA701053); and Programs Supported by Ningbo Natural Science Foundation (nos. 2014A610148 and 2015A610107).

References

[1] F. Ganz, D. Puschmann, P. Barnaghi, and F. Carrez, "A practical evaluation of information processing and abstraction techniques for the internet of things," *IEEE Internet of Things Journal*, vol. 2, no. 4, pp. 340–354, 2015.

[2] M. Görges, G. A. Dumont, C. L. Petersen, and J. M. Ansermino, "Using machine-to-machine/'Internet of Things' communication to simplify medical device information exchange," in *Proceedings of the International Conference on the Internet of Things (IoT '14)*, pp. 49–54, Cambridge, Mass, USA, October 2014.

[3] http://www.gartner.com/newsroom/id/3291817.

[4] R. Roman, P. Najera, and J. Lopez, "Securing the Internet of things," *Computer*, vol. 44, no. 9, pp. 51–58, 2011.

[5] http://ahmedbanafa.blogspot.com/2015/03/internet-of-things-iot-security-privacy.html.

[6] A. P. Johnson, R. S. Chakraborty, and D. Mukhopadhyay, "A PUF-enabled secure architecture for FPGA-based IoT applications," *IEEE Transactions on Multi-Scale Computing Systems*, vol. 1, no. 2, pp. 110–122, 2015.

[7] D. Mukhopadhyay, "PUFs as promising tools for security in Internet of things," *IEEE Design & Test*, vol. 33, no. 3, pp. 103–115, 2016.

[8] J. Das, K. Scott, S. Rajaram, D. Burgett, and S. Bhanja, "MRAM PUF: a novel geometry based magnetic PUF with integrated CMOS," *IEEE Transactions on Nanotechnology*, vol. 14, no. 3, pp. 436–443, 2015.

[9] Y. Zhang, P. Wang, Y. Li, X. Zhang, Z. Yu, and Y. Fan, "Model and physical implementation of multi-port PUF in 65 nm CMOS," *International Journal of Electronics*, vol. 100, no. 1, pp. 112–125, 2013.

[10] M. T. Rahman, F. Rahman, D. Forte, and M. Tehranipoor, "An aging-resistant RO-PUF for reliable key generation," *IEEE Transactions on Emerging Topics in Computing*, vol. 4, no. 3, pp. 335–348, 2016.

[11] D. Lim, J. W. Lee, B. Gassend, G. E. Suh, M. Van Dijk, and S. Devadas, "Extracting secret keys from integrated circuits," *IEEE Transactions on Very Large Scale Integration (VLSI) Systems*, vol. 13, no. 10, pp. 1200–1205, 2005.

[12] D. Suzuki and K. Shimizu, "The glitch PUF: a new delay-PUF architecture exploiting glitch shapes," in *Proceedings of the 12th International Conference on Cryptographic Hardware and Embedded Systems (CHES '10)*, pp. 366–382, San Diego, Calif, USA, 2010.

[13] M. Majzoobi and F. Koushanfar, "Time-bounded authentication of FPGAs," *IEEE Transactions on Information Forensics and Security*, vol. 6, no. 3, pp. 1123–1135, 2011.

[14] Y. Gao, D. C. Ranasinghe, S. F. Al-Sarawi, O. Kavehei, and D. Abbott, "Memristive crypto primitive for building highly secure physical unclonable functions," *Scientific Reports*, vol. 5, Article ID 12785, 2015.

[15] T. Machida, D. Yamamoto, M. Iwamoto, and K. Sakiyama, "A new arbiter PUF for enhancing unpredictability on FPGA," *The Scientific World Journal*, vol. 2015, Article ID 864812, 13 pages, 2015.

[16] M. Wan, Z. He, S. Han, K. Dai, and X. Zou, "An invasive-attack-resistant PUF based on switched-capacitor circuit," *IEEE Transactions on Circuits and Systems. I. Regular Papers*, vol. 62, no. 8, pp. 2024–2034, 2015.

[17] D. P. Sahoo, P. H. Nguyen, D. Mukhopadhyay, and R. S. Chakraborty, "A case of lightweight PUF constructions: cryptanalysis and machine learning attacks," *IEEE Transactions on Computer-Aided Design of Integrated Circuits and Systems*, vol. 34, no. 8, pp. 1334–1343, 2015.

[18] G. T. Becker, A. Wild, and T. Güneysu, "Security analysis of index-based syndrome coding for PUF-based key generation," in *Proceedings of the IEEE International Symposium on Hardware-Oriented Security and Trust (HOST '15)*, pp. 20–25, IEEE, Washington, DC, USA, May 2015.

[19] Y. J. Peng, Y. G. He, J. R. Guo, S. H. Wang, and B. F. Cao, "Study for signal reflections in transmission lines," *Modern Electronic Technology*, vol. 30, no. 21, pp. 179–184, 2007.

[20] C. D. Wu, *A Study On Signal Integrity of High-Speed Digital Design*, Xi'an Electronic and Science University, Xi'an, China, 2005.

[21] L. Chen, "Analysis of reflection of signal integrity of transmission line," *Industry and Minc Automation*, vol. 40, no. 3, pp. 49–52, 2014.

[22] G. Z. Wen and J. D. Tan, "Impedance matching on transmission line," *Modern Electronics Technique*, vol. 29, no. 10, pp. 140–142, 2006.

[23] X. Y. Chen, K. Li, T. Dan, and M. D. Chen, "Model and calculation of microstrip multi capacitor load impedance matching," *Information and Electronic Engineering*, vol. 2, no. 2, pp. 106–108, 2004.

[24] THS3001 Datasheet, http://www.ti.com/lit/ds/symlink/ths3001-die.pdf.

[25] S. B. Tong and C. Y. Hua, *Analog Electronic Technology Foundation*, Higher Education Press, Beijing, China, 2006.

[26] X. Su, "Analysis and simulation of impedance matching in transmission line of high speed circuits," *Coal Technology*, vol. 30, no. 10, pp. 38–40, 2011.

[27] P. Jiang, *The Research of Board-Level Signal Integrity, Power Integrity and Electromagnetic Interference*, Inner Mongolia University, Hohhot, China, 2015.

[28] C. B. Zheng, *Analysis and Design of PCB Signal Integrity*, 2008.

[29] OPA847 Datasheet, http://www.ti.com/lit/ds/symlink/opa847.pdf.

[30] https://en.wikipedia.org/wiki/Autocorrelation.

[31] K. Yang, Q. Dong, D. Blaauw, and D. Sylvester, "14.2 A physically unclonable function with BER < 10^{-8} for robust chip authentication using oscillator collapse in 40 nm CMOS," in *Proceedings of the 2015 IEEE International Solid-State Circuits Conference (ISSCC '15)*, pp. 1–3, San Francisco, Calif, USA, 2015.

[32] Y. Su, J. Holleman, and B. P. Otis, "A digital 1.6 pJ/bit chip identification circuit using process variations," *IEEE Journal of Solid-State Circuits*, vol. 43, no. 1, pp. 69–77, 2008.

6

The Design and Life Test of a Multifunction Power Amplifier for Space Application

Xiuqin Xu, Hui Xu, Yongheng Shang, Zhiyu Wang, Yang Wang, Liping Wang, Hao Luo, Zhengliang Huang, and Faxin Yu

School of Aeronautics and Astronautics, Zhejiang University, Hangzhou 310027, China

Correspondence should be addressed to Yongheng Shang; yh_shang@zju.edu.cn

Academic Editor: Gerard Ghibaudo

A new multifunction power amplifier (MFPA) is designed and fabricated for the application of point-to-point K-Band backhaul TR module. A DC temperature life test was performed to model the up-limit temperature effect of the designed MFPA under space application. After 240 hours of 100°C life test, the test results illustrate that the designed MFPA has only slight power degradation at the saturation region without change of the linear gain. The general performance of the designed MFPA satisfies the requirement of the application scenario.

1. Introduction

As one of the most popular III-V binary compound semiconductor devices, GaAs monolithic microwave integrated circuits (MMIC) have attracted many attentions with the advantages of high electron mobility, high cutoff frequency, low noise figure, and good output power and performed superior capabilities for commercial, military, and space applications [1–5]. Regarding the stringent requirements from the market, the most distinct change of GaAs MMIC is the shrinkage of the device size. It brings on high device reliability which becomes a priority for GaAs MMIC device designers. In order to ensure that the reliability of the designed devices fits the proposed application, the Joint Electron Device Engineering Council (JEDEC) was formed, and many standards with respect to semiconductor device fabrication and test were published [6–9] along with other standards for specific applications, such as mil-std-886F. In general, with respect to the application scenario, different test methods are used. The high gate-drain voltage stress used for large signal test is provided to evaluate hot electron effect [10–12]. The test of hydrogen effects which caused the degradation of maximum drain current and surface corrosion is introduced in [13–15].

In this paper, a multifunction power amplifier (MFPA) for the application of point-to-point K-Band backhaul satellite communication is proposed. Due to the space application environment, only the thermal reliability is considered in here at the present stage. The high temperature life test results are reported, under quiescent DC stress with the bias voltages of $V_{DS} = 5\,V$, $V_{GS} = -1\,V$, 100°C of temperature, and 240 hours of time period.

In the following, Section 2 introduces the TR module and MFPA circuit design; Section 3 states the test related setup and test results; following that is the discussion in Section 4; finally, the conclusion is given in Section 5.

2. Circuit Design

The proposed MFPA is used for a point-to-point K-Band backhaul TR module. Each of the designed TR modules includes 4 channels; each channel consists of a vector modulator (VM) for phase and amplitude modulation, a driver amplifier (DRV), an MFPA, a low noise amplifier (LNA), and a CMOS chip used to receive the control signal from the beam forming computer and provide the bias to different chips with respect to the control signal. At the transmission/receiving mode of the MFPA, the RF signal is delivered/collected to/from the left polarized antenna unit at the end of each channel via a driver amplifier (DRV) and

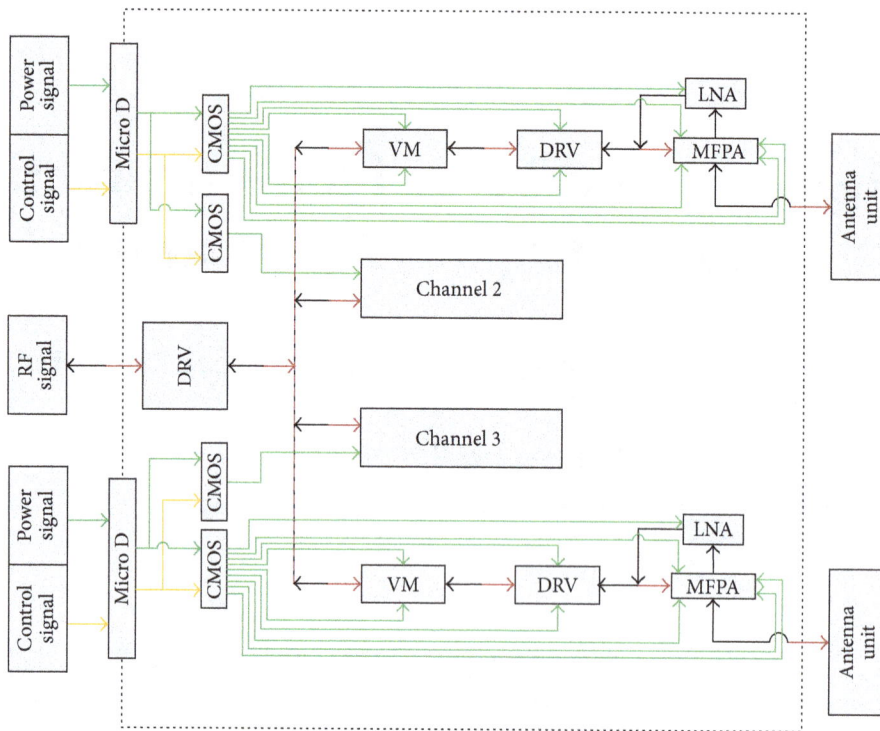

FIGURE 1: Schematic diagram of the designed TR module hosting the proposed MFPA.

a power splitter/combiner. The schematic design of the TR module is illustrated in Figure 1.

The low temperature cofired ceramic (LTCC) technology is applied to the housing of the TR module [16–18]. It has 16 layers, which are used to cope with different signals and the ground plane. The schematic diagram of the layer distribution is illustrated in Figure 2. As shown, the RF signal is transmitted on layer 12 via a 1 : 4 power splitter (or 4 : 1 power combiner under receive mode). The RF layer is isolated by layers 10, 11, 13, and 14 to prevent the RF interference to DC signal and provide large enough ground for the RF signal. The DC signals are placed on the top, from layer 1 to layer 8. Via holes are used to send the DC signals to the corresponding pads in the RF layer for bias and power supply purpose. The photography of the designed LTCC house cavity is shown in Figure 3.

For the proposed MFPA, it consists of a single pole double throw (SPDT) switch, a transmission channel, and a receiving channel, activating at the transmission mode and the receiving mode of the TR module, respectively. The transmission channel consists of a two-stage amplifier with a $2 \times 40\,\mu m$ field effect transistor (FET) as its first stage and a $2 \times 60\,\mu m$ FET as its second stage. The receiving channel only consists of a 50 Ω transmission line. The SPDT switch consists of a $2 \times 80\,\mu m$ FET and is employed to switch between the transmission and receiving channel. The layout of the designed MFPA is illustrated in Figure 4.

A $0.25\,\mu m$ GaAs pHEMT technology is used for MFPA fabrication. Typical parameters of the MFPA are measured. The drain current of the MFPA is 360 mA/mm at $V_{GS} = 0$ V and reaches its maximum, 490 mA/mm, at $V_{GS} = 0.5$ V. The

channel-on-resistance is 0.94 Ω·mm, and the transconductance is 410 mS/mm. The photograph of the MFPA is shown in Figure 5.

3. Test Setup and Life Test

In order to evaluate the designed MFPA, a three-port test fixture is designed to host the device under test (DUT), shown in Figure 6. In order to retrieve the insertion loss of the DUT and remove the dissipations due to the three microstrip feeding lines and the bonding wires, following test procedures are applied to calibrate the test fixture.

Firstly, the two well-aligned microstrip feed lines connecting port 1 and port 3, respectively, are fabricated by mounting a single microstrip transmission line which connects both ports and removing the middle section of it, whose length is the same as the DUT. Then, one pair of $25\,\mu m$ gold bonding wires is used to link the two feed lines. After measuring the insertion loss IL_{31} between port 1 and port 3, the bonding wires are removed and another pair of bonding wires is used to link the feed lines connecting port 1 and port 2. Again, the insertion loss IL_{21} between port 1 and port 2 is measured. Finally, the DUT is assembled, and the S-parameters of the test system are measured. The insertion loss of the DUT is then achieved by compensating S21 and S31 with IL_{21} and IL_{31}.

Before performing the life test, the DUT assembled with the test fixture is sealed in a nitrogen environment using a Ni/Au plated Kovar lid and a parallel seam sealer as an individual packaged sample, shown in Figure 7. Then, all the

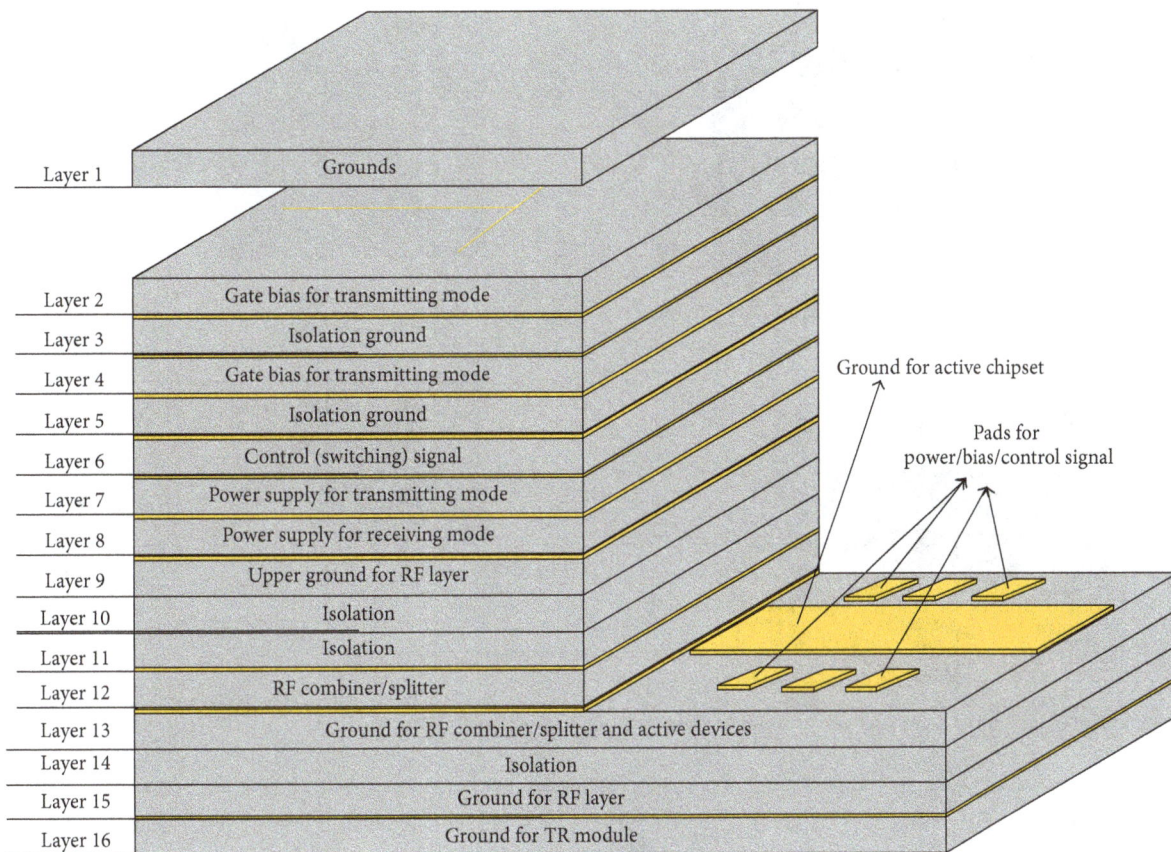

Layer 1	Grounds
Layer 2	Gate bias for transmitting mode
Layer 3	Isolation ground
Layer 4	Gate bias for transmitting mode
Layer 5	Isolation ground
Layer 6	Control (switching) signal
Layer 7	Power supply for transmitting mode
Layer 8	Power supply for receiving mode
Layer 9	Upper ground for RF layer
Layer 10	Isolation
Layer 11	Isolation
Layer 12	RF combiner/splitter
Layer 13	Ground for RF combiner/splitter and active devices
Layer 14	Isolation
Layer 15	Ground for RF layer
Layer 16	Ground for TR module

Ground for active chipset

Pads for power/bias/control signal

FIGURE 2: The housing cavity of the MFPA.

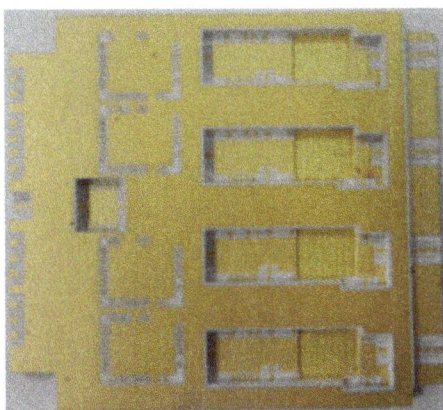

FIGURE 3: The realization of the LTCC housing cavity.

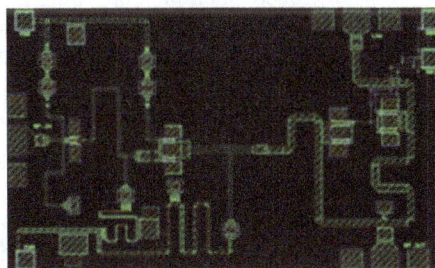

FIGURE 4: Layout of the proposed MFPA circuit.

packaged samples are fixed onto a mother board as illustrated in Figure 8.

After the packaging, the life test is performed based on a high temperature stress with DC bias. The temperature is set to 100°C and lasts for 240 hours. The same biases as normal operation are set as $V_{DS} = 5$ V and $V_{GS} = -1$ V. The output power and gain against input power before and after the temperature stress are illustrated in Figure 9. When the input power is low, only tiny changes have been observed after 240 hours of temperature stress. As the input power increases, the curves before and after the temperature stress tend to deviate from each other. This deviation reveals the same effect with the one introduced in [19], the saturation gain of which is degraded without change of the linear gain after the RF life test under the condition of 150°C channel temperature, 5 V drain voltage, and 2000-hour test time.

The insertion loss and isolation against frequency before and after temperature stress are illustrated in Figure 10. The isolation and insertion loss are similar before and after the life test. In conclusion, the test results demonstrate that the designed MFPA is insensitive to the life test which means that it is qualified to be used in space applications.

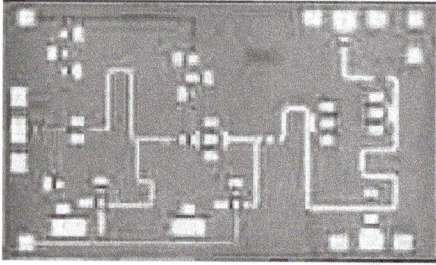

FIGURE 5: Photo of the fabricated MFPA (chip size: 2.5 mm × 1.4 mm).

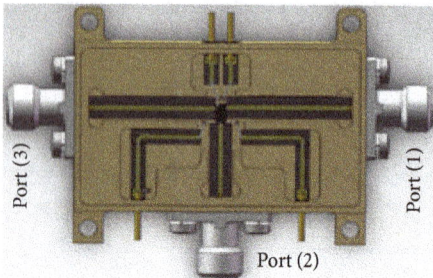

FIGURE 6: The designed test fixture for the proposed MFPA.

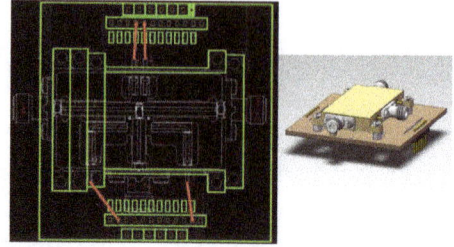

FIGURE 7: The illustration of packaged DUT.

FIGURE 8: The mother board of the packaged DUT.

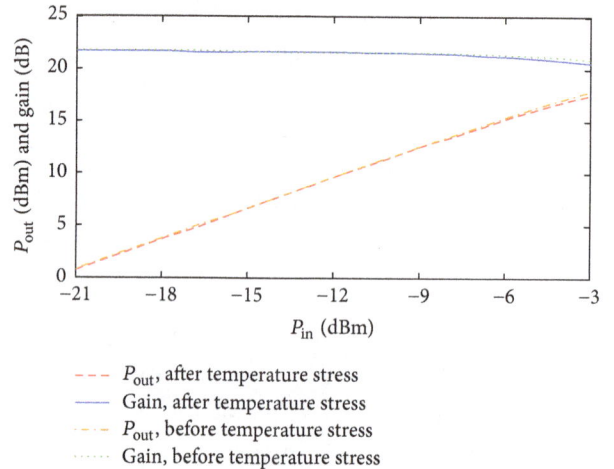

P_{out}, after temperature stress
Gain, after temperature stress
P_{out}, before temperature stress
Gain, before temperature stress

FIGURE 9: The output power and gain against input power before (blue color traces) and after (red color traces) temperature stress under transmission mode.

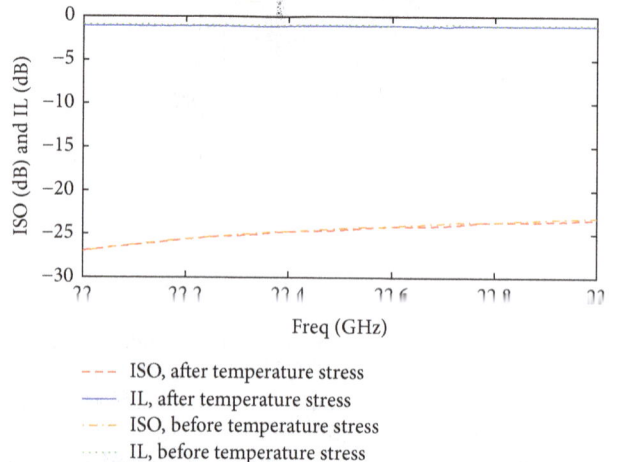

ISO, after temperature stress
IL, after temperature stress
ISO, before temperature stress
IL, before temperature stress

FIGURE 10: The insertion loss and isolation against frequency before (blue color traces) and after (red color traces) temperature stress.

4. Discussion

As illustrated in Figure 9, there is a small degradation at the light saturation (compression region) region. This indicates that the device has some sort of change within the device physically. Such phenomenon could be caused by several GaAs-related degradations such as "hot electron" phenomenon [20–23], metal-semiconductor induced interdiffusion [15, 24], or humidity and hydrogen related degradation [13, 14]. However, in case of hot electron, the holes produced by the impact ionization effects can be trapped in the passivation layer. This leads to the degradation of output power, but according to the authors in [10, 22], the hot electron effect degrades the output power not only in the saturation region but also in the linear region. Therefore, this has ruled out the hot electron to be the major effect that causes the saturation degradation behavior shown in Figure 9.

In case of metal-semiconductor induced interdiffusion, it mainly occurs when the device is under high temperature. Such interdiffusion phenomenon changes the surface state between the metal-semiconductor interface and the buffer layer above the channel layer. It further leads to the change of the barrier height of the metal-semiconductor junction, the channel carrier density, and gate resistivity. All these effects act as the degradation of the drain current and output power. At the linear region, due to the effective number of carriers participating in the drain, current is less than the overall number of the carriers. Therefore, this metal-semiconductor interdiffusion has much less effect on the effective carrier density, which is the reason that the linear region is unchanged after the stress.

With respect to humidity and hydrogen related degradation, the main signatures are the reduction of drain current, which leads to the reduction of output power, especially when the DUT is stressed with high temperature condition combined with humidity and hydrogen effects [25, 26]. The possibility of humidity and hydrogen within the hermetic package could cause the adhesion phenomenon with respect to the passivation layer and the damage of the surface state of the semiconductor layer. For the case of our packaging, the sealing process and the housing cavity are checked to avoid the humidity and hydrogen related effects. However, there is no guarantee or evidence to support the absence of such degradation during this life test. More experiments used by the authors in [19] can be used in the near future to estimate the resistivity of humidity and hydrogen related degradation effect for the designed MFPA. Furthermore, the humidity and hydrogen occur on the ground when the device module is prepared. Therefore, the techniques for reducing such effects are highly depending on the module sealing techniques and the components designed techniques. In case of humidity, the following method can be applied to minimize the effects of humidity: (1) Preheating is the most used method to reduce the possibility of humidity for the components and the module cavity. (2) Sealing environmental control is another efficient way of controlling the humidity condition by using temperature control and protection gas. (3) Sealing finishing is to take care of all the interconnections between all the pins and the module cavity and ensure that there is no leakage. For hydrogen effect, the following techniques can be employed to reduce the effect: (1) sealing material selection, to choose the material which has a low hydrogen absorption concentration such as Al; (2) components process technique, to improve the hydrogen resistance by using carefully designed process technique such as using Al as the gate material for GaAs chip; (3) module cavity structure, to optimize the design of the structure in order to let the hydrogen flow within the cavity (this is due to the fact that the flowing hydrogen will have no harm to the components); (4) using of hydrogen absorber, to reduce the hydrogen level under the safe level by placing hydrogen absorber material within the module cavity [15, 27, 28].

In summary, the observed degradation of the output power at saturation region is possibly caused by the metal-semiconductor interdiffusion and/or the humidity and hydrogen related degradation. The true nature of this phenomenon can be further revealed with more tests such as longer temperature life test and thermal shock. The observation techniques such as transmission electron microscopy (TEM) can be used to analyze the change of the pHEMT unit of the designed MFPA with respect to the test conditions and the changing of electrical characteristic, in order to find the relations between the stress condition, the changing of physical, and electrical characteristics of the designed MFPA fully.

5. Conclusion

This paper introduced the design and life test of a K-band MFPA for the application of point-to-point K-Band backhaul TR module. The design of the TR module using LTCC technology and the MFPA using 0.25 μm GaAs MMIC process is provided. After the 240 hours of life test, only slight degradation at saturation region was observed. This indicated that the designed MFPA are suitable for the proposed spatial application scenario. The possible reasons of the degradation phenomenon were discussed in detail. Future work with additional test methods, such as temperature cycling, thermal shocking, and long life test (over 1000 hours), will be performed to further evaluate the designed MFPA.

Competing Interests

The authors declare that they have no competing interests.

Acknowledgments

This work was supported by the National Natural Science Foundation of China under Grant no. 61401395, the Scientific Research Fund of Zhejiang Provincial Education Department under Grant no. Y201533913, Zhejiang Provincial Natural Science Foundation of China under Grant no. LY14F020024, and the Fundamental Research Funds for the Central Universities under Grant nos. 2016QNA4025 and 2016QNA81002.

References

[1] S. Kayali, G. Ponchak, and R. Shaw, *GaAs MMIC Reliability Assurance Guideline for Space Applications*, Jet Propulsion Laboratory, California Institute of Technology Pasadena, Pasadena, Calif, USA, 1996.

[2] Y. C. Chou, D. Leung, R. Grundbacher et al., "Gate metal interdiffusion induced degradation in space-qualified GaAs PHEMTs," *Microelectronics Reliability*, vol. 46, no. 1, pp. 24–40, 2006.

[3] M. Dammann, A. Leuther, F. Benkhelifa, T. Feltgen, and W. Jantz, "Reliability and degradation mechanism of AlGaAs/InGaAs and InAlAs/InGaAs HEMTs," *Physica Status Solidi (A): Applied Research*, vol. 195, no. 1, pp. 81–86, 2003.

[4] H. L. Hartnagel, "III-V compounds for high-temperature operation," *Materials Science and Engineering B*, vol. 46, no. 1–3, pp. 47–51, 1997.

[5] N. Balkan, *Hot Electron in Semiconductors: Physics and Devices*, Clarendon Press, Oxford, UK, 1998.

[6] W. J. Roesch, "The ROCS workshop and 25 years of compound semiconductor reliability," *Microelectronics Reliability*, vol. 51, no. 2, pp. 188–194, 2011.

[7] JEP, "Guidelines for GaAs MMIC life testing," JEP 118, EIA, JEDEC Publication, 1993.

[8] *Reliability Qualification of Power Amplifier Modules*, JESD 237, EIA, JEDEC Publication, 2014.

[9] JEP, "Foundry process qualification guidelines," JEP 001A, EIA, JEDEC Publication, 2014.

[10] M. Borgarino, R. Menozzi, Y. Baeyens, P. Cova, and F. Fantini, "Hot electron degradation of the DC and RF characteristics of AlGaAs/InGaAs/GaAs PHEMT's," *IEEE Transactions on Electron Devices*, vol. 45, no. 2, pp. 366–372, 1998.

[11] H.-K. Huang, C.-P. Chang, M.-P. Houng, and Y.-H. Wang, "Current-dependent hot-electron stresses on InGaP-gated and AlGaAs-gated low noise PHEMTs," *Microelectronics Reliability*, vol. 46, no. 12, pp. 2038–2043, 2006.

[12] J. C. M. Hwang, "Gradual degradation under RF overdrive of MESFETs and PHEMTs," in *Proceedings of the 17th Annual IEEE Gallium Arsenide Integrated Circuit Symposium*, pp. 81–84, San Diego, Calif, USA, November 1995.

[13] A. R. Reisinger, S. B. Adams, and A. A. Immorlica, "Outgassing of hydrogen in an enclosed cavity and ramifications on the reliability of GaAs devices," in *Proceedings of the GaAs Reliability Workshop*, pp. 77–95, Anaheim, Calif, USA, October 1997.

[14] D. C. Eng and J. Scarpulla, "Hydrogen sensitivity of GaAs HEMT amplifiers: the effect of bias mode," in *Proceedings of the GaAs Reliability Workshop*, pp. 89–93, Monterey, Calif, USA, October 1999.

[15] T. Hisaka, H. Sasaki, Y. Nogami et al., "Corrosion-induced degradation of GaAs PHEMTs under operation in high humidity conditions," *Microelectronics Reliability*, vol. 49, no. 12, pp. 1515–1519, 2009.

[16] K. Malecha, T. Maeder, C. Jacq, and P. Ryser, "Structuration of the low temperature co-fired ceramics (LTCC) using novel sacrificial graphite paste with PVA-propylene glycol-glycerol-water vehicle," *Microelectronics Reliability*, vol. 51, no. 4, pp. 805–811, 2011.

[17] D. Nowak and A. Dziedzic, "LTCC package for high temperature applications," *Microelectronics Reliability*, vol. 51, no. 7, pp. 1241–1244, 2011.

[18] T. Maeder, Y. Fournier, J.-B. Coma, N. Craquelin, and P. Ryser, "Integrated SMD pressure/flow/temperature multisensor for compressed air in LTCC technology: thermal flow and temperature sensing," *Microelectronics Reliability*, vol. 51, no. 7, pp. 1245–1249, 2011.

[19] T. Hisaka, Y. Nogami, H. Sasaki et al., "Degradation mechanism of HEMT under large signal operation," in *Proceedings of the IEEE 25th Annual Technical Digest Gallium Arsenide Integrated Circuit (GaAs IC) Symposium*, pp. 67–70, San Diego, Calif, USA, November 2003.

[20] C. Canali, P. Cova, E. De Bortoli et al., "Enhancement and degradation of drain current in pseudomorphic AlGaAs/InGaAs HEMT's induced by hot-electrons," in *Proceedings of the IEEE International Reliability Physics Symposium*, pp. 205–211, Las Vegas, Nev, USA, April 1995.

[21] G. Meneghesso, E. De Bortoli, P. Cova, and R. Menozzi, "On temperature and hot electron induced degradation in AlGaAs/GaAs PM-HEMT's," in *Proceedings of the Workshop on High Performance Electron Devices for Microwave and Optoelectronic Applications*, pp. 136–141, 1995.

[22] R. Menozzi, M. Borgarino, P. Cova, Y. Baeyens, and F. Fantini, "The effect of hot electron stress on the DC and microwave characteristics of AlGaAs/InGaAs/GaAs PHEMTs," *Microelectronics Reliability*, vol. 36, no. 11-12, pp. 1899–1902, 1996.

[23] R. Menozzi, P. Cova, C. Canali, and F. Fantini, "Breakdown walkout in pseudomorphic HEMT's," *IEEE Transactions on Electron Devices*, vol. 43, no. 4, pp. 543–546, 1996.

[24] D. J. Cheney, E. A. Douglas, L. Liu et al., "Degradation mechanisms for GaN and GaAs high speed transistors," *Materials*, vol. 5, no. 12, pp. 2498–2520, 2012.

[25] T. Hisaka, Y. Aihara, Y. Nogami et al., "Degradation mechanisms of GaAs PHEMTs in high humidity conditions," in *Proceedings of the Reliability of Compound Semiconductors Workshop (ROCS '04)*, pp. 81–88, October 2004.

[26] W. J. Roesch, "Compound semiconductor activation energy in humidity," *Microelectronics Reliability*, vol. 46, no. 8, pp. 1238–1246, 2006.

[27] W. Guo and Q. Ge, "Investigation on the moisture resistance of Plastic IC," *Electronics and Packaging*, vol. 5, no. 4, pp. 16–19, 2005.

[28] W. Wu, H. Bai, Y. Liu, X. Sun, and Y. Song, "The generation and control of hydrogen in sealed devices," *Electronics and Packaging*, vol. 9, no. 8, pp. 34–37, 2009.

7

Microwave Impedance Spectroscopy and Temperature Effects on the Electrical Properties of Au/BN/C Interfaces

Hazem K. Khanfar,[1] A. F. Qasrawi,[2] and Yasmeen Kh. Ghannam[2]

[1]*Department of Telecommunication Engineering, Arab-American University, Jenin, State of Palestine*
[2]*Department of Physics, Arab-American University, Jenin, State of Palestine*

Correspondence should be addressed to Hazem K. Khanfar; hazem.khanfar@aauj.edu

Academic Editor: Gerard Ghibaudo

In the current study, an Au/BN/C microwave back-to-back Schottky device is designed and characterized. The device morphology and roughness were evaluated by means of scanning electron and atomic force microscopy. As verified by the Richardson–Schottky current conduction transport mechanism which is well fitted to the experimental data, the temperature dependence of the current-voltage characteristics of the devices is dominated by the electric field assisted thermionic emission of charge carriers over a barrier height of ~0.87 eV and depletion region width of ~1.1 μm. Both the depletion width and barrier height followed an increasing trend with increasing temperature. On the other hand, the alternating current conductivity analysis which was carried out in the frequency range of 100–1400 MHz revealed the domination of the phonon assisted quantum mechanical tunneling (hopping) of charge carriers through correlated barriers (CBH). In addition, the impedance and power spectral studies carried out in the gigahertz-frequency domain revealed a resonance-antiresonance feature at frequency of ~1.6 GHz. The microwave power spectra of this device revealed an ideal band stop filter of notch frequency of ~1.6 GHz. The ac signal analysis of this device displays promising characteristics for using this device as wave traps.

1. Introduction

Metal-semiconductor-metal (MSM) devices have been in the focus of scientists for decades due to their wide range of applications. They have been used for plasmonic hot-electron photodetection applications [1] like solar blind ultraviolet photodetectors [2]. They are also used as a bistable resistor for digital imaging applications [3]. In addition, the MSM devices are employed as varactors, which can reach cutoff frequency up to 308 GHz [4]. One more accountable add for the applications of these devices is the use as nanoplasmonic waveguides which can find applications in integrated nanophotonic circuits [5].

For ultraviolet applications, wide band gap materials are preferably used to form MSM devices. As a fast-response flexible ultraviolet photodetector employing a metal-semiconductor-metal structure InGaZnO photodiode is fabricated [6]. The flexible photodetector shows relatively good photoresponse characteristics before and after bending and retains good folding reproducibility after repeated bending up to 500 cycles. More importantly, it shows a fast speed with response and recovery times of 0.8 ms and 2.0 ms, 33.8 ms, being much faster than that of the reported flexible ultraviolet detectors [6]. Boron nitride is also used in the fabrication of MSM devices. Gold-hexagonal boron nitride-gold tunnel junctions are used for conversion of electrons to free-space photons, mediated by resonant slot antennas [7]. They achieved polarized, directional, and resonantly enhanced light emission from inelastic electron tunneling [7]. Also, boron nitride based MSM are highly promising for realizing highly sensitive solid-state thermal neutron detectors with expected advantages resulting from semiconductor technologies, including compact size, light weight, ability to integrate with other functional devices, and low cost [4]. On the other hand, in our previous work [8], an Ag/BN/Ni microwave rejection-band filter was

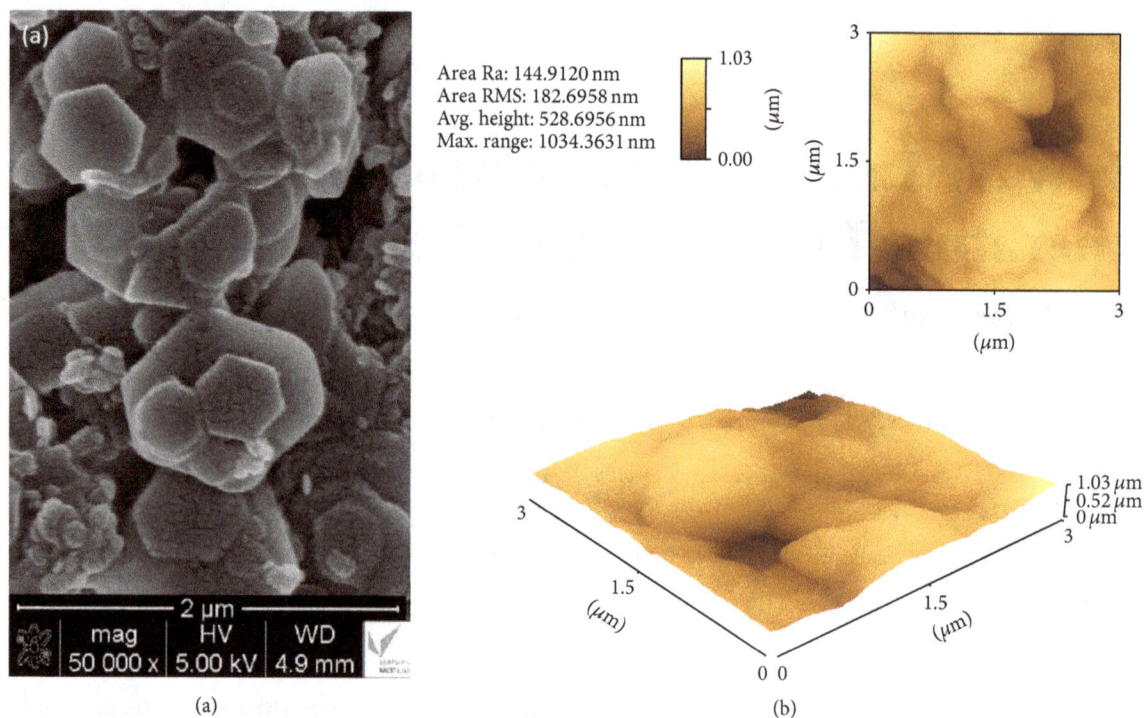

Area Ra: 144.9120 nm
Area RMS: 182.6958 nm
Avg. height: 528.6956 nm
Max. range: 1034.3631 nm

(a)   (b)

FIGURE 1: (a) The scanning electron microscopic image for the BN films being magnified 5000 times; (b) the atomic force microscopy images for the BN surface.

designed and characterized. The device was characterized by current-voltage (I-V) characteristics, differential resistance, frequency, and power dependence in the frequency range 1.0–3.0 GHz [8]. The above-mentioned technological applications of boron nitride indicate its novelty for microwave sensing. Because the BN exhibits a very high work function of 9.64 eV owing to its wide band gap it easily establishes Schottky barrier with most metals. In addition, recalling that the Ag/BN/Ni device is novel as it was able to screen very weak microwave signals which suits mobile amplifiers at 2.9 GHz [8], here in this work we aim to fabricate new class of these MSM devices for the purpose of optimizing wide range of microwave signals sensing. Particularly, an Au/BN/C MSM band stop filter that works at high temperatures up to 100°C is fabricated and characterized. The band stop filter is designed so that it can stop signals at gigahertz frequencies even in high temperature media. The device is characterized by (I-V) characteristics with different temperature in the range of 306–373 K. On the other hand, the device is characterized using an impedance material analyzer and using a 3.0 GHz signal generator and 3.0 GHz spectrum analyzer. The AC current conduction properties of the Au/BN/C device are also explored through the analysis of the frequency dependent electrical conductivity in the frequency range of 100–1800 MHz.

2. Experimental Details

The 1.0 μm gold substrate was thermally evaporated under a vacuum pressure of 5×10^{-5} mbar onto glass slides. The

Alfa Aesar 43773-BN paste is used as a wide band gap semiconductor in the design of Au/BN/C MSM device. The paste, which was prepared at the Alfa Aesar chemical firm (product number 043773) included a typical solution of BN nanopowders (50%) solved in 40% water and 10% aluminum phosphate (AlPO4). Further details about the properties of this material were previously given [9]. The BN liquid was poured over the gold substrate, shacked gently, and left to dry for 24 hours. As determined by a digital micrometer, the thickness of the BN layer was ~400 μm. The scanning electron microscopy images which were recorded with the help of Jeol JSM 7600F microscope and represented in Figure 1(a) display well-shaped hexagonal structured grains of size in the range of 0.60–0.36 μm. On the other hand, the atomic force microscopy (AFM) images for the same samples which are displayed in Figure 1(b) and evaluated for a surface area of 9.0 μm^2 have shown an average surface roughness (R_a) of 145 nm and root mean square average of height deviation (R_{MS}) of ~183 nm. The deviation in the surface height is small as compared to the sample thickness and represents only about 0.046% of the total thickness. This result indicates that the metallic layer (carbon) which is located on the top surface of the BN will represent homogenous and stable electrode area. The Au/BN layer was heat treated at 530 K for 15 minutes to guarantee the absence of the moisture between the grains or through the BN flakes. The contact area of conducting carbon was $\approx 3.14 \times 10^{-2}$ cm^2. The I-V characteristics data were recorded at different temperatures in the range of 306–373 K using Keithley 230 programmable voltage source and Keithley 6485 picoammeter. The temperature was varied in

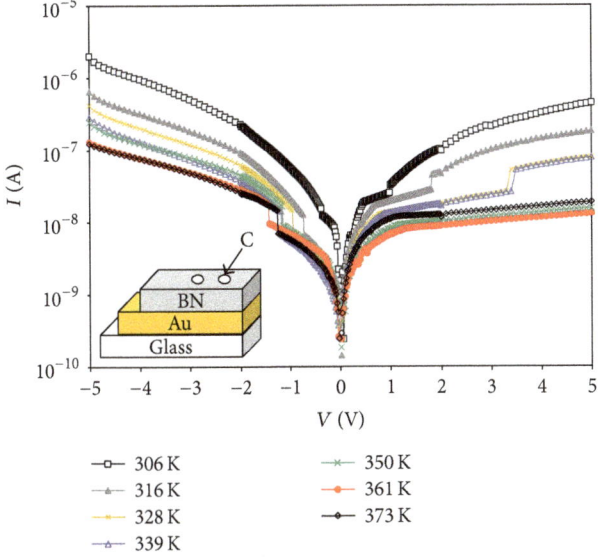

FIGURE 2: The I-V characteristics and geometrical design of Au/BN/C device.

room atmosphere to allow observation of device natural operational conditions. For this reason, the temperature stability with atmosphere was controlled by the most accurate temperatures for at least 10-minute period of time. After each heating cycle within the mentioned range, the current-voltage characteristic curve was compared to the previous cycle to guarantee operation stability. On the other hand, a 4291B RF Impedance/Material Analyzer was used to register the dielectric data. Also, an N9310A RF Signal Generator (9 kHz–3 GHz) with Instek GSP 830 3 GHz Spectrum Analyzer is used to collect the power spectrum data.

3. Results and Discussion

As the metal work functions of the gold and carbon are 5.34 eV and 5.1 eV, respectively, and the BN is of p-type (as determined by the hot probe technique) with barrier height of 9.64 eV [9], the Au/BN represent a Schottky diode and the C/BN represent another Schottky diode. The presentation of both Schottky-type metals on the front and rare part of the BN makes the Au/BN/C device structure represent a metal-semiconductor-metal (MSM) back-to-back Schottky device. Figure 2 shows the I-V characteristics of the Au/BN/C MSM device being registered in the temperature range of 300–375 K. As the curves, which are plotted on a logarithmic scale reads ($I \propto e^{qV/kT}$), increasing the temperature decreases the current values. As, for example, at a forward biasing voltage of 2.0 V and temperature of 306 K, the current value is 0.99 nA, when the temperature is increased to 339 K, the current becomes 0.17 nA and it reaches a value of 0.12 nA at $T = 373$ K. In addition, the forward biasing of device exhibits a temperature dependent switching property. The switching voltage (V_S) from low injected current to high injected current increases with increasing temperature. Namely, V_S increase from 1.0 V, 1.8 V, 3.4 V, to 3.5 V as the temperature

increases from 306, 316, 328, to 339 K, respectively. At higher temperatures above 339 K, the switching feature was not detectable due to measuring range limitations. In the reverse bias direction V_S exhibited negative values of 0.38, 0.68, 0.94, 1.08, 1.18, 1.42, and 1.22 V as the temperature is raised from 306, 316, 328, 339, 350, 361, to 373 K, respectively. Following the previous investigation [8, 10], the I-V characteristics of the forward bias current were analyzed in accordance with the Richardson–Schottky model that assumes direct current electrical conduction through electric field assisted thermionic emission. The current depends on the square root of the voltage and can be presented by the following equation [10]:

$$I = AA^{**}T^2 \exp\left(-\frac{\phi_b}{kT}\right) \exp\left(\frac{-e\beta_s \sqrt{V}}{(kT\sqrt{w_o})}\right). \quad (1)$$

The analysis of the I-V with the help of this equation reveals the depletion width (w_o) from the slope and the value of the field independent barrier height (ϕ_b) to the electron motion from the intercept of the $\ln(I) - \sqrt{V}$ variation through the respective relations:

$$w_o = \left(\frac{-e\beta_s}{kT\left(d\left(\ln\left(I\right)\right)/d\left(\sqrt{V}\right)\right)}\right), \quad (2)$$

$$\phi_b = -\frac{kT}{e}\left[\frac{\ln\left(I\left(V=O\right)\right)}{AA^{**}T^2}\right]. \quad (3)$$

In (2) and (3), A is the device area, $A^{**} = 90$ [11] is Richardson constant, $\beta_s = \sqrt{e/(4\pi\varepsilon_o\varepsilon_r)} = 1.39\times10^{-4}$ (Vcm)$^{1/2}$ is Richardson–Schottky coefficient, and $\varepsilon_r = 7.5$ [11] is the optical dielectric constant of the material. The temperature dependence of the values of the depletion width (w_o) and ϕ_b are shown in Figures 3(a) and 3(b). As it is readable from Figure 3(a), the junction width changes from 1.14 μm to 2.26 and reaches 3.3 μm as the temperature increased from 306 to 339 and reaches 361 K, respectively. Consistently, the values of the barrier height which are displayed in Figure 3(b) increased from 0.872 to 0.913 and reached 0.93 eV as the temperature increased from 306 to 339 and reached 361 K, respectively. The rate of change of the depletion width and energy barrier height with temperature in accordance with the linear fitting of the data (shown by solid lines in Figure 3) are $w_o = 0.037T - 10.20$ (μm) and $\varphi_o(T) = 7.08 \times 10^{-4}T + 0.68$ (eV). The increase in the barrier height and in the depletion width of the Au/BN/C devices with increasing temperature could be ascribed to many physical reasons like the increase in the electron-hole recombination rate, the thermal expansion, and the lattice constants variations with temperature. The hexagonal BN is reported [12] to exhibit a temperature dependent lattice constants of $c = 6.6516$ Å $+ 2.74 \times 10^{-4}T$ (Å°C^{-1}) and $a = 2.50424$ Å $- 7.42 \times 10^{-6}T$ (Å°C^{-1}) $+ 4.79 \times 10^{-9}T^2$ (Å°C^{-2}). The lattice parameters along the c-axis and along the a-axis of the BN polycrystals increased from 6.7338 to 6.7505 Å and from 2.5024 to 2.5384 Å as the temperature increases from 300 to 360 K for each unit cell. On the other hand, the excitonic energy band gap of the hexagonal BN is reported

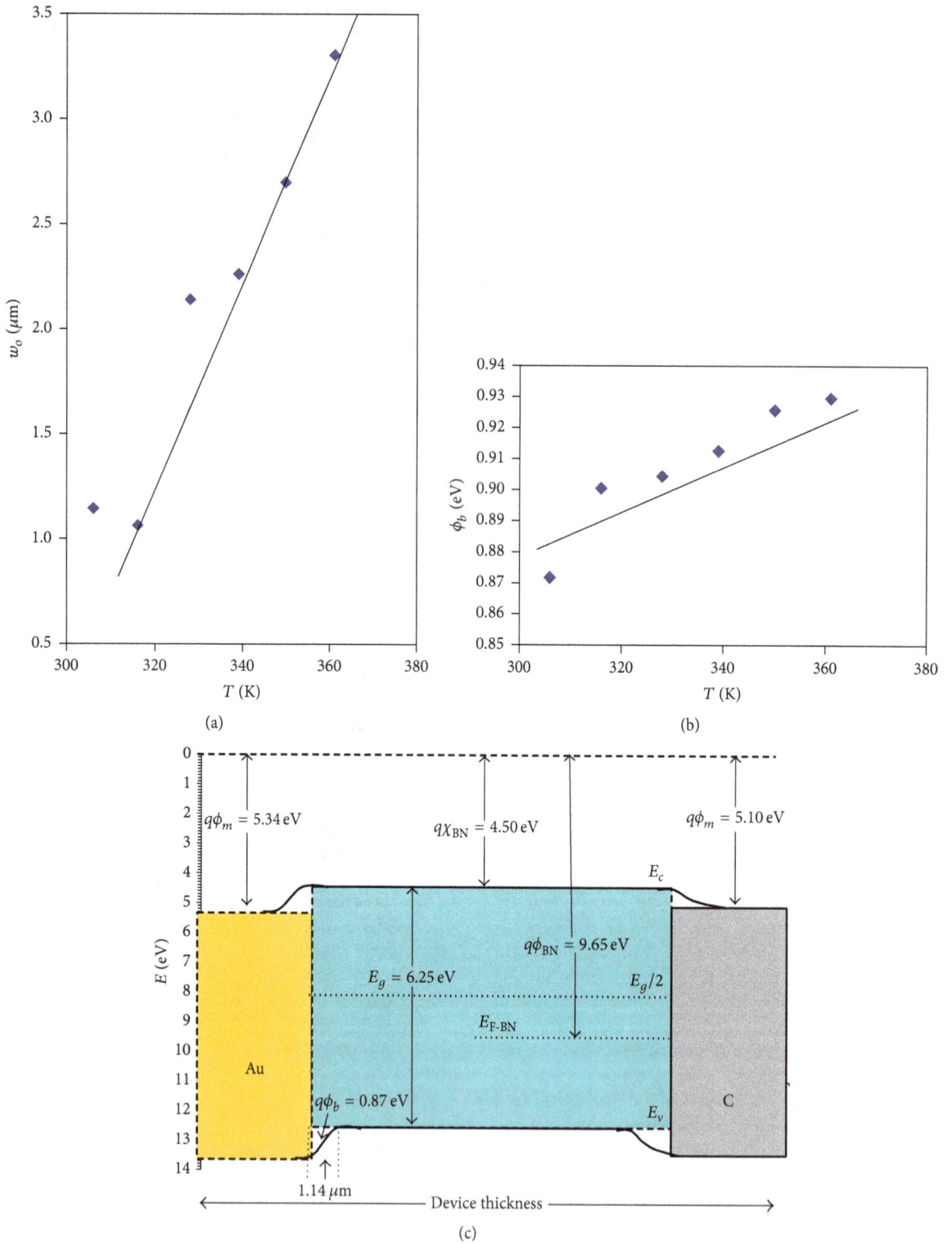

FIGURE 3: The temperature dependence of (a) the depletion region width and (b) the barrier height of the Au/BN/C device. (c) The energy band diagram of the device. In the diagram ϕ_m, $q\chi$, $q\phi_b$, E_c, E_v, $E_{F\text{-}BN}$, and E_g are the metal work function, electron affinity, barrier height, conduction band energy level, valence band energy level, the Fermi level of BN, and the energy band gap, respectively.

to exhibit a negative temperature coefficient of ~4.3 × 10⁻⁴ eV/°K [13]. The latter value indicates an expected effect on the device parameters (φ_b, w_o) associated with the narrowing of the energy band gap of BN. Although the shrinkage in the energy band gap may be of less effect, it indicates an increase in the number of charge carriers that are transported from the valence to the conduction band which means more electron-hole recombination ratios and wider depletion region. Similar characteristics which involve increasing the barrier height with increasing temperature were observed for Au/n-GaAs/n-Ge diodes [14]. The increase in the barrier height with increasing temperature was ascribed to the electron-hole recombination dynamics and to the electric field dependence of the barrier height (similar to our case). In another study, the increase in the barrier height with increasing temperature was also assigned to the existence of double Gaussian distribution of energy barriers represented by the screened mean barrier height in the sample. The double Gaussian distribution of energy barriers results from the barrier height inhomogeneity at the metal/semiconductor (MS) interface.

The energy band diagram of the device is shown in Figure 3(c). The working principle of the Au/BN/C device which represents two Schottky barriers connected back to back is that the biasing of any polarity will put one Schottky barrier in the reverse direction (Au side as a cathode) and the other in the forward direction (C-side as an anode) [10]. While the depletion width of the forward biased diode narrows, the depletion width of the reverse biased diode widens. Thus, the measured width which is presented in Figure 3(a) indicated the effective depletion region width of the device [10]. At particular temperature, 306 K, for example, the appearance of the large reverse current which lowers the rectification ratio and makes the I-V characteristic appear as if it was ohmic is explained by means of the current transport mechanism through the two diodes which have depletion widths W_{D1} and W_{D2}. When one of the diodes is forward biased, its depletion width (w_{D1}) narrows, causing large forward and low reverse current for low applied voltages. The narrow barrier makes the charge transport via tunneling process preferable [10]. In this case the current conduction mechanism is governed by electric field assisted tunneling of charged particles through the narrow barrier consistent with (1). On the other side of the device, the second diode is reverse biased and its depletion width (w_{D2}) widens with increasing voltage. This behavior also lowers the values of the reverse current compared to the forward one. However, as the voltage is further increased, the depletion region of second diode reaches the first region leading to the flat band condition, in which the ohmic nature of contact dominates [10]. Another point that is worth of consideration is that when the reach-through voltage is reached, the exponentially increasing current with increasing applied voltage is also affected by electron-hole generation-recombination, avalanche breakdowns, and image force lowering potentials processes [10] leading to the observed high reverse current.

It is also worth noting that the current-voltage characteristics of the device exhibit low rectification ratio associated with high resistance values (R_s > 10 MΩ). The rectification

ratio varies in the range of 5–11. This behavior which was also observed for some organic/inorganic heterojunctions was assigned to the recombination losses caused by surface trap states [15]. The leakage current is observed to be affected by the distribution of defects and the quality of the interfaces. In addition, the lattice mismatch between the hexagonal and the face centered cubic gold film and diamond structured carbon film could be a main reason as the force confinement of electrons and holes in different spaces.

Figure 4 shows the ac signal analysis of Au/BN/C MSM device being recorded with the help of a 1.8 GHz impedance/material analyzer under no biasing conditions. The parallel capacitance spectra of the device which are illustrated in Figure 4(a) slightly decreased with increasing frequency up to 1594 MHz. For larger frequencies there appear resonance (series) antiresonance (parallel) oscillatory modes at two different frequency regions. The series resonance (f_s) appears at 1620 and 1647 MHz, while the parallel resonance (f_p) is apparent at 1611 and at 1629 MHz. The resonance-antiresonance modes of oscillation in the Au/BN/C device are of importance as they can be used as wave traps, which are sometimes inserted in series with antennas of radio receivers to block the flow of alternating current at the frequency of an interfering station, while allowing other frequencies to pass [16].

The resonance-antiresonance modes of oscillation in the Au/BN/C device can be modeled as two parallel arms: the motional arm and the static arm. The motional arm is represented by a series RLC circuit while the static arm is represented by a shunt capacitor. The reactive impedance (X_m) of the motional arm can be either positive or negative depending on the frequency; also at specific frequency series resonance (f_s) the reactive impedance is minimum ($X_m \approx 0$). The antiresonance f_p (resonance of high impedance) occurs at a particular frequency when the impedance of the motional arm is very high compared to the impedance of the static arm and the device can be modeled as shunt capacitor. At this frequency (f_p) the total admittance (Y) of the device is minimum ($Y = Y_{motional} + Y_{static} \approx 0$) which indicates that the reactive impedance of the static arm and the reactive impedance of the motional arm are equivalent in magnitude and opposite in signs. On the other hand, Figure 4(b) shows the parallel resistance of the device versus the frequency. The resistance decreases from 6365.42 Ω at 100 MHz to 448.14 at 1800 MHz. The trend of the impedance looks like the capacitive part of the device and can be ascribed to the same dynamics. The total effect which is known as the impedance ($-Z-$) spectra of the resistive and reactive parts (inductive and capacitive) is plotted in Figure 4(c). It is noticed that as the frequency increases, $|Z|$ decreases. Namely, $|Z|$ decreases from 548.59 Ω at 100 MHz to 33.94 Ω at 1800 MHz. This indicates that the capacitive part is more dominating than the inductive part of the impedance load.

To give more deep significance to the impedance spectral data, the alternating current conductivity (σ_{ac}) is calculated from the conductance spectra ($1/R_p$). The conductivity spectrum which is displayed in Figure 4(d) exhibits an increasing trend of variation with increasing frequency. In the frequency range of 100–1400 MHz, the $\sigma_{ac} - f$ variation

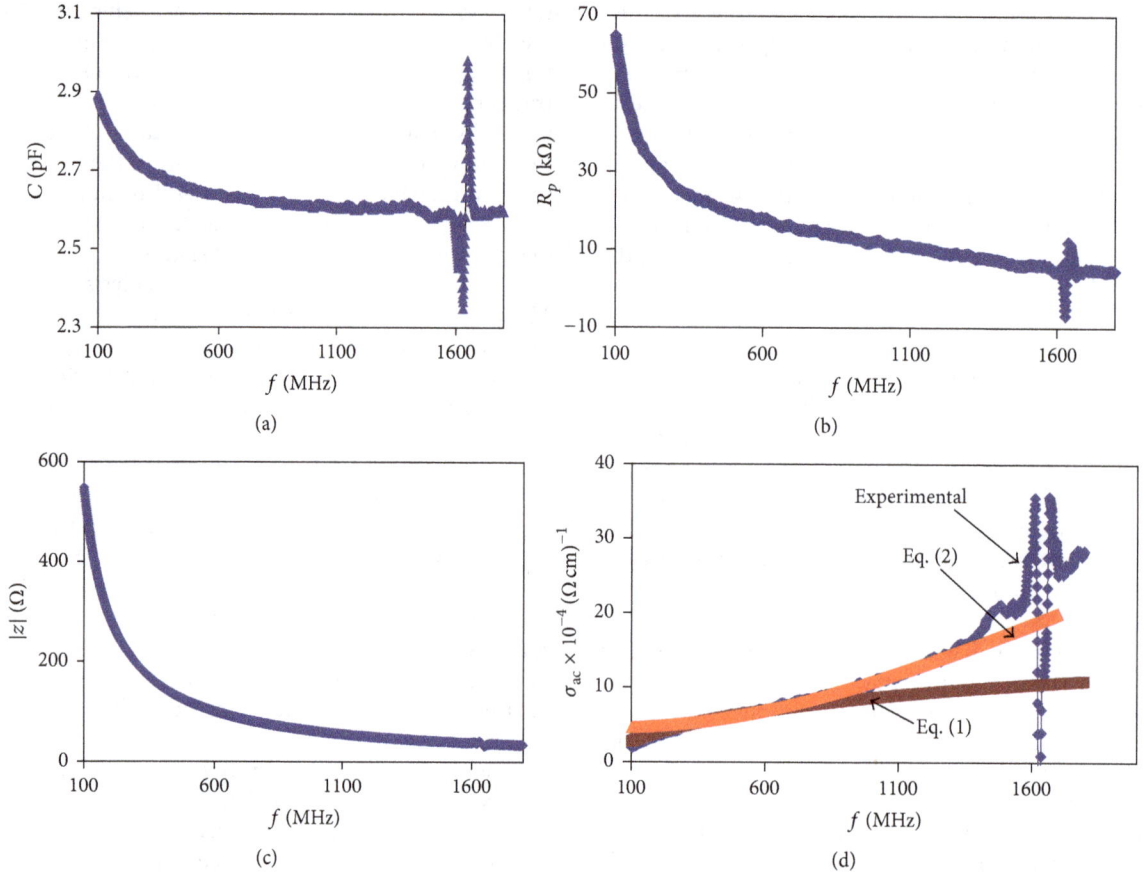

FIGURE 4: (a) The capacitance, (b) the parallel resistance, (c) the total impedance, and (d) the conductivity spectra for the Au/BN/C device.

could be represented by the relation $\sigma_{ac} \propto w^s$. The angular frequency (w) exponent (s) for the Au/BN/C device is 0.79. The increase of σ_{ac} with frequency in the frequency range of 100–1400 MHz is an indication of the domination of the correlated barrier hopping mechanism (CBH) in the samples [17, 18]. In accordance with this model the s value takes the form $s = 1 - 6kT/(W_M + kT\ln(w\tau_o))$ with W_M and τ_o being the maximum barrier height at infinite intersite separation (binding energy of carrier in its localized site) and the relaxation time, respectively. The CBH model assumes that the electrons hop between pairs of localized states at the Fermi level and relates the conductivity to the density of states ($N(E_F)$) at the Fermi level through the relation [17, 18]

$$\sigma_{ac} = \left(\frac{\pi}{3}\right)e^2kTw\xi^5\left(N(E_F)\right)^2\left(\ln\left(\frac{v_{ph}}{w}\right)\right)^4. \quad (4)$$

In this relation, e is the electronic charge, $\xi = 10\,\text{Å}$ is the typical localization length, and $v_{ph} = 10^{10}\,\text{Hz}$ is the phonon frequency. The fitting of the experimental data of conductivity in accordance with (1) that reveals the exact frequency exponent parameter ($s = 0.79$) which is shown in Figure 4(d) as brown colored allowed determining the value of $N(E_F)$ as $5.40 \times 10^{18}\,\text{eV}^{-1}\text{cm}^{-3}$ in the frequency range of 100–850 MHz. The corresponding maximum barrier heights at infinite intersite separation W_M and τ_o are found to be

0.97 eV and 50 ps, respectively. The values of W_M and τ_o are in good consistency with the literature data [17, 18].

Recalling that the AC conductivity increases with increasing frequency as long as the frequency of the field is lower than the charge carrier jump frequency in solids, one may reproduce the experimental data in a wider range of frequency (100–1400 MHz) assuming the existence of the high $\sigma_{ac}(H)$ and low $\sigma_{ac}(L)$ limits to the CBH conductivity in accordance with the relation

$$\sigma_{ac} = \sigma_{ac}(H) + \frac{\sigma_{ac}(L) - \sigma_{ac}(H)}{1 + (w\tau_o)^2}. \quad (5)$$

The fitting curve which is shown as red colored in Figure 4(d) can be reproduced by the substitution of $\sigma_{ac}(L) = 4.50 \times 10^{-4}\,(\Omega\text{cm})^{-1}$ $\sigma_{ac}(H) = 7.00 \times 10^{-3}\,(\Omega\text{cm})^{-1}$. The excellent consistency between the experimental and theoretical data ensures the correctness of the domination of the CBH conduction mechanism in the Au/BN/C devices.

It is worth notifying that the impedance spectra which were discussed in this work related to no electric field biasing. The correlated barrier hopping conduction mechanism is a phonon assisted quantum mechanical tunneling process that takes place between defect states and/or in the depletion regions of the device at the Au/BN and C/BN interfaces. The electric field assisted tunneling conduction mechanism which was determined from the direct current analysis

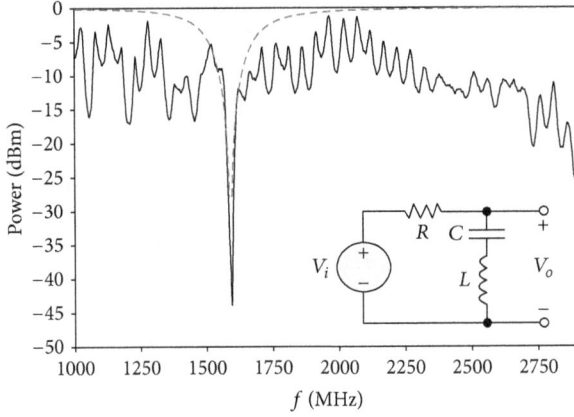

FIGURE 5: Gain-frequency dependence of the Au/BN/C device. Dashed plot is the transfer function of the modeled RLC circuit.

revealed a room temperature barrier height of 0.87 eV. This value is less than the maximum barrier height at infinite intersite separation which is determined as 0.97 eV indicating the correctness of the evaluation procedure of the conduction mechanism. It is also important to notice that, for applied frequencies larger than 1400 MHz, the dynamics of the conductance changes abnormally as a result of the resonance-antiresonance phenomenon which is observed near 1600 MHz. This series-parallel resonance case at this frequency is also observed in many wide gap glassy materials including the glass itself (measured in our laboratory). However, because the glass is coated with metal and metal is directly connected to the fixture of the impedance analyzer, the resonance case is mostly due to the BN itself. Considering the peak which appears at 1600 MHz in the negative parallel resistance mode to be due to the BN, the negative resistance property and the peak itself could have originated from the gate and applied biases as was also observed for the boron nitride-graphene heterostructure [19]. The gate behaves like Fabry-Perot interference which is suppressed by the electron-phonon scattering and the applied biases that represent an in-plane wave vector matching when the Dirac points of electrodes align. It is reported that the hexagonal BN layers can induce an asymmetry in the I-V characteristics which can be modulated by the applied bias [19]. In another work which concerns the negative resistance and negative capacitance in Au/ZnO/n-GaAs Schottky barrier diodes, the increment of negative capacitance (values by high frequency at forward biases) was assigned to the series resistance, interface states, and interfacial layer.

The appearance of the biasing voltage dependent resonance peak in the dielectric spectra which was also observed in the frequency range of 1 K–1 MHz for the graphene oxide-doped praseodymium barium cobalt oxide nanoceramics was attributed to the particular distribution of interface states located between Au and interfacial layer and to the interfacial polarization [20].

As a complementary work, Figure 5 shows the output gain versus input frequency being recorded with the help of a 3.0 GHz power spectrum and 3.0 GHz signal generator;

the output gain demonstrates the band reject filter with central frequency of 1595 MHz. The value of the critical frequency (known as notch frequency) is consistent with that detected from the capacitance spectra at the resonance-antiresonance frequency range. This consistency between the two measuring techniques indicates that the Au/BN/C device is ideal for use as wave traps in the microwave frequency range. The two terminal simple devices which exhibit the behavior of a band reject filter can be modeled using RLC circuit as shown in the inset of Figure 5. The transfer function of the modeled RLC filter circuit is given by [21]

$$G = \frac{\eta^2 + (w^*)^2}{\eta^2 + 2\gamma w^* \eta + (w^*)^2}, \tag{6}$$

where $w^* = 2\pi f^*$ is the angular frequency, $f^* = 1/(2\pi\sqrt{LC})$ is the center frequency, $\eta = 2\pi i f$, and $\gamma = (R/2)\sqrt{C/L}$ is the damping factor. The modeling of the device which was carried out using the MATLAB program is presented by the dashed line in Figure 5. The line represents the output of the RLC circuit for fitting parameters of $R = 20.0\,\Omega$, $C = 1.0\,\text{pF}$, and $L = 9.96\,\text{nH}$. The center frequency (f^*) of this circuit is 1595 MHz with a damping factor of 100.22×10^{-3}. The fitting parameters values of this device appear to be ideal for use in microwave technology. Such types of filters find applications in telecommunication networks.

It is important to mention that in all parametric calculations the fitting procedures were carried out by a special high-convergence minimization program that makes use of regression and residual sums of squares (R^2), coefficient of determination, and residual mean squares statistical analysis. The errors in the data were evaluated to be 4–12%. Typical best fits for the experimental data and its related modeling are illustrated in Figure 5. All the other calculated slopes (shown in Figures 3(a) and 3(b)) were restricted to give a residual sum of squares $R^2 = 0.97$.

4. Conclusions

In this work a new class of BN based devices is produced and characterized. The device was found to be properly operated at temperatures up to 100°C. The temperature effect of this device is explored by means of the current-voltage characteristics which revealed an increasing barrier height and depletion width with increasing temperature. The impedance spectroscopy which was analyzed in the frequency range of 100–1800 MHz allowed determining the current conduction mechanism by correlated barrier hopping. The AC signal analysis of the Au/BN/C device displayed resonance-antiresonance physical phenomena at frequency of ~1.6 GHz. The device wave trapping property was verified by the input of an ac signal in the suitable frequency domain and found to behave as an ideal narrow band stop filter with promising features for use in microwave communication technology.

Competing Interests

The authors declare that there is no conflict of interests regarding the publication of this paper.

References

[1] K. Wu, Y. Zhan, C. Zhang, S. Wu, and X. Li, "Strong and highly asymmetrical optical absorption in conformal metal-semiconductor-metal grating system for plasmonic hot-electron photodetection application," *Scientific Reports*, vol. 5, Article ID 14304, 2015.

[2] W. Zhang, J. Xu, W. Ye et al., "High-performance AlGaN metal-semiconductor-metal solar-blind ultraviolet photodetectors by localized surface plasmon enhancement," *Applied Physics Letters*, vol. 106, no. 2, 2015.

[3] M. J. Kumar, M. Maram, and P. P. Varma, "Schottky biristor: a metal-semiconductor-metal bistable resistor," *IEEE Transactions on Electron Devices*, vol. 62, no. 7, pp. 2360–2363, 2015.

[4] D. M. Geum, S. H. Shin, S. M. Hong et al., "Metal-semiconductor–metal varactors based on InAlN/GaN heterostructure with cutoff frequency of 308 GHz," *IEEE Electron Device Letters*, vol. 36, no. 4, pp. 306–308, 2015.

[5] Y. Fang and M. Sun, "Nanoplasmonic waveguides: towards applications in integrated nanophotonic circuits," *Light: Science and Applications*, vol. 4, article e294, 2015.

[6] H. T. Zhou, L. Li, H. Y. Chen, Z. Guo, S. J. Jiao, and W. J. Sun, "Realization of a fast-response flexible ultraviolet photodetector employing a metal-semiconductor-metal structure InGaZnO photodiode," *RSC Advances*, vol. 5, no. 107, pp. 87993–87997, 2015.

[7] M. Parzefall, P. Bharadwaj, A. Jain, T. Taniguchi, K. Watanabe, and L. Novotny, "Antenna-coupled photon emission from hexagonal boron nitride tunnel junctions," *Nature Nanotechnology*, vol. 10, no. 12, pp. 1058–1063, 2015.

[8] H. K. Khanfar, "Fabrication and characterization of Ag/BN/Ni microwave rejection-band filters," *IEEE Transactions on Electron Devices*, vol. 61, no. 6, pp. 2154–2157, 2014.

[9] S. E. Al Garni and A. F. Qasrawi, "Design and characterization of (Al, C)/p-Ge/p-BN/C isotype resonant electronic devices," *Physica Status Solidi (A) Applications and Materials Science*, vol. 212, no. 8, pp. 1845–1850, 2015.

[10] S. Sze and K. K. Ng, *Physics of Semiconductor Devices*, John Wiley & Sons, Inc., Hoboken, NJ, USA, 2006.

[11] E. W. S. Caetano, V. N. Freire, G. A. Farias, and E. F. Da Silva Jr., "Optical properties of zincblende GaN/BN cylindrical nanowires," *Applied Surface Science*, vol. 234, no. 1-4, pp. 50–53, 2004.

[12] O. Madelung, *Semiconductors: Data Handbook*, Springer, Berlin, Germany, 2004.

[13] X. Z. Du, C. D. Fyre, J. H. Edgar, J. Y. Lin, and H. X. Jiang, "Temperature dependence of the energy bandgap of two-dimensional hexagonal boron nitride probed by excitonic photoluminescence," *Journal of Applied Physics*, vol. 115, no. 5, Article ID 053503, 2014.

[14] M. K. Hudait, P. Venkateswarlu, and S. B. Krupanidhi, "Electrical transport characteristics of Au/n-GaAs Schottky diodes on n-Ge at low temperatures," *Solid-State Electronics*, vol. 45, no. 1, pp. 133–141, 2001.

[15] M. Soylu, M. Gülen, and S. Sönmezoğlu, "Temperature-dependent model for hole transport mechanism in a poly(1.8-diaminocarbazole)/Si structure," *Philosophical Magazine*, vol. 96, no. 25, pp. 2600–2614, 2016.

[16] D. M. Pozar, *Microwave Engineering*, Wiley Interscience, 3rd edition, 2005.

[17] A. A. A. Darwish, M. M. El-Nahass, and A. E. Bekheet, "AC electrical conductivity and dielectric studies on evaporated nanostructured InSe thin films," *Journal of Alloys and Compounds*, vol. 586, pp. 142–147, 2014.

[18] R. Murti, S. K. Tripathi, N. Goyal, and S. Prakash, "Random free energy barrier hopping model for ac conduction in chalcogenide glasses," *AIP Advances*, vol. 6, no. 3, Article ID 035010, 2016.

[19] Y. Zhao, Z. Wan, X. Xu, S. R. Patil, U. Hetmaniuk, and M. P. Anantram, "Negative differential resistance in boron nitride graphene heterostructures: physical mechanisms and size scaling analysis," *Scientific Reports*, vol. 5, Article ID 10712, 2015.

[20] A. Kaya, S. Alialy, S. Demirezen, M. Balbaşı, S. Yerişkin, and A. Aytimur, "The investigation of dielectric properties and ac conductivity of Au/GO-doped PrBaCoO nanoceramic/n-Si capacitors using impedance spectroscopy method," *Ceramics International*, vol. 42, no. 2, pp. 3322–3329, 2016.

[21] H. G. Dimopoulos, *Analog Electronic Filters: Theory, Design and Synthesis*, Springer, Dordrecht, Netherlands, 2012.

8

Design of High-Voltage Switch-Mode Power Amplifier based on Digital-Controlled Hybrid Multilevel Converter

Yanbin Hou, Wanrong Sun, Aifeng Ren, and Shuming Liu

School of Electronic Engineering, Xidian University, Xi'an 710071, China

Correspondence should be addressed to Yanbin Hou; ybhou@mail.xidian.edu.cn

Academic Editor: Wei-Zen Chen

Compared with conventional Class-A, Class-B, and Class-AB amplifiers, Class-D amplifier, also known as switching amplifier, employs pulse width modulation (PWM) technology and solid-state switching devices, capable of achieving much higher efficiency. However, PWM-based switching amplifier is usually designed for low-voltage application, offering a maximum output voltage of several hundred Volts. Therefore, a step-up transformer is indispensably adopted in PWM-based Class-D amplifier to produce high-voltage output. In this paper, a switching amplifier without step-up transformer is developed based on digital pulse step modulation (PSM) and hybrid multilevel converter. Under the control of input signal, cascaded power converters with separate DC sources operate in PSM switch mode to directly generate high-voltage and high-power output. The relevant topological structure, operating principle, and design scheme are introduced. Finally, a prototype system is built, which can provide power up to 1400 Watts and peak voltage up to ±1700 Volts. And the performance, including efficiency, linearity, and distortion, is evaluated by experimental tests.

1. Introduction

Class-A, Class-B, and Class-AB amplifiers are usually termed as linear amplifiers for their switching devices operate in linear mode, while Class-D amplifiers are also known as switching amplifiers, in which the switching devices are either fully turned on or completely turned off, operating in switch mode. This means that when the switch is conducting (turned on) there is virtually no voltage across the switch and that when the switch is not conducting (turned off) there is no current flowing through the switch. So the ideal efficiency of Class-D amplifier is 100% in theory [1]. In fact, the commonly used semiconductor switching devices, such as IGBT (Insulated Gate Bipolar Transistor) and MOSFET (Metal Oxide Semiconductor Field Effect Transistor), have saturation voltage drop or on-state resistance when they are turned on. Although part of power has been consumed as heat by switches, efficiencies over 85% can be achieved in most situations for Class-D amplifier. By comparison, linear amplifier has a theoretical efficiency of 78% at the utmost [2].

In a typical Class-D amplifier, input signal is converted into a series of pulses through pulse width modulation (PWM), which serve as control signals for driving power switches. The switching devices together with DC sources realize power and voltage amplification. At the output stage of Class-D amplifier, low-frequency amplified signal is retrieved across the load by filtering out high-frequency carrier wave. Due to its single half-bridge or full-bridge topology, PWM-based switching amplifier usually outputs a peak voltage of dozens or hundreds of Volts [3]. Step-up transformer seems to be a ready-made solution to boost up output voltage, since existing PWM-based switching amplifiers can provide enough output power. However, the noisy and heavy midfrequency step-up transformer will make the whole amplifier system bulky and expensive. What is worse, the insertion of step-up transformer between amplifier and load will introduce additional power loss, thus reducing the overall system efficiency. One of our outside partners is confronted with such a dilemma. They built an analog PWM-based switching amplifier and used a bulky step-up transformer to produce the rated voltage of 1000 Volts and rated power of 1000 Watts across a 1000-Ohm resistor over the operating frequency range of 100–1000 Hertz. But the average system efficiency is about 70%, presenting a challenge to cool. What

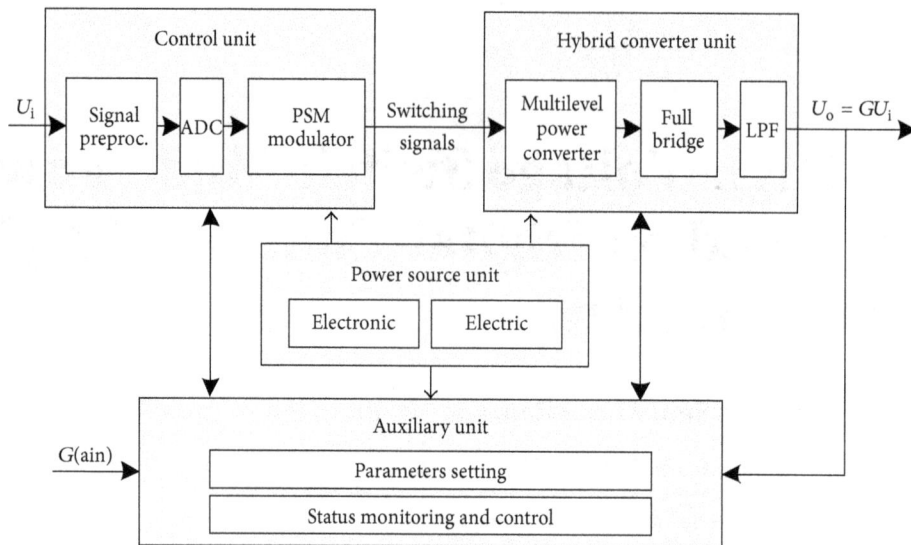

FIGURE 1: Block diagram of high-voltage switching amplifier.

is more, the bulky step-up transformer has a weight of over 50 kilograms, almost two-thirds of the gross weight. Our target is to rebuild a switching amplifier with main requirements as follows:

Output voltage up to 1000 Vrms.

Output power up to 1000 W.

Overall efficiency above 85%.

Output distortion below 2.5%.

Gross weight less than 80 Kg.

The requirements about high efficiency and light weight exclude the possibility of using step-up transformer to boost up voltage. From the view of power electronics, the desired switching amplifier can be considered as a programmable high-voltage power supply operating in switch mode under the control of input signal. NPC- (neutral point clamped-) based and cascaded H-bridge (full-bridge) multilevel converters are two commonly used topologies to directly produce high-voltage output [4, 5]. The former uses a single high-voltage DC source and multiple cascaded semiconductor switches together with clamping diodes or flying capacitors, usually involving complicated voltage balancing measures [6–8]. The latter adopts modular design, of which each converter consists of a low-voltage DC source and four semiconductor switches. Thus it will demand a great number of semiconductor switches and control signals when employing multiple cascaded full-bridge converters to generate high-voltage output.

In order to directly generate high-voltage output and reduce power semiconductor switches as well, a new circuit architecture, called hybrid multilevel converter, is proposed in this paper. The corresponding control scheme is designed on the basis of pulse step modulation (PSM) by adopting digital signal processing technology. Section 2 introduces the design scheme, Section 3 gives the performance tests, and

Section 4 is the conclusion. Experimental results show that this new system can provide up to 1400 Watts and ±1700 Volts peak voltage with high efficiency and low distortion as expected.

2. Design of High-Voltage Switching Amplifier

Figure 1 shows the functional block diagram of high-voltage switch-mode power amplifier developed in this paper.

As shown in Figure 1, this switching amplifier consists of four units, which are control unit, hybrid converter unit, power source unit, and auxiliary unit, respectively. The workflow is as follows.

Small analog input signal is firstly preprocessed and then digitalized by a 12-bit analog-to-digital converter (ADC). According to the ADC data and user settings, PSM modulator generates a set of switching signals, which are further isolated and driven before applying to hybrid converter unit. Under the control of corresponding switching signals, multilevel power converter transforms energy from separate low-voltage DC sources to augment modulated switching signals and full bridge takes charge of the switchover between positive and negative phase. At the final output stage, a low-pass filter is used to retrieve the expected amplified output waveform.

For the sake of efficiency and reliability, the whole system was developed utilizing digital signal processing technology. PSM modulator was implemented with a high-performance floating-point digital signal processor (DSP) and a high-volume field programmable gate array (FPGA). Auxiliary unit adopted digital technologies at the most extent, like human-machine interface (HMI) for setting gain and other parameters, measuring output voltage and current, monitoring temperatures, and so on.

The following sections introduce key points in detail to develop a ±1700 Vp/1400 W switch-mode power amplifier, of

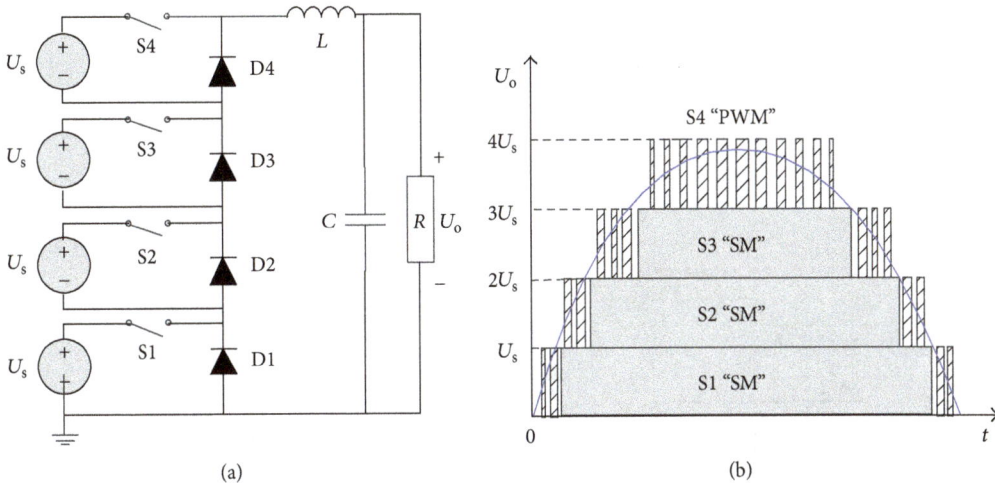

FIGURE 2: PSM-based HVPS. (a) Schematic and (b) principle.

which efficiency is expected to be no less than 85% with low distortion in the midfrequency range (100–1000 Hz).

2.1. Pulse Step Modulation. Brown Boveri Corporation presented pulse step modulation (PSM) and applied it broadly in short-wave transmitter and high-voltage power supply (HVPS) [9–11]. Figure 2 illustrates the schematic circuit and work principle of PSM-based HVPS to quadruple its output voltage of each converter.

As shown in Figure 2(a), a PSM-based HVPS basically consists of multiple (four in this example) power converters connected in series, which work in switch mode under the control of input reference signal (here taking semi-sine wave as an example). Each power converter has a separate low-voltage DC source U_s, a semiconductor switch S, and a free-wheeling diode D [12]. The diode is to provide a low impedance path when the corresponding converter is switched off, allowing current from other working converters to pass through. Regardless of voltage drops across diodes and switches, total output voltage U_o is simply equal to several times of unit step voltage U_s, depending on the number of switched-on power inverters. Obviously, actual output voltage is just a roughly stepwise approximation to the expected waveform, which is referred to as step modulation (SM). To further refine output, pulse width modulation (PWM) and LC components are adopted to smooth the transition from one step to the other. Taking Figure 2(a) as an example, let switches S1~S3 operate in SM mode and let switch S4 operate in PWM mode; the filtered output voltage U_o across the load is plotted as a blue solid line in Figure 2(b).

As stated above, PSM consists of coarse SM and fine PWM. Figure 2(b) shows that each SM operates at a rather low frequency the same as input reference signal. It is also noticed that PWM need to switch much faster than SM to achieve an accurate representation of the input signal. Following Nyquist theorem, PWM need to run at least twice the maximum input frequency, but actual design generally adopts a much higher ratio (typically 10 to 50) for reducing distortion [13]. In theory, high-frequency PWM can improve output waveform. However, switch loss increases as switching becomes faster. Generally, PWM will take a compromise between output waveform and switch loss, depending on the practical application.

2.2. Hybrid Converter Unit. As illustrated in Figure 2, PSM-based HVPS can only produce unipolar voltage. In order to operate as an amplifier, circuit needs to be further modified. Conventional method is that every converter adopts full-bridge architecture, employing four switches and two groups switching signals to produce $0~\pm U_s$. However, it will increase system cost and control complexity if many full-bridge converters are used to produce high voltage.

In this paper, a hybrid multilevel converter was designed, in which information on amplitude and phase of input signal is separately modulated. As shown in Figure 3, eight cascaded power converters (PC1~PC8) produce desired amplitude while full bridge (composed of Q1~Q4) controls phase. Compared with cascaded full-bridge converters, this hybrid architecture employs much fewer switches and control signals.

Power converters PC1~PC8 are identical, except for corresponding to different switching signals. Each one consists of separated 220 V/3 A DC source U_s, semiconductor switch S, and free-wheeling diode D. In consideration of switching speed and power loss, semiconductor switches S1~S8 all use power MOSFETs, which have small on-state resistance. Free-wheeling diodes D1~D8 adopt fast recovery diodes. Intensive simulations and tests confirm that MOSFET and diode suffer the maximum potential difference of U_s in all situations, so many power MOSFETs available can be used, such as IRF840, 2SK1507, and FMV11N60E. 2SK1507 with a typical drain-source on-state resistance of $R_{DS(on)} = 0.85\,\Omega$ was adopted. Diodes adopted MUR860 with a maximum instantaneous forward voltage drop $V_F = 1.5$ V. Switches S1~S7 and S8 are

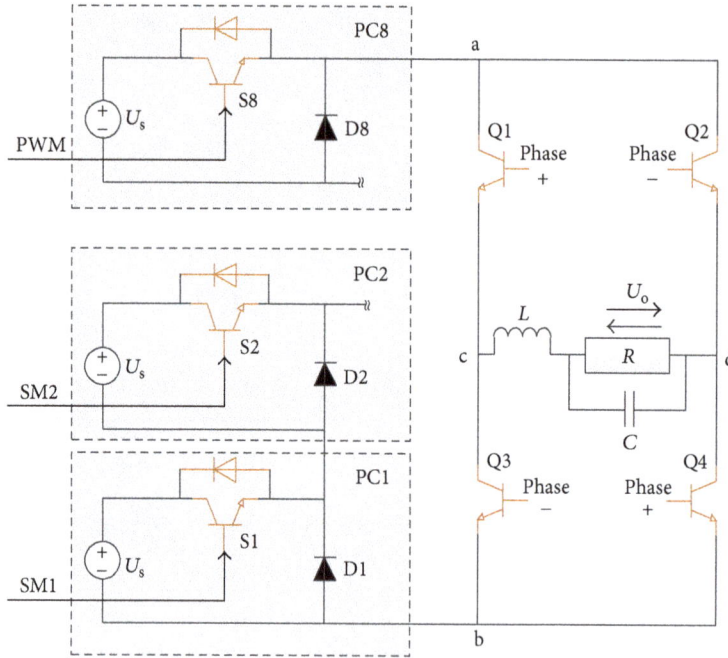

FIGURE 3: Schematic of hybrid multilevel converter.

under the control of switching signals SM1~SM7 and PWM, respectively.

The full bridge is built with four discrete IGBTs Q1~Q4, under the control of switching signals Phase+/Phase−. When the input signal is nonnegative, Q2 and Q3 are switched off simultaneously; meanwhile Q1 and Q4 are switched on simultaneously and vice versa. Although ZVS (Zero Voltage Switching) is applied, dead time is also inserted between commutations to avoid potential shoot-through risk. IGBTs adopted IXYS IXBH12N300 with a maximum saturation voltage drop $V_{CE(sat)} = 3.2$ V.

Inductor L, capacitor C, and resistive load R make up a Butterworth low-pass filter (LPF) to recover the amplified output signal from high-voltage modulated waveform. Given load resistance and cut-off frequency, inductance and capacitance can be obtained by

$$L = \frac{R}{\left(\sqrt{2}\pi f_C\right)},$$

$$C = \frac{1}{\left(2\sqrt{2}\pi R f_C\right)},$$

$$(1)$$

where f_C denotes the cut-off frequency, f_C in Hertz, L in Henrys, C in Faradays, and R in Ohms, respectively.

The LPF filter has a −3 dB cut-off frequency of 5 kHz, corresponding to $R = 1$ kΩ, $L = 44$ mH, and $C = 18$ nF.

2.3. Digital PSM Modulator.

Unlike carrier-based PWM that adopts analog modulation [14], we designed digital PSM

modulator to generate SM and PWM switching signals by the following formula:

$$M(n) = \text{fix}\left(\frac{G|U_i(n)|}{U_s}\right)$$

$$d(n) = \text{rem}\left(\frac{G|U_i(n)|}{U_s}\right)$$

$$\text{Phase+} = \begin{cases} \text{ON,} & \text{if } U_i(n) \geq 0 \\ \text{OFF,} & \text{if } U_i(n) < 0 \end{cases} \qquad (2)$$

$$\text{Phase−} = \begin{cases} \text{ON,} & \text{if } U_i(n) < 0 \\ \text{OFF,} & \text{if } U_i(n) \geq 0 \end{cases}$$

$$n = t_1, t_2, \ldots, t_n,$$

where $|U_i(n)|$ is the absolute amplitude of sampled input signal at moment t_n. G is predefined voltage gain; U_s is DC source voltage. Function fix$(\cdot)$ rounds the data towards zero, resulting in an integer $M(n)$, which is the number of power converters switched on between t_n and t_{n+1}. Function rem$(\cdot)$ retrieves the remainder after division, so $d(n)$ is a decimal, representing the "on" time of a power converter relative to sampling interval; that is, $d(n)$ determines the duty cycle of PWM switching signal (e.g., 0.50 is equal to 50%). Phase+ and Phase− record the phase information about sampled input signal, serving as the switching signals for full bridge.

In other words, during sampling interval between t_n and t_{n+1}, $M(n)$ power converters work in SM mode, and one power converter operates in PWM mode with a duty cycle of $d(n)$. Obviously, the switching frequency of PWM is equal to the sampling frequency of ADC. Here we adopted unipolar centre-aligned PWM, whose advantages and implementation could be referred to [15].

2.4. Software Simulation. The workflow of the above switching amplifier is demonstrated through simulation. For example, a sine wave of 1 kHz and ±1 Vp needs to be amplified by a voltage gain of $G = 800$. It is assumed that sampling frequency of ADC is 40 kSa/s, so that there will be 40 sampled points in a cycle, as shown in Figure 4(a). According to (2), if $U_s = 220$ Vdc, it needs that three converters work in SM mode and one converter operates in PWM mode. Digital PSM modulator generates switching signals, as plotted in Figure 4(b). The unused converters are always switched off, which are not plotted in Figure 4(b). Figure 4(c) gives the modulated amplitude after amplification, measured between points a and b (marked in Figure 3). Figure 4(d) shows the output voltages after adding phase information and filtering, measured between points c and d (marked in Figure 3), in which the blue solid line represents modulated waveform after amplification, and the red dotted line is the final output voltage across load after LPF demodulation.

In this simulation, zero-crossing points of input signal are all sampled by ADC. However, this may be not true in most actual situations. So ZVS will be applied to force all SM and PWM signals to be off at the moments of commutation, for example, when $t = 0.5, 1.0, 1.5,$ and 2.0 ms in Figure 4(b).

It is important to note that IGBTs usually take more time to be switched on or off compared to MOSFETs due to finite switching speed. In order to protect full bridge at the moment of voltage phase alternation, proper dead time is added to ensure that short circuit will not happen between points a and b (marked in Figure 3), which means that both phase control signals in Figure 4(b) are switched off during the dead time.

3. Experimental Tests and Results

3.1. Prototype System. Based on the above introduction, a high-voltage switch-mode power amplifier is developed, as shown in Figure 5.

From top to down, there are main distribution box, electronic case for control unit and auxiliary unit, electric case for hybrid converter unit, and two 4-channel AC-DC 220 V power sources in a 19-inch standard cabinet (600 mm × 600 mm × 1050 mm, W × D × H). A resistor of 1000 Ohms with cooling equipment is used as dummy load, laid outside the cabinet for facilitating heat dissipation.

3.2. Total Harmonic Distortion. Many factors in switching amplifier can cause distortion, such as power switches, nonlinearity of output LPF, DC source voltage fluctuations, dead time, and modulation technologies. As one of the most common and important features, total harmonic distortion (THD) is usually used to evaluate system distortion. Given that input signal is a sine wave, THD is most commonly defined as the ratio of the rms amplitude of a set of harmonic

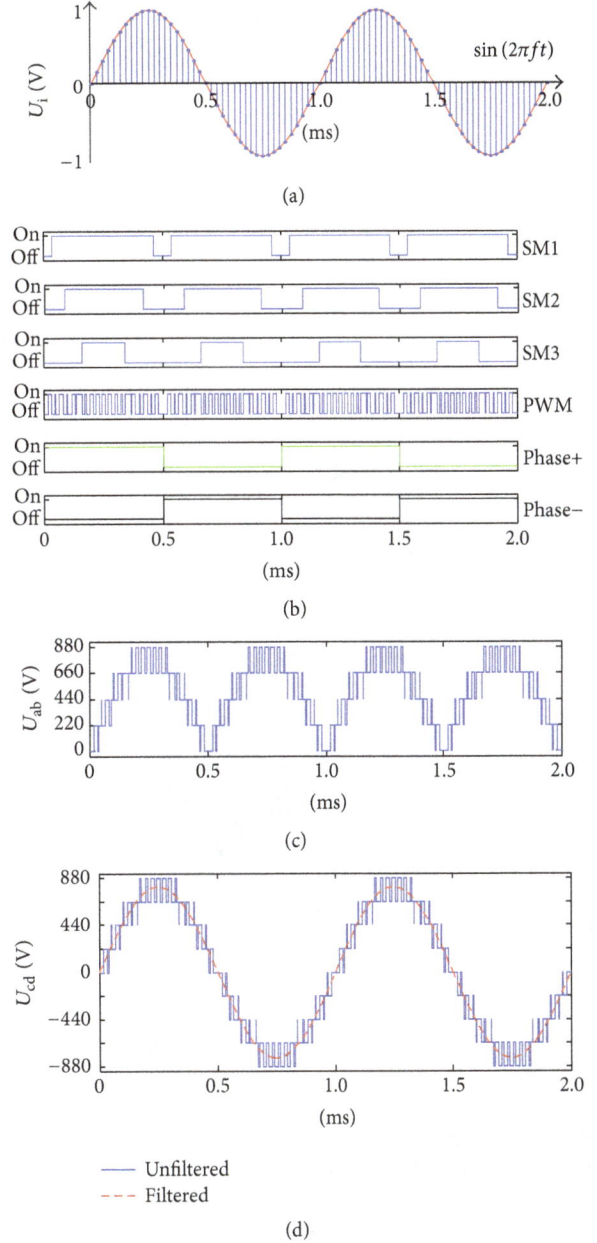

FIGURE 4: Simulation of switching power amplifier. (a) Sampled input signal, (b) switching signals for power converters and full bridge, and ((c) and (d)) voltage waveforms measured between a-b and c-d points, respectively (points a-d marked in Figure 3).

frequencies to that of fundamental frequency, which can be formulated as follows [16]:

$$\text{THD} = \frac{\sqrt{U_2{}^2 + U_3{}^2 + \cdots + U_n{}^2}}{U_1}, \quad (3)$$

where U_1 denotes the rms voltage of fundamental frequency and U_n is the rms voltage of the nth harmonic frequency.

Sine wave of 1 Vrms with frequency sweeping from 100 to 1000 Hz was used as input signal to investigate the prototype system. Figure 6 shows the screenshots of output waveforms

TABLE 1: THD values of power amplifier system.

Freq. (Hz)	Output power (W)			
	300	650	960	1300
100	1.52%	1.47%	1.06%	1.04%
200	1.59%	1.53%	1.12%	1.05%
300	1.65%	1.52%	1.15%	1.04%
400	1.73%	1.58%	1.23%	1.05%
500	1.68%	1.58%	1.26%	1.09%
600	1.69%	1.60%	1.27%	1.19%
800	1.71%	1.66%	1.33%	1.30%
1000	1.77%	1.76%	1.40%	1.34%

TABLE 2: Efficiency of power amplifier system.

P_c (W)	P_o (W)	Efficiency
350	300	85.7%
712	650	91.3%
1022	960	93.9%
1361	1300	95.5%

FIGURE 5: Photo of power amplifier system.

at frequencies of 100, 500, and 1000 Hz. On the oscilloscope display, the upper channel shows output voltage and the lower one is its corresponding fast Fourier transform (FFT). It is noticed that the FFT spectrum is plotted as magnitude in dB relative to 1 Vrms; the corresponding rms voltage is retrieved by

$$U_n = 10^{dB_n/20}, \qquad (4)$$

where dB_n denotes the voltage gain relative to 1 Vrms at the nth harmonic frequency and U_n is the rms output voltage of the nth harmonic frequency.

Table 1 lists THD values measured at four different output power values. Without regard to individual measurement errors, it can be found that THD value increases as signal frequency rises and that THD value decreases as output power increases. It is not difficult to explain this phenomenon. With input signal sweeping from low frequency to high frequency, the sampled points in a single cycle reduce, resulting in output waveform not so smooth as before. When output power increases, signal-to-noise ratio is improved to a certain degree.

It is noticed that measured output power in Table 1 is an average of output power values measured from 100 to 1000 Hz with a step size of 100 Hz, keeping input signal amplitude

constant. Averaging is aimed at diminishing measurement error for high voltage. A typical example is shown in Figure 6. With the same input signal and amplification gain, peak and rms output voltages fluctuated irregularly at different frequencies.

3.3. The Maximum Output. To measure the maximum output power, sine wave of 1 kHz is used as input signal and its amplitude increases until output distortion reaches THD = 10%. Figure 7 shows output voltage when amplifier outputs the maximum power of about 1500 W.

As shown in Figure 7, waveform appears apparent clipping at the maximum output power value. For the sake of practicability, the maximum output voltage is limited to 3400 Vpp (a critical point of appearing voltage clipping), and the corresponding maximum output power is hence lowered to about 1400 W, estimated by

$$P_o = \frac{1}{8R}U_{pp}^2 = \frac{1}{8 \times 1000} \times (3400 \text{ V})^2 = 1445 \text{ Watts}. \quad (5)$$

3.4. Efficiency and Linearity. The efficiency corresponding to the output power in Table 1 is listed in Table 2. P_c represents the whole power consumed by entire amplifier system; P_o is the output power measured on load. It is noticed that P_o is an average of output power values measured from 100 to 1000 Hz with a step size of 100 Hz, as the same reason explained in Section 3.2.

As for an amplifier, nonlinearity caused by gain variation will affect the waveform shape of analog output with respect to the corresponding analog input. Figure 8 plots three groups of output versus input signal amplitude with a constant voltage gain of $G = 1000$.

As shown in Figure 8, the output approximately follows a linear relationship relative to the input. The average voltage gains measured at frequencies of 400, 800, and 1000 Hz are 1027, 1036, and 1021, respectively. That is to say, this new developed high-voltage switch-mode power amplifier has a statistical linearity error less than 4%.

3.5. Automatic Protection. The running status of power amplifier system is under real-time surveillance, including output voltage, output current, output power, and system temperature. Once any alarm about overvoltage, overcurrent, or overtemperature is triggered, amplifier system will be shut down within 1 ms, as illustrated in Figure 9. In the oscilloscope screenshot, Channel 1 records alarm signal, and Channel 2 is recording switching signal. If a failure occurs, falling edge of alarm signal is first detected and low level

(a)

(b)

(c)

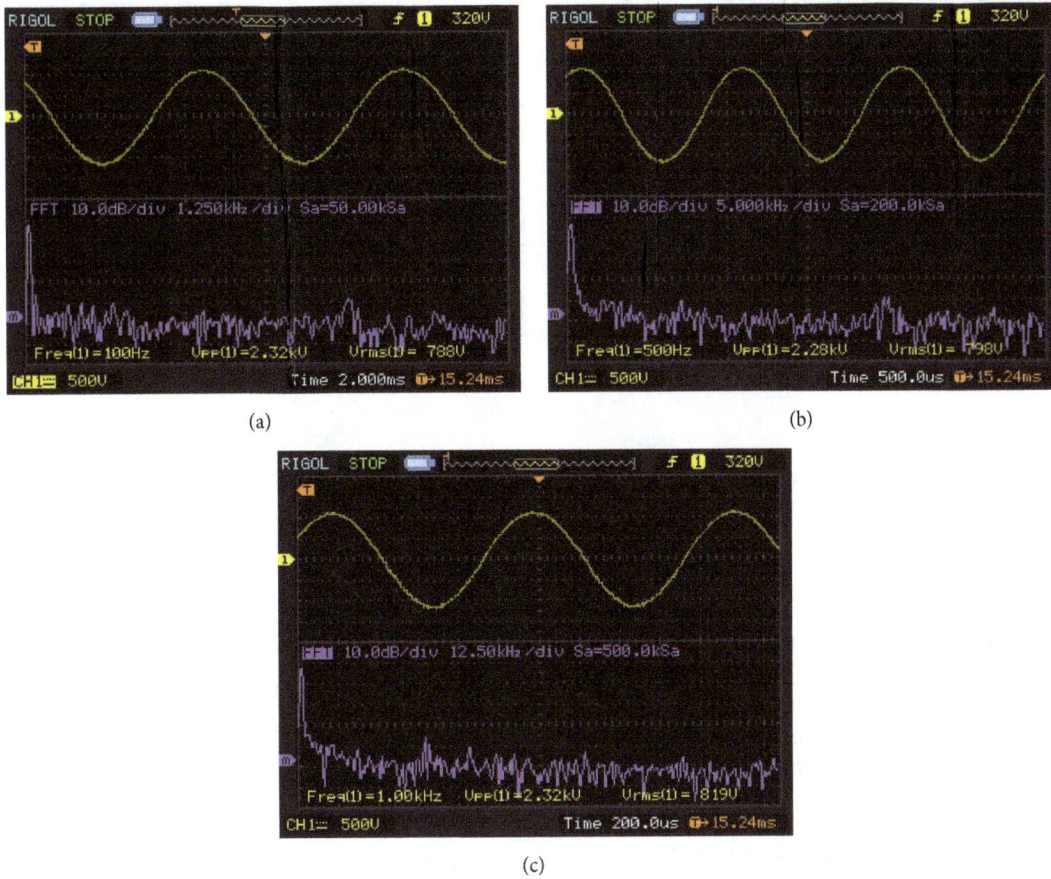

FIGURE 6: Oscilloscope screenshots of output voltage at frequencies of (a) 100 Hz (500 V/div, 2 ms/div), (b) 500 Hz (500 V/div, 500 us/div), and (c) 1000 Hz (500 V/div, 200 us/div).

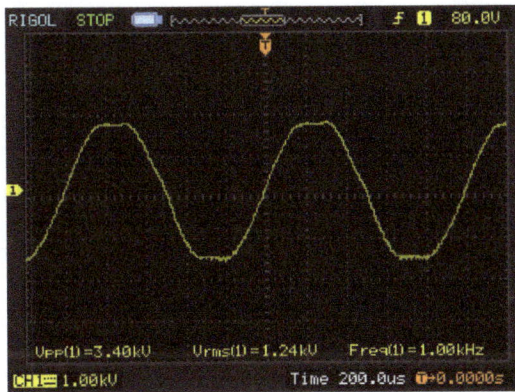

FIGURE 7: High-voltage output waveform of clipping with THD = 10% (1 kV/div, 200 us/div).

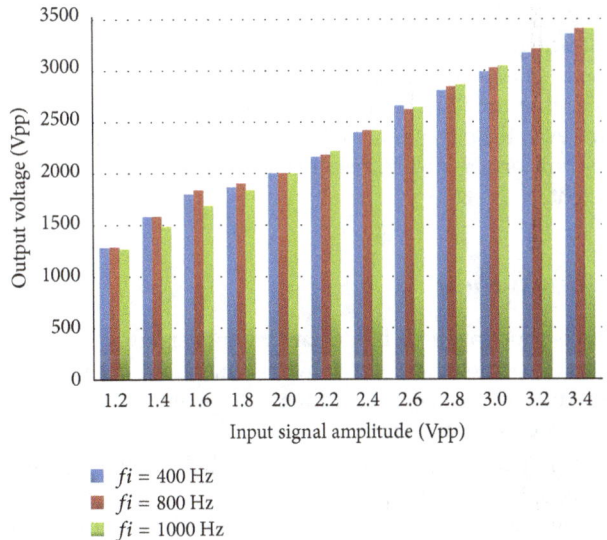

FIGURE 8: Linearity of power amplifier system.

of alarm signal is then reconfirmed within about 600 us. If failure does happen, control unit will switch off all power converters one by one within 40 us.

3.6. Overall Comparison. Table 3 gives an overall comparison between the new digital PSM-based switching amplifier and the existing analog PWM-based switching amplifier. The specification and performance of the latter are provided by our partner off campus, including modulation technology, circuit topology, output voltage and power, distortion,

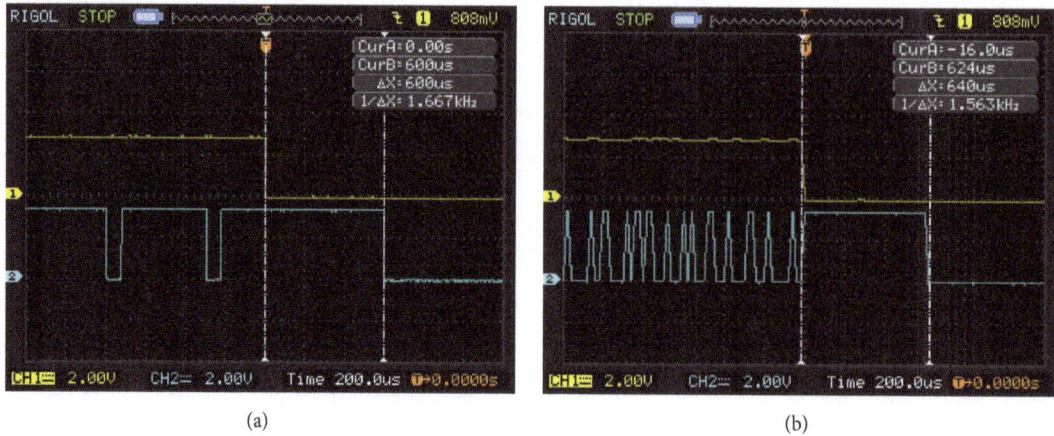

(a)   (b)

FIGURE 9: Oscilloscope screenshots of automatic protection. (a) CH1: failure alarm signal and CH2: first closed SM1 signal. (b) CH1: failure alarm and CH2: last closed PWM signal (2 V/div, 200 us/div).

efficiency, and weight. For convenience, the two switching amplifiers are labelled as PSM-based amplifier and PWM-based amplifier, respectively.

In PWM-based switching amplifier, a high-speed analog comparator compares a triangular carrier wave of 300 KHz with the input signal to generate PWM switching signals. But, in PSM-based switching amplifier as described in Section 2.4, the corresponding PWM switching signal is 40 KHz. High PWM switching frequency means that more sampled points are used to approximate the input signal, thus yielding smooth waveform and low distortion. Furthermore, PWM-based switching amplifier only adopts fast power MOSFETs in the half-bridge stage, while PSM-based switching amplifier adopts IGBTs with much higher blocking voltage in the full-bridge stage. But IGBT switches slower than MOSFET due to tail-current effect at turn-off. Therefore, PWM-based switching amplifier can generate better output waveform. However, the dependence of PWM-based amplifier on step-up transformer lowers overall efficiency and increases gross weight.

4. Conclusion

In this paper, a switch-mode power amplifier based on digital pulse step modulation and hybrid multilevel converter is designed to directly produce high-voltage output without the aid of step-up transformer.

Compared with available PWM-based switching amplifier coupled with a step-up transformer, this new developed system can provide output power up to 1400 Watts and peak voltage up to ±1700 Volts with encouraging efficiency and portability. The major advantages are its hybrid architecture and digital modulation. When directly generating high-voltage output, a lot of control signals and power switches are cut down by modulating the information on amplitude and phase of input signal separately.

However, there is further work to improve this design.

TABLE 3: Comparisons of two power amplifier systems.

Items	PSM-based amplifier	PWM-based amplifier
Modulation	Digital PSM	Analog PWM
Circuit topology	Multilevel	Half bridge
Power supplies	DC 220 V	DC ±80 V
Step-up transformer	No need	Need
Maximum output	±1700 Vp	±1500 Vp
	1400 W	1200 W
THD*	1.38%	0.85%
Efficiency*	94.2%	82.5%
Gross weight	50 Kg	80 Kg
User interface	Touch LCD	LED indicators

*The test condition is outputting 1000 Vrms sine at 1 KHz.

Firstly, switching frequency needs to be increased for more practical applications. However, MOSFETs in high-voltage converter cannot switch at a frequency as high (300 kHz or above) as they do in PWM-based audio amplifier, because this will lead to serious overshoot and EMI problems. Thus, more elaborate modulation scheme needs to be considered, such as phase-shifting modulation technologies.

Secondly, more intelligent strategies should be adopted. For example, redundant power converters can timely replace the failed ones without shutdown. If there is no usable power converter, amplifier will appropriately lower output voltage to prevent serious distortion.

Competing Interests

The authors declare that there are no competing interests regarding the publication of this paper.

Acknowledgments

This research is supported by the Fundamental Research Funds for the Central Universities under Grant no. JB140205.

References

[1] M. Bloechl, M. Bataineh, and D. Harrell, "Class D switching power amplifiers: theory, design, and performance," in *Proceedings of the IEEE SoutheastCon 2004*, pp. 123–146, Greensboro, NC, USA, March 2004.

[2] M. Berkhout and L. Dooper, "Class-D audio amplifiers in mobile applications," *IEEE Transactions on Circuits and Systems*, vol. 57, no. 5, pp. 992–1002, 2010.

[3] D. Self, *Audio Power Amplifier Design Handbook*, Focal Press, Boston, Mass, USA, 5th edition, 2009.

[4] F. J. Liu, *Modern Inverter Technology and Application*, Publishing House of Electronics Industry, Beijing, China, 2006 (Chinese).

[5] V. Sala, R. Salehi, M. Moreno-Eguilaz, M. Salehifar, and L. Romeral, "Clamping diode caused distortion in multilevel NPC Full-Bridge audio power amplifiers," in *Proceedings of the 38th Annual Conference on IEEE Industrial Electronics Society (IECON '12)*, pp. 4941–4948, IEEE, Montreal, Canada, October 2012.

[6] W. W. He, P. Palmer, X. Q. Zhang, M. Snook, and Z. H. Wang, "IGBT series connection under active voltage control," in *Proceedings of the 14th European Conference on Power Electronics and Applications (EPE '11)*, pp. 1–9, IEEE, Birmingham, UK, September 2011.

[7] P. Palmer, W. W. He, X. Q. Zhang, J. Zhang, and M. Snook, "IGBT series connection under Active Voltage Control with temporary clamp," in *Proceedings of the 38th Annual Conference on IEEE Industrial Electronics Society (IECON '12)*, pp. 465–470, Montreal, Canada, October 2012.

[8] T. Lu, Z. M. Zhao, S. Q. Ji et al., "Design of voltage balancing control circuit for series connected HV-IGBTs," in *Proceedings of the 16th International Conference on Electrical Machines and Systems (ICEMS '13)*, pp. 515–518, IEEE, Busan, South Korea, October 2013.

[9] W. Schminke, "The merits of modern technology for today's high power short-wave transmitters," *IEEE Transactions on Broadcasting*, vol. 34, no. 2, pp. 126–133, 1988.

[10] J. Alex and W. Schminke, "A high voltage power supply for negative ion NBI based on PSM technology," in *Proceedings of the 17th IEEE/NPSS Symposium on Fusion Engineering (SOFE '97)*, vol. 2, pp. 1063–1066, IEEE, San Diego, Calif, USA, October 1997.

[11] L. Y. Yao, Y. Q. Wang, X. H. Mao, Y. L. Wang, and Q. Li, "A fully digital controller of high-voltage power supply for ECRH system on HL-2A," *IEEE Transactions on Plasma Science*, vol. 40, no. 3, pp. 793–797, 2012.

[12] J. Alex and W. Schminke, "Fast switching, modular high-voltage DC/AC-power supplies for RF-amplifiers and other applications," in *Proceedings of the 16th IEEE/NPSS Symposium on Fusion Engineering (SOFE '95)*, pp. 936–939, IEEE, Champaign, Ill, USA, October 1995.

[13] M. J. Hawksford, "Modulation and system techniques in PWM and SDM switching amplifiers," *Journal of the Audio Engineering Society*, vol. 54, no. 3, pp. 107–139, 2006.

[14] M. H. Rashid, *Power Electronics Handbook—Devices, Circuits, and Applications*, Butterworth-Heinemann, Boston, Mass, USA, 3rd edition, 2011.

[15] B. Lei, G.-C. Xiao, and X.-L. Wu, "Comparison of performance between bipolar and unipolar double-frequency sinusoidal pulse width modulation in a digitally controlled H-bridge inverter system," *Chinese Physics B*, vol. 22, no. 6, Article ID 060509, pp. 281–288, 2013.

[16] D. Shmilovitz, "On the definition of total harmonic distortion and its effect on measurement interpretation," *IEEE Transactions on Power Delivery*, vol. 20, no. 1, pp. 526–528, 2005.

9

Three-Input Single-Output Voltage-Mode Multifunction Filter with Electronic Controllability based on Single Commercially Available IC

Supachai Klungtong,[1] **Dusit Thanapatay,**[1] **and Winai Jaikla**[2]

[1]*Department of Electrical Engineering, Faculty of Engineering, Kasetsart University, Bangkok 10900, Thailand*
[2]*Department of Engineering Education, Faculty of Industrial Education, King Mongkut's Institute of Technology Ladkrabang, Bangkok 10520, Thailand*

Correspondence should be addressed to Winai Jaikla; winai.ja@hotmail.com

Academic Editor: Jiun-Wei Horng

This paper presents a second-order voltage-mode filter with three inputs and single-output voltage using single commercially available IC, one resistor, and two capacitors. The used commercially available IC, called LT1228, is manufactured by Linear Technology Corporation. The proposed filter is based on parallel RLC circuit. The filter provides five output filter responses, namely, band-pass (BP), band-reject (BR), low-pass (LP), high-pass (HP), and all-pass (AP) functions. The selection of each filter response can be done without the requirement of active and passive component matching condition. Furthermore, the natural frequency and quality factor are electronically controlled. Besides, the nonideal case is also investigated. The output voltage node exhibits low impedance. The experimental results can validate the theoretical analyses.

1. Introduction

Analog filter is widely utilized in numerous applications such as communication, sound system, instrumentation, and control system. The biquadratic or second-order filter is the important building block. Also, this filter is the basic block to design high order filter. In particular, the second-order multifunction filter which provides many filter responses in the same circuit has gained significant attention and has become an interesting research topic [1, 2]. The multiple-inputs single-output (MISO) universal filter is the interesting one and has been continuously proposed. In case of voltage-mode MISO filter, the selection of output filter response by switching on or off the input voltages should be done without the matching condition of passive and active elements. Moreover, the additional double gain amplifier should not be required.

Attention has been paid to the use of active building block in synthesis and design of electronic circuits for analog signal processing [3–6]. The active building block based circuits require a minimum number of active elements (most of them use only single active building block). Thus, the new active

building blocks have been continuously introduced especially the electronically controllable active building block. Most of them are designed and constructed from BJT or CMOS transistors. Practically, these devices should be fabricated into the chip for the best way to test their performances. However, their performances and applications are often proved via simulation by only using computer program due to the investment cost reason. Although the new active building block can be constructed from commercially available IC, for example, in [7, 8] using AD844 and in [9] using OPA860 and EL2082, they still require more than one commercially available IC. Despite the fact that some circuits can use single AD844 as active building block, the AD844 based circuits are not electronically controlled.

In the literature, a number of multiple-input single output voltage-mode multifunction filters based on different active building blocks have been reported in [10–42] and the references cited. However, the proposed filter in [10–16, 18–23, 27, 29, 30, 32, 34, 38, 39, 41, 42] uses more than one active building block. The natural frequency and quality factor of the filters in [10–17, 19, 21–25, 27, 28, 30, 32, 35, 38, 39] are not electronically tuned. The matching condition for selection of output filter response is required for the circuit in [14, 15, 17, 21, 23, 26, 28, 33, 35–38]. The active building block used in [19, 24, 27, 30, 32, 36, 38] is not commercially available IC. The output voltage node does not exhibit low impedance for the filter in [11, 12, 14, 15, 17, 18, 20–22, 24, 25, 29, 31, 33, 34, 36–38, 40]. The proposed filter in [37, 41] requires double input signal. Additionally, only the proposed filters in [11–18, 21, 22, 25, 26, 31, 35, 39, 41, 42] are supported by the experimental measurements.

The three-input single-output voltage-mode biquad filter emphasizing the use of single commercially available IC, LT1228 from Linear Technology Inc., is present in this paper. The proposed filter consists of single LT1228, single resistor, and two capacitors which are suitable for off-the-shelf implementation. The selection of output filter response can be done without the requirement of any passive and active component matching condition. The natural frequency and quality factor can be electronically adjusted. The experimental results of proposed filter agree well with the theoretical expectation.

2. Principle of Operation

2.1. Active Building Block: LT1228.
LT1228 is commercially manufactured by Linear Technology Inc. [43]. It is the combination of transconductance amplifier (OTA) and current feedback amplifier (CFA). The symbolic representation of LT1228 is shown in Figure 1(a). Let us denote the name of each port as v_+, v_-, z, x, and w. In ideal consideration, impedance at ports v_+, v_-, and z exhibits high and the impedance at ports x and w exhibits low. Figure 1(b) shows the equivalent circuit and pin configuration is illustrated in Figure 1(c). Ideally, the port relation can be described by the following matrix:

$$\begin{pmatrix} I_{v_+} \\ I_{v_-} \\ I_z \\ V_x \\ V_w \end{pmatrix} = \begin{pmatrix} 0 & 0 & 0 & 0 & 0 \\ 0 & 0 & 0 & 0 & 0 \\ g_m & -g_m & 0 & 0 & 0 \\ 0 & 0 & 1 & 0 & 0 \\ 0 & 0 & 0 & R_T & 0 \end{pmatrix} \begin{pmatrix} V_+ \\ V_- \\ V_z \\ I_x \\ I_w \end{pmatrix}, \quad (1)$$

where R_T is the transresistance gain. In an ideal case, R_T is typically very large and can be considered as infinite value. g_m of LT1228 is controlled by DC bias current I_B as follows:

$$g_m = 10 I_B. \quad (2)$$

2.2. Proposed Filter.
The structure of three-input and single-output voltage-mode filter which is composed of single commercially available IC, single resistor, and two capacitors is presented in Figure 1. This filter is based on parallel RLC circuit. The input voltage v_{in1} is applied at v_+ terminal which is ideally high impedance, v_{in2} is applied through C_1, and v_{in3} is applied through R. The output voltage v_o is at x terminal which exhibits low impedance. In routine

analysis, the output voltage of the proposed filter can be given as

$$v_o = \frac{v_{in1}\left(g_m/C_1\right)s + v_{in2}s^2 + v_{in3}\left(g_m/C_1C_2R\right)}{s^2 + \left(g_m/C_1\right)s + g_m/C_1C_2R}. \quad (3)$$

From (3), the natural frequency is given as

$$\omega_0 = \sqrt{\frac{g_m}{C_1C_2R}}. \quad (4)$$

Subsequently, the quality factor is given as

$$Q = \sqrt{\frac{C_1}{C_2 g_m R}}. \quad (5)$$

It is evident from (4) and (5) that the natural frequency and quality factor can be electronically tuned via g_m.

It is found from (3) that the derivation of five filter responses can be done as follows:

(i) If the input voltage is applied at node v_{in3} while nodes v_{in1} and v_{in2} are grounded, the noninverting low-pass filter is achieved.

(ii) If the input voltage is applied at node v_{in2} while nodes v_{in1} and v_{in3} are grounded, the noninverting high-pass filter is achieved.

(iii) If the input voltage is applied at node v_{in1} while nodes v_{in2} and v_{in3} are grounded, the noninverting band-pass filter is achieved.

(iv) If the input voltage is applied at nodes v_{in2} and v_{in3} while node v_{in1} is grounded, the noninverting band-reject filter is achieved.

(v) If the input voltage is applied at nodes v_{in2} and v_{in3} while the inverting input voltage is applied at node v_{in1}, the noninverting all-pass filter is.

It is found from the above statement that the output filter response can be selected without the active and passive matching condition. Moreover, the all-pass filter response does not require the double gain amplifier circuit unlike the MISO filters in [37, 41]. However, the inverting unit gain amplifier circuit is required for the all-pass function [44].

3. Effect of Parasitic Elements

Practically, the influence of parasitic element in LT1228 will affect the performances of the proposed filter. High impedance ports V_+, V_-, and z and the parallel RC appeared. The parasitic resistance and capacitance are, respectively, named as R_+, C_+, R_-, C_-, R_z, and C_z. At low impedance port x, the series resistance appears. This resistance is denoted as R_x. Also, the transresistance gain (R_T) is considered as R_T paralleled with C_T. These important parasitic impedances most affect the performance of the proposed circuit. Taking them into account, the output voltage of the circuit in Figure 2 is obtained as

$$v_o = \frac{g_m\left(Y_T + sC_2\right)v_{in1} + \left(Y_T + sC_2\right)sC_1 v_{in2} + \left[R_x Y_T\left(Y_z + sC_1\right) + g_m\right]G v_{in3}}{\left\{\left(sC_2 + G\right)\left[R_x Y_T\left(Y_z + sC_1\right) + g_m\right]\right\} + \left[\left(Y_T + sC_2\right)\left(Y_z + sC_1\right)\right]}, \quad (6)$$

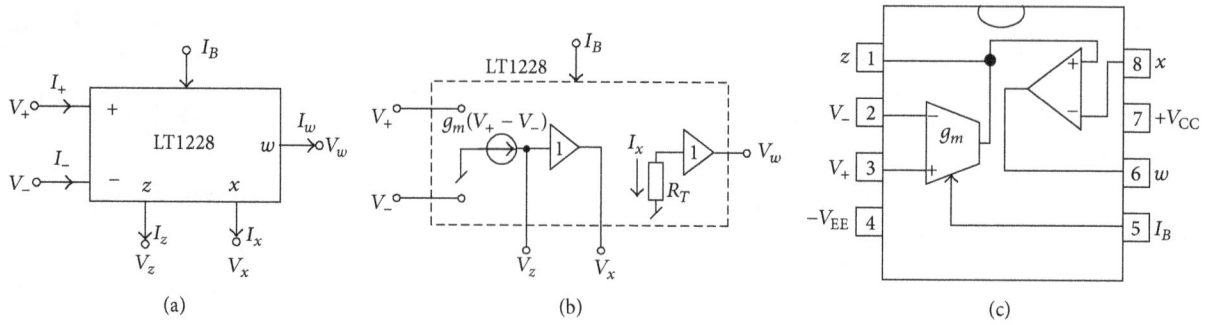

FIGURE 1: LT1228. (a) Electrical symbol of LT1228. (b) Equivalent circuit. (c) Pin configuration.

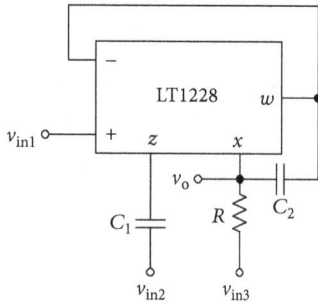

FIGURE 2: Proposed filter.

where $Y_T = sC_T + G_T$, $Y_z = sC_z + G_z$, $G_z = 1/R_z$, $G_T = 1/R_T$, and $G = 1/R$. If the operational frequency $f_{op} \ll 1/C_T R_T$, the output voltage in (6) becomes

$$v_o = \frac{sC_2 g_m v_{in1} + s^2 C_1 C_2 v_{in2} + g_m G v_{in3}}{s^2 + s\left((G_z + g_m)/(C_1 + C_z)\right) + g_m G/(C_1 + C_z) C_2}. \quad (7)$$

From (7), the natural frequency is given as

$$\omega_0 = \sqrt{\frac{g_m}{(C_1 + C_z) C_2 R}}. \quad (8)$$

Subsequently, the quality factor is given as

$$Q = \frac{1}{G_z + g_m} \sqrt{\frac{(C_1 + C_z) g_m}{C_2 R}}. \quad (9)$$

4. Experimental Results

In order to evaluate the performances of the proposed three-input single-output voltage-mode multifunction filter in Figure 2, the experiment was done by using LT1228. The power supply voltage of the LT1228 was ±5 V. An experimental setup was made by taking $C_1 = C_2 = 1$ nF, $R = 1$ kΩ, $I_B = 100$ μA. A resistor of 2 kΩ in series with the x terminal (pin 8 of LT1228) was connected as recommended in datasheet [43]. With above component values, the natural frequency and quality factor as analyzed in (4) and (5) become $f_0 = 159.15$ kHz and $Q = 1$. For this test, the sinusoidal voltage with 60 mV$_{p-p}$ was applied as input voltage. The frequency response of the LP, HP, BP, and BR function is reported in Figure 3 and the phase and gain response of AP function is shown in Figure 4. It is obvious that the proposed filter can provide five filter responses as described in Section 2. The theoretical

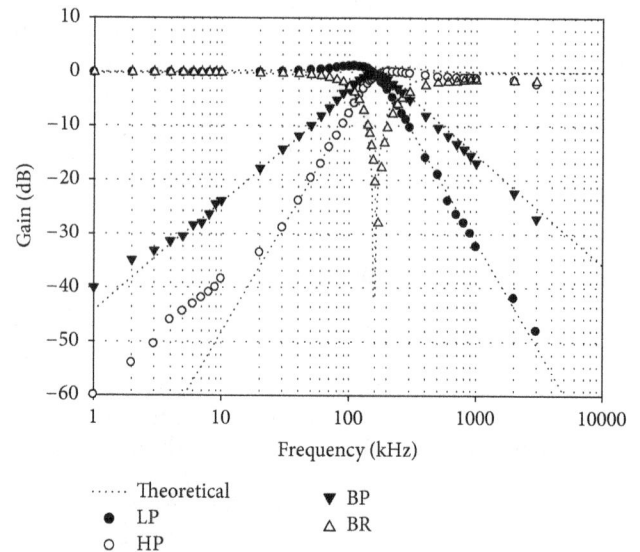

FIGURE 3: Experimental gain response of the proposed filer.

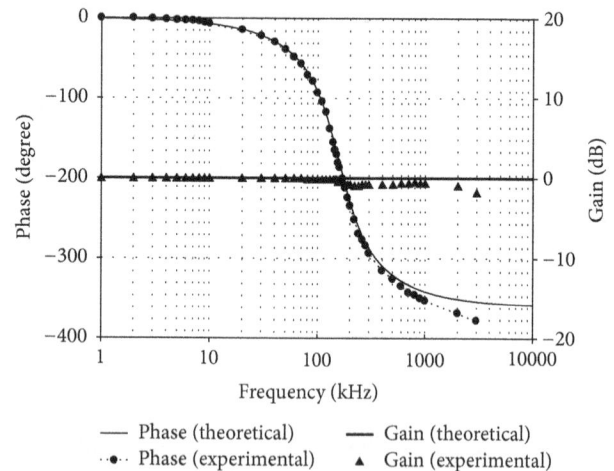

FIGURE 4: Experimental phase and gain response of all-pass function.

and experimental gain responses are slightly different during low and high frequency due to the effect of parasitic elements of LT1228 as studied in Section 2. The experimental natural frequency is about 155 kHz. The deviation of theoretical and experimental natural frequency is about 2.6%. The time-domain responses of output voltage in LP, HP, BP, BR, and AP functions are, respectively, shown in Figures 5, 6, 7, 8, and 9

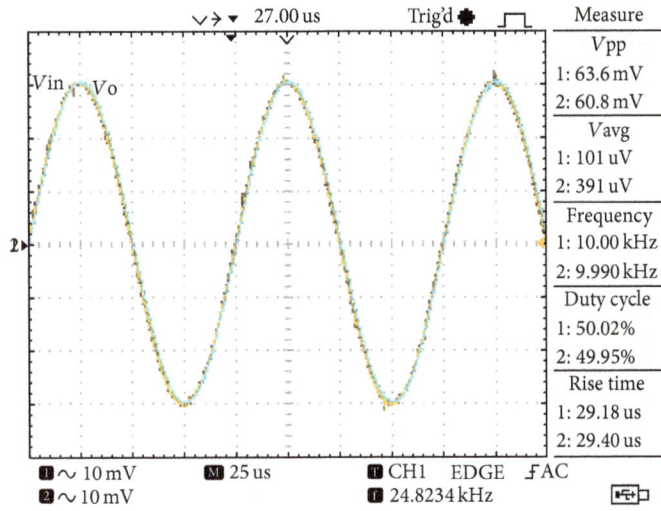

(a)

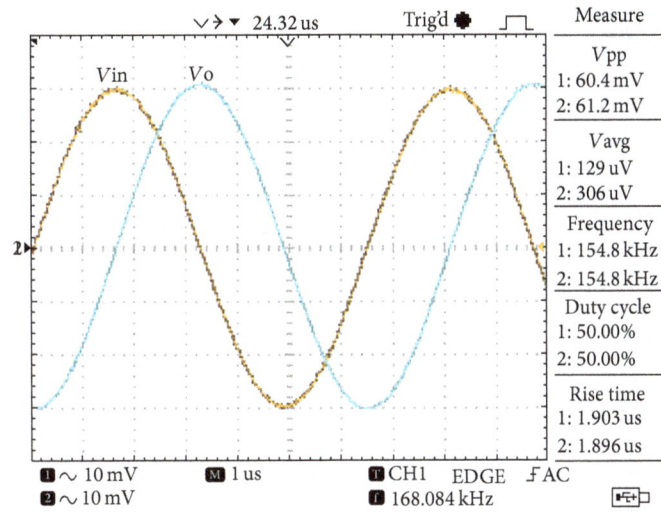

(b)

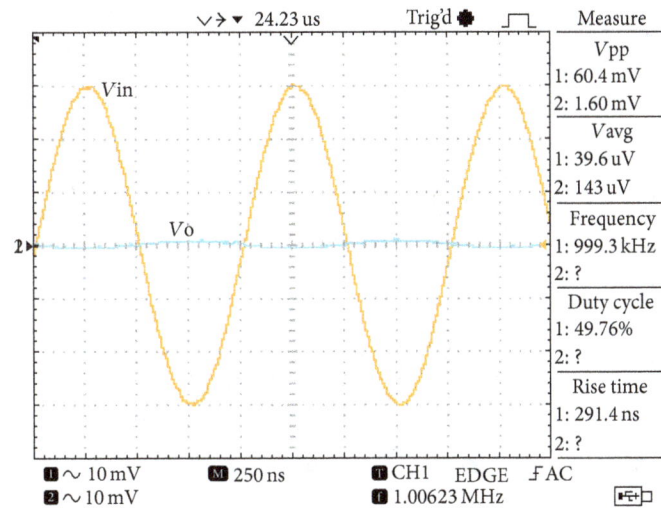

(c)

FIGURE 5: The measured input and output waveforms of low-pass filter at (a) 10 kHz, (b) 155 kHz, and (c) 1 MHz.

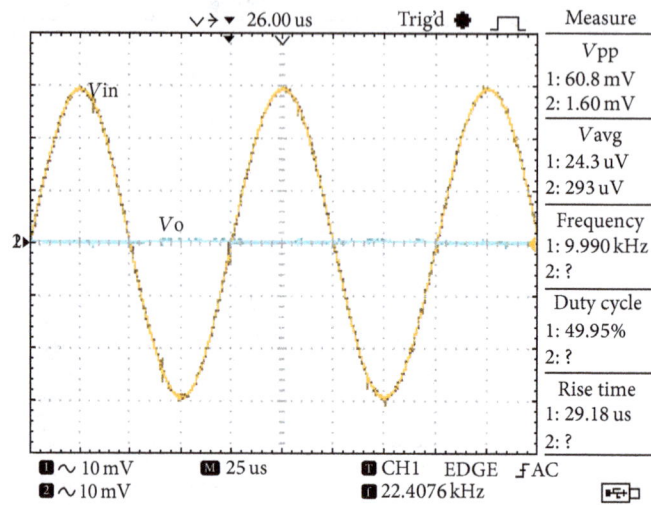

(a)

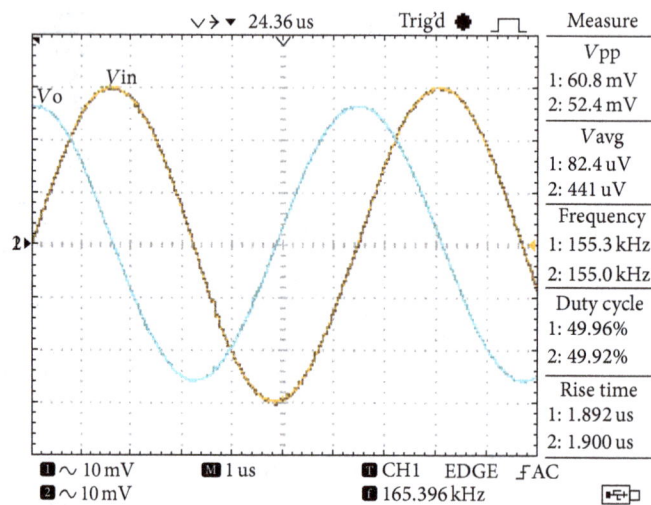

(b)

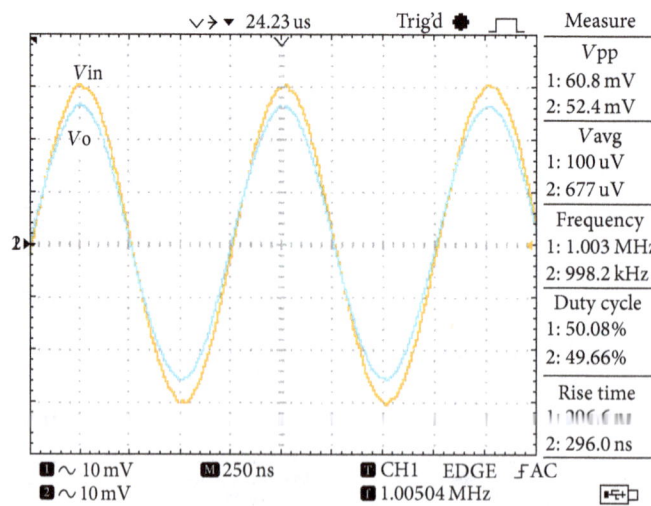

(c)

FIGURE 6: The measured input and output waveforms of high-pass filter at (a) 10 kHz, (b) 155 kHz, and (c) 1 MHz.

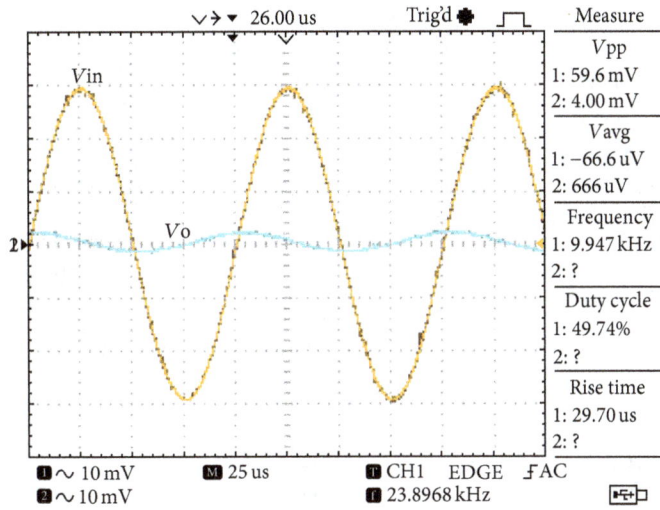

(a)

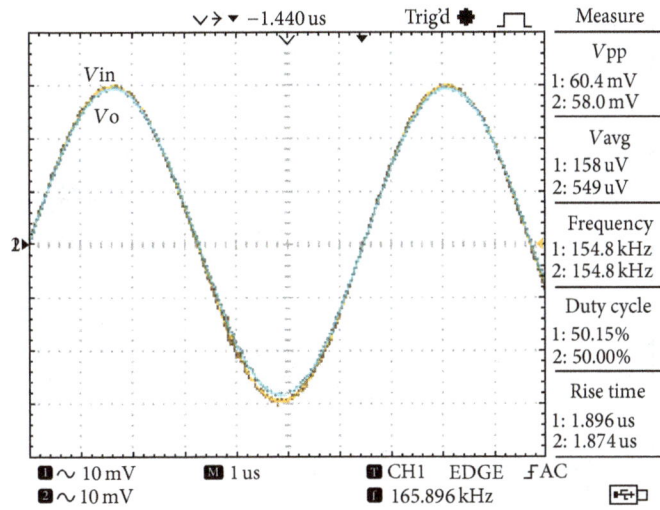

(b)

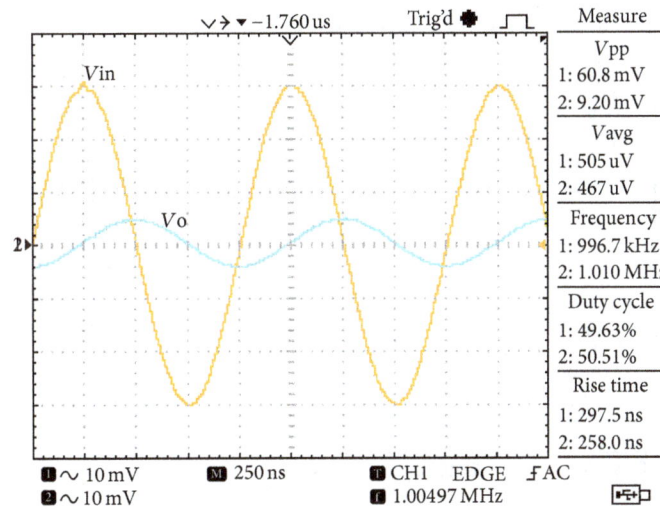

(c)

FIGURE 7: The measured input and output waveforms of band-pass filter at (a) 10 kHz, (b) 155 kHz, and (c) 1 MHz.

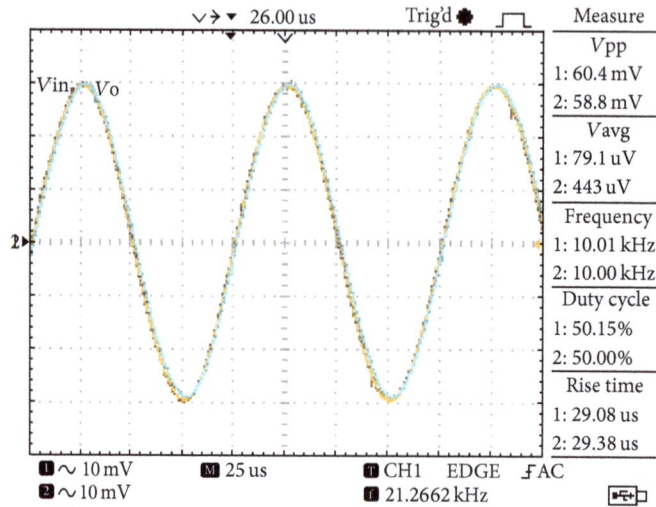

(a)

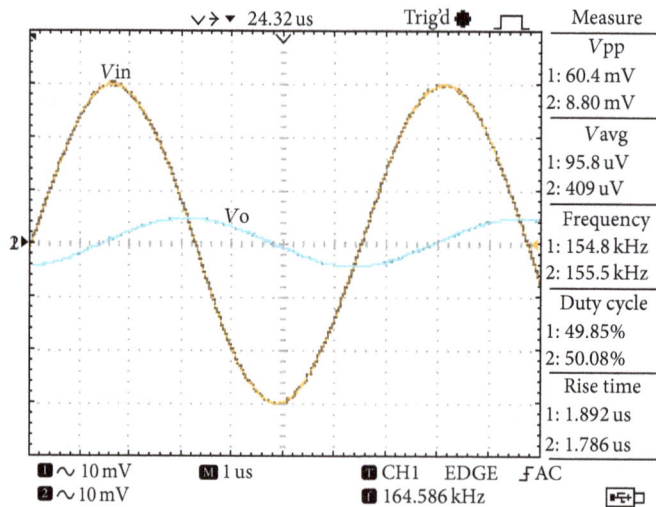

(b)

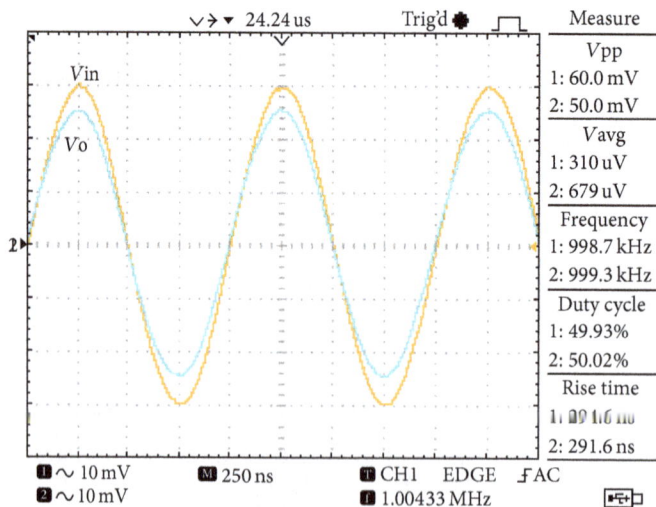

(c)

FIGURE 8: The measured input and output waveforms of band-reject filter at (a) 10 kHz, (b) 155 kHz, and (c) 1 MHz.

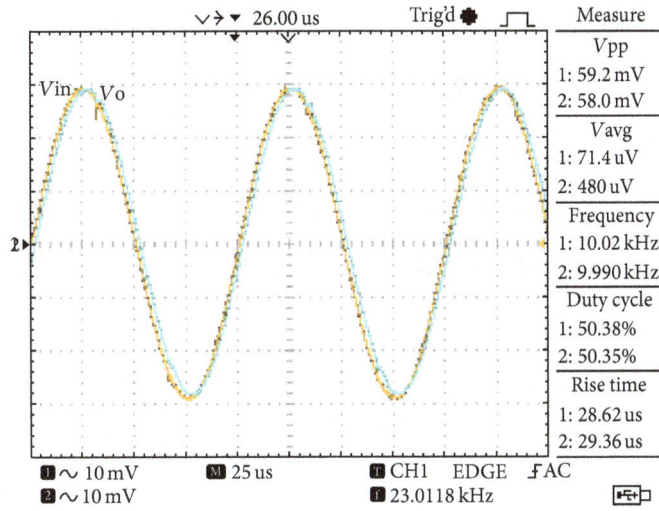

(a)

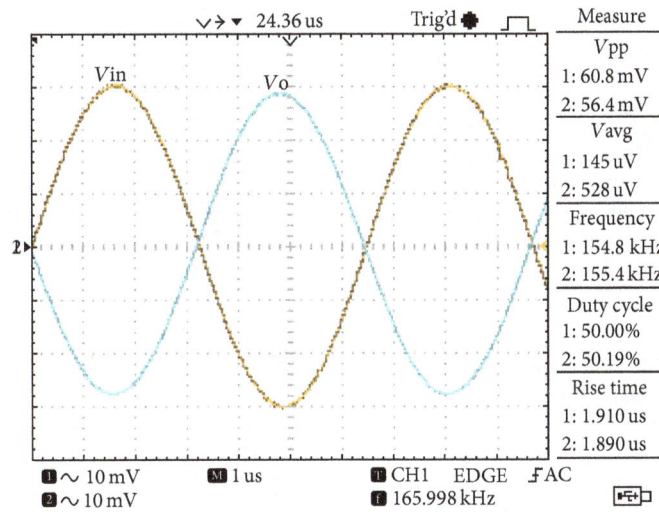

(b)

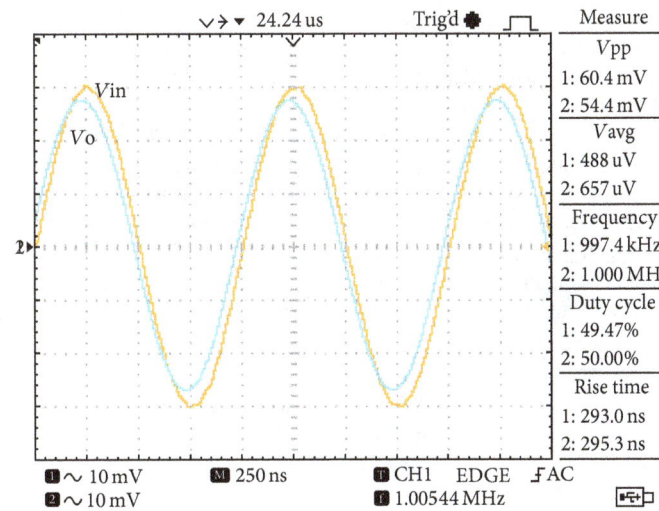

(c)

FIGURE 9: The measured input and output waveforms of all-pass filter at (a) 10 kHz, (b) 155 kHz, and (c) 1 MHz.

when three frequencies, 10 kHz, 155 kHz, and 1 MHz with 60 mV$_{p\text{-}p}$ were applied at the input voltage.

5. Conclusion

In this contribution, the three-input single-output voltage-mode filter is presented. The proposed filter uses only single commercially available IC, LT1228 as active element. The natural frequency and quality factor can be tuned electronically by changing the bias current of LT1228. The selection of output filter response can be done without requirement of the matching condition of passive and active component. Also, the selection of all-pass filter response can be done without the requirement of double gain amplifier. Using only single commercially available IC, the proposed filter is suitable for off-the-shelf implementation. The workability of the proposed filter is demonstrated by experimental results.

Conflicts of Interest

The authors declare that they have no conflicts of interest.

Acknowledgments

The authors are very grateful to Kasetsart University's Graduate School for funding used for publications in international academic journals.

References

[1] N. Afzal and D. Singh, "Reconfigurable mixed mode universal filter," *Active and Passive Electronic Components*, vol. 2014, Article ID 769198, 14 pages, 2014.

[2] P. Beg and S. Maheshwari, "Generalized filter topology using grounded components and single novel active element," *Circuits, Systems, and Signal Processing*, vol. 33, no. 11, pp. 3603–3619, 2014.

[3] W. Mekhum and W. Jaikla, "Three input single output voltage-mode multifunction filter with independent control of pole frequency and quality factor," *Advances in Electrical and Electronic Engineering*, vol. 11, no. 6, pp. 494–500, 2013.

[4] J.-W. Horng, C.-M. Wu, and N. Herencsar, "Three-input-one-output current-mode universal biquadratic filter using one differential difference current conveyor," *Indian Journal of Pure and Applied Physics*, vol. 52, no. 8, pp. 556–562, 2014.

[5] P. Uttaphut, "New current-mode quadrature sinusoidal oscillator using single DVCCTA as active element," *Przegląd Elektrotechniczny*, vol. 1, no. 9, pp. 231–234, 2016.

[6] W. Jaikla, A. Noppakarn, and S. Lawanwisut, "New gain controllable resistor-less current-mode first order allpass filter and its application," *Radioengineering*, vol. 21, no. 1, pp. 312–316, 2012.

[7] S. Maheshwari and M. S. Ansari, "Catalog of realizations for DXCCII using commercially available ICs and applications," *Radioengineering*, vol. 21, no. 1, pp. 281–289, 2012.

[8] S. Maheshwari, "Current conveyor all-pass sections: brief review and novel solution," *The Scientific World Journal*, vol. 2013, Article ID 429391, 6 pages, 2013.

[9] R. Sotner, J. Jerabek, N. Herencsar, J.-W. Horng, K. Vrba, and T. Dostal, "Simple oscillator with enlarged tunability range based

on ECCII and VGA utilizing commercially available analog multiplier," *Measurement Science Review*, vol. 16, no. 2, pp. 35–41, 2016.

[10] C. Chang and M.-S. Lee, "Universal voltage-mode filter with three inputs and one output using three current conveyors and one voltage follower," *Electronics Letters*, vol. 30, no. 25, pp. 2112–2113, 1994.

[11] J.-W. Horng, C.-C. Tsai, and M.-H. Lee, "Novel universal voltage-mode biquad filter with three inputs and one output using only two current conveyors," *International Journal of Electronics*, vol. 80, no. 4, pp. 543–546, 1996.

[12] J.-W. Horng, M.-H. Lee, H.-C. Cheng, and C.-W. Chang, "New CCII-based voltage-mode universal biquadratic filter," *International Journal of Electronics*, vol. 82, no. 2, pp. 151–155, 1997.

[13] C.-M. Chang, "Multifunction biquadratic filters using current conveyors," *IEEE Transactions on Circuits and Systems II: Analog and Digital Signal Processing*, vol. 44, no. 11, pp. 956–958, 1997.

[14] C.-M. Chang and S.-H. Tu, "Universal voltage-mode filter with four inputs and one output using two CCII s," *International Journal of Electronics*, vol. 86, no. 3, pp. 305–309, 1999.

[15] J.-W. Horng, "High-input impedance voltage-mode universal biquadratic filter using three plus-type CCIIs," *IEEE Transactions on Circuits and Systems II: Analog and Digital Signal Processing*, vol. 48, no. 10, pp. 996–997, 2001.

[16] J.-W. Horng, "Voltage-mode multifunction filter using one current feedback amplifier and one voltage follower," *International Journal of Electronics*, vol. 88, no. 2, pp. 153–157, 2001.

[17] J.-W. Horng, C.-K. Chang, and J.-M. Chu, "Voltage-mode universal biquadratic filter using single current-feedback amplifier," *IEICE Transactions on Fundamentals of Electronics, Communications and Computer Sciences*, vol. 85, no. 8, pp. 1970–1973, 2002.

[18] J.-W. Horng, "High input impedance voltage-mode universal biquadratic filter using two OTAs and one CCII," *International Journal of Electronics*, vol. 90, no. 3, pp. 185–191, 2003.

[19] C.-M. Chang and H.-P. Chen, "Universal capacitor-grounded voltage-mode filter with three inputs and a single output," *International Journal of Electronics*, vol. 90, no. 6, pp. 401–406, 2003.

[20] J.-W. Horng, "Voltage-mode universal biquadratic filter using two OTAs," *Active and Passive Electronic Components*, vol. 27, no. 2, pp. 85–89, 2004.

[21] J.-W. Horng, "High input impedance voltage-mode universal biquadratic filters with three inputs using plus-type CCIIs," *International Journal of Electronics*, vol. 91, no. 8, pp. 465–475, 2004.

[22] J. W. Horng, "Voltage-mode universal biquadratic filters using CCIIs," *IEICE Transactions on Fundamentals of Electronics, Communications and Computer Sciences*, vol. 87, pp. 406–409, 2004.

[23] N. A. Shah and M. A. Malik, "Voltage/current-mode universal filter using FTFN and CFA," *Analog Integrated Circuits and Signal Processing*, vol. 45, no. 2, pp. 197–203, 2005.

[24] C.-M. Chang and H.-P. Chen, "Single FDCCII-based tunable universal voltage-mode filter," *Circuits, Systems, and Signal Processing*, vol. 24, no. 2, pp. 221–227, 2005.

[25] N. A. Shah, M. F. Rather, and S. Z. Iqbal, "A novel voltage-mode universal filter using a single CFA," *Active and Passive Electronic Devices*, vol. 1, pp. 183–188, 2005.

[26] M. Sagbas and M. Koksal, "Voltage-mode three-input single-output multifunction filters employing minimum number of components," *Frequenz*, vol. 61, no. 3-4, pp. 87–93, 2007.

[27] W.-Y. Chiu and J.-W. Horng, "High-input and low-output impedance voltage-mode universal biquadratic filter using DDCCs," *IEEE Transactions on Circuits and Systems II: Express Briefs*, vol. 54, no. 8, pp. 649–652, 2007.

[28] S. Kilinç, A. Ü. Keskin, and U. Çam, "Cascadable voltage-mode multifunction biquad employing single OTRA," *Frequenz*, vol. 61, no. 3-4, pp. 84–86, 2007.

[29] M. Kumngern, M. Somdunyakanok, and P. Prommee, "High-input impedance voltage-mode multifunction filter with three-input single-output based on simple CMOS OTAs," in *Proceedings of the International Symposium on Communications and Information Technologies (ISCIT '08)*, pp. 426–431, October 2008.

[30] H.-P. Chen and Y.-Z. Liao, "High-input and low-output impedance voltage-mode universal biquadratic filter using FDCCIIs," in *Proceedings of the 9th International Conference on Solid-State and Integrated-Circuit Technology (ICSICT '08)*, pp. 1794–1798, October 2008.

[31] N. Herencsar, J. Koton, and K. Vrba, "Single CCTA–based universal biquadratic filters employing minimum components," *International Journal of Computer and Electrical Engineering*, vol. 1, no. 3, pp. 307–310, 2009.

[32] H.-P. Chen, "Voltage-mode FDCCII-based universal filters," *AEU—International Journal of Electronics and Communications*, vol. 62, no. 4, pp. 320–323, 2008.

[33] W. Tangsrirat, "Novel current-mode and voltage-mode universal biquad filters using single CFTA," *Indian Journal of Engineering and Materials Sciences*, vol. 17, no. 2, pp. 99–104, 2010.

[34] A. Ranjan and S. K. Paul, "Voltage mode universal biquad using CCCII," *Active and Passive Electronic Components*, vol. 2011, Article ID 439052, 5 pages, 2011.

[35] I. Myderrizi, S. Minaei, and E. Yuce, "DXCCII-based grounded inductance simulators and filter applications," *Microelectronics Journal*, vol. 42, no. 9, pp. 1074–1081, 2011.

[36] W. Tangsrirat and O. Channumsin, "Voltage-mode multifunctional biquadratic filter using single DVCC and minimum number of passive elements," *Indian Journal of Pure and Applied Physics*, vol. 49, no. 10, pp. 703–707, 2011.

[37] J. Satansup and W. Tangsrirat, "Single VDTA-based voltage-mode electronically tunable universal filter," in *Proceedings of the 27th International Technical Conference on Circuits/Systems, Computers and Communications*, Sapporo, Japan, July 2012.

[38] J.-W. Horng, C.-H. Hsu, and C.-Y. Tseng, "High input impedance voltage-mode universal biquadratic filters with three inputs using three CCs and grounding capacitors," *Radioengineering*, vol. 21, no. 1, pp. 290–296, 2012.

[39] J. K. Pathak, A. K. Singh, and R. Senani, "New Voltage Mode Universal Filters Using Only Two CDBAs," *ISRN Electronics*, vol. 2013, Article ID 987867, 6 pages, 2013.

[40] K. L. Pushkar, D. R. Bhaskar, and D. Prasad, "A new MISO-type voltage-mode universal biquad using single VD-DIBA," *ISRN Electronics*, vol. 2013, Article ID 478213, 5 pages, 2013.

[41] W. Ninsraku, D. Biolek, W. Jaikla, S. Siripongdee, and P. Suwanjan, "Electronically controlled high input and low output impedance voltage mode multifunction filter with grounded capacitors," *AEU—International Journal of Electronics and Communications*, vol. 68, no. 12, pp. 1239–1246, 2014.

[42] S. Sangyaem, S. Siripongdee, W. Jaikla, and F. Khateb, "Five-inputs single-output voltage mode universal filter with high input and low output impedance using VDDDAs," *International Journal for Light and Electron Optics*, vol. 128, pp. 14–25, 2017.

[43] http://www.linear.com/product/LT1228.

[44] S. Siripongdee and W. Jaikla, "Electronically controllable grounded inductance simulators using single commercially available IC: LT1228," *AEU—International Journal of Electronics and Communications*, 2017.

10

A New Capacitor-Less Buck DC-DC Converter for LED Applications

Munir Al-Absi, Zainulabideen Khalifa, and Alaa Hussein

Electrical Engineering Department, Faculty of Engineering, King Fahd University of Petroleum and Minerals, Dhahran, Saudi Arabia

Correspondence should be addressed to Munir Al-Absi; mkulaib@kfupm.edu.sa

Academic Editor: Mingxiang Wang

In this paper, a new capacitor-less DC-DC converter is proposed to be used as a light emitting diode (LED) driver. The design is based on the utilization of the internal capacitance of the LED to replace the smoothing capacitor. LED lighting systems usually have many LEDs for better illumination that can reach multiple tens of LEDs. Such configuration can be utilized to enlarge the total internal capacitance and hence minimize the output ripple. Also, the switching frequency is selected such that a minimum ripple appears at the output. The functionality of the proposed design is confirmed experimentally and the efficiency of the driver is 85% at full load.

1. Introduction

Light emitting diodes (LEDs) are starting to experience widespread usage in many lighting applications. LEDs are replacing florescent lighting because of the LEDs' advantages compared to the florescent lamps. These advantages are mainly lower power consumption and longer life expectancy. However, commercial LED drivers limit the life expectancy of the LED lighting system to around one-fifth of the lifetime of the LED itself. The main reason of the driver short lifetime is the smoothing capacitor at the output. This is due to leakage in this capacitor, and hence this causes degradation in the driver performance with time. Several works on electrolytic capacitor-less LED drives have been presented to maximize the overall lifetime of the LED system and the recent states of the art are given in [1–7]. In [1], a current injection approach is used. In [2–7] several single-stage topologies using multiple switches or using shared switch techniques are presented. Most of the works presented require relatively complicated power circuits or current controlled technique to reduce the size of the energy storage capacitor. These topologies lead to larger area and higher cost. A new design of capacitor-less driver is presented in [0]. The design used a storage capacitor C_d and a two-winding dual inductor.

The major intention of this paper is to build upon the results obtained in [9] and present the mathematical model and experimental results to confirm the functionality of the design. The rest of the paper is organized as follows: Section 2 describes the proposed design. Mathematical analysis and experimental results are given in Section 3. Section 4 concludes the paper.

2. The Proposed Design

The proposed design is based on the well-known buck converter shown in Figure 1, where the output voltage is the voltage across the load resistance R_L and C_0. Vpulse represents the controlling pulses generated from the control circuit. The DC output voltage is given by

$$V_{O(DC)} = \frac{D\left(V_{in} - V_{ds}\right) - D'V_d}{1 + r_L/R_L}, \tag{1}$$

where V_{ds} is the drain-to-source voltage of the MOS transistor used for switching, r_L is the inductor resistance, V_d is the diode voltage drop, D is the ON duty cycle of the control pulse, and D' is the OFF duty cycle of the pulse.

The inductor L and the smoothing capacitor C_0 will average the pulses passing through Q1 causing ripples on the load. The ripple voltage will be affected by the duty cycle, the switching frequency, the inductance, the internal resistance of

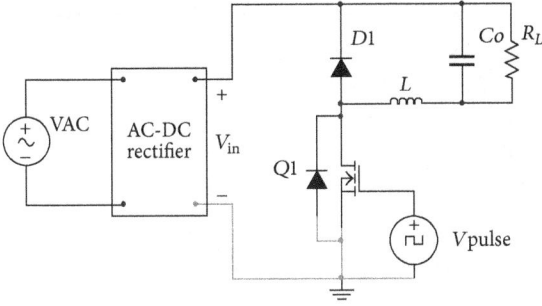

FIGURE 1: Buck DC-DC converter.

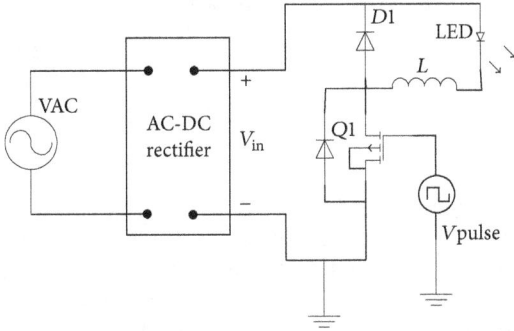

FIGURE 2: The proposed capacitor-less buck DC-DC converter.

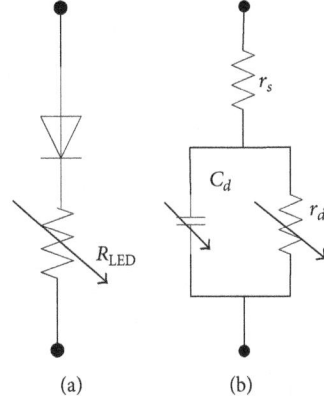

(a) (b)

FIGURE 3: LED model in conduction mode for (a) DC mode and (b) AC mode.

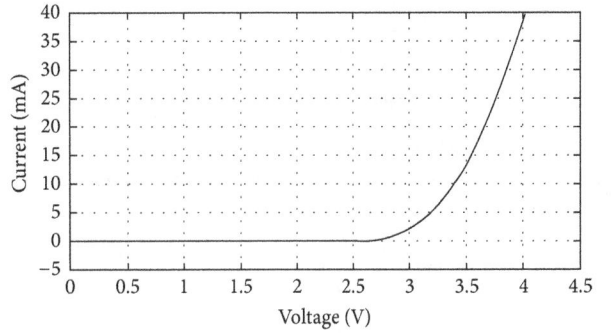

FIGURE 4: The V-I characteristics curve of a single white LED.

the smoothing capacitor ESR, and the value of the smoothing capacitor.

The approximate voltage ripples assuming linear models and a small ripple voltage are given by [10]

$$\Delta V_r = \frac{V_{in} - (V_o + V_{ds} + V_{rL})}{Lf_s} \times D \left(\frac{1}{8C_o f_s} + \text{ESR} \right), \quad (2)$$

where f_s is the switch frequency of Vpulse, L is the inductor, V_{r_L} is the voltage across the inductor resistance, and ESR is the capacitor series resistor.

The proposed design is a modified version of the design in Figure 1 and is shown in Figure 2, where the load is an array of LEDs, as it is the case in all commercially available LED lamps. The internal capacitance of the LED array will act as a smoothing capacitor if a proper switching frequency and duty cycle are chosen, and hence no external smoothing capacitor is needed.

3. Mathematical Analysis and Experimental Results

3.1. Mathematical Analysis. It is well known that the LED in conduction mode can be modeled using a resistor and an ideal diode for DC mode and a capacitor and a resistor in parallel for AC mode as shown in Figures 3(a) and 3(b), respectively. The resistance r_s represents the constant series contact resistance and quasineutral region resistance of the LED, r_d represents the small signal resistance of the LED at certain DC current, and C_d represents the diffusion

capacitance at a certain DC current. In conduction mode, r_d is the reciprocal of the conductance which is equal to the DC current divided by the thermal voltage. This indicates that as the DC current increases, the value of the resistance r_d will decrease. Moreover, the value of C_d also is a function of the conductance and its value will increase as the current increases [11].

With reference to Figure 2, the DC output voltage across the LEDs is the same as in (1) with R_L replaced by R_{LED}. The LED equivalent circuits shown in Figure 3 are used in this analysis. The DC output voltage is given by

$$V_{O(DC)} = \frac{D(V_{in} - V_{ds}) - D'V_d}{1 + r_L/R_{LED}}, \quad (3)$$

where r_L is inductor resistance. The value of R_{LED} depends on the current passing through the LED, and it can be deduced from the I-V characteristics curve of the LED shown in Figure 4. It is clear form Figure 4 that as the current increases, the value of R_{LED} will decrease.

To find the effective capacitance of the LED, the ripple current is given by

$$\Delta I_{pp} = \frac{V_{in} - (V_o + V_{ds} + V_{rL})}{Lf_s} \times D, \quad (4)$$

where ΔI_{pp} is the ripple current through the inductor L. From Figure 2 and the model of Figure 3, the output voltage ripple

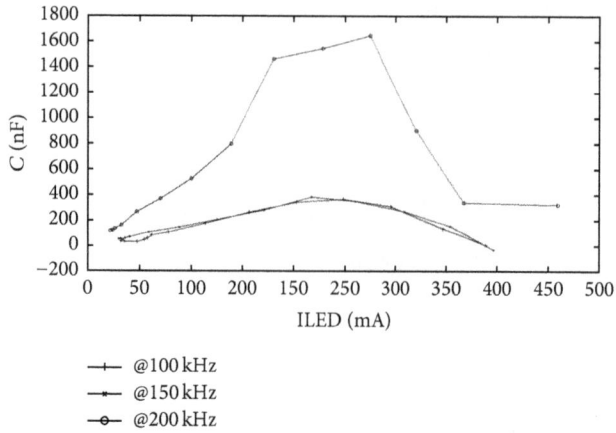

—+— @100 kHz
—•— @150 kHz
—◦— @200 kHz

FIGURE 5: Plot of the effective capacitance C_d versus the load current.

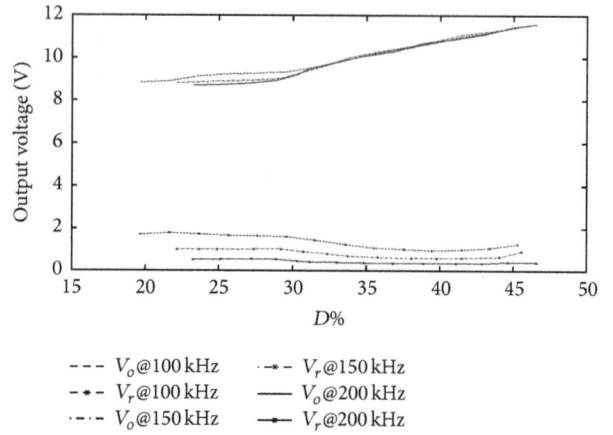

--- V_o@100 kHz   ·–•– V_r@150 kHz
--•-- V_r@100 kHz   —— V_o@200 kHz
·–·– V_o@150 kHz   —•— V_r@200 kHz

FIGURE 6: The output DC (V_o) and ripple (V_r) voltage versus the duty cycle for different frequencies.

is given by

$$\Delta V_r = \Delta I_{pp} \times R_{\text{LOAD}} = \Delta I_{pp}\,(z + r_s), \qquad (5)$$

where $z = (r_d \times 1/8 f_s C_d)/(r_d + 1/8 f_s C_d) = r_d/(1 + 8 r_d f_s C_d)$ and $1/8 f_s C_d$ is the impedance of the diffusion capacitor [10].

Combining (4) and (5), the output voltage ripple is given by

$$\Delta V_r = \Delta I_{pp}\left(r_s + \frac{1}{g_d + 8 f_s C_d}\right), \qquad (6)$$

where $g_d = 1/r_d$.

Rewriting (6) to find the effective capacitance C_d,

$$C_d = \frac{1}{8 f_s}\left(\frac{1}{\Delta V_r/\Delta I_{pp} - r_s} - g_d\right). \qquad (7)$$

Plots of the effective capacitance as a function of the LED current for different frequencies are shown in Figure 5. It is evident from the figure that the effective capacitance at 200 kHz is high since the impedance of the capacitance is much smaller than that of the dynamic resistance.

In the AC model of Figure 3, the behavior of r_d and C_d gives an indication that as the DC current increases, the ripple voltage will decrease, which is another parameter that can be controlled and affect the ripple voltage. This fact is supported by the experimental results we have carried out and it is explained in the next section.

It is important to point out that the value of C_d is linearly changing with the DC current only in strong conduction mode [12]. However, during the OFF period in the switching buck converter pulse, the LED internal resistance will draw the stored charge and the output voltage will decrease. If the OFF period is long enough, the value of the diffusion capacitor will be very small causing a sharp drop in the output voltage that might cause flicker in the LED light.

3.2. Experimental Results. The circuit shown in Figure 2 was connected in the laboratory using off-the-shelf components to test the proposed design experimentally. The LED used is the sum of 3 series packages of 11 parallel LEDs per

package giving a total of 33 LEDs. The output voltage is measured across the LED packages. The components used are as follows: L is an inductor of 470 uH, Q1 is an N-MOS power transistor BUZ71, Vpulse is the switching control pulse with an amplitude of 10 V, and $D1$ is a silicon fast switching diode 1N914. The inductor's series resistance was measured and its value was approximately 4 Ω. It was assumed that the AC source was rectified and provided a DC output called V_{in} with nominal voltage of 35 V. The LED's I-V characteristics shown in Figure 4 have been used to extract the value of R_{LED} for different DC current values.

The behavior of the circuit was studied by varying the duty cycle of Vpulse from 18% to 44% at three different frequencies (100 kHz, 150 kHz, and 200 kHz). The maximum duty cycle was set to 44% because this duty cycle will produce the maximum LED current. The DC output and ripple voltage are plotted in Figure 6. As is clear from the figure, as the duty cycle increases, the DC output voltage increases. The ripple voltage decreased with the increase in frequency.

From Figure 6, the DC voltage is changing linearly with the duty cycle for $D > 30\%$. Also, it is clear that, for duty cycle greater than 30%, the error is less than 3%. The deviation between theoretical and experimental results is shown in Figure 7. It is evident from the plot that a designer should select the switching pulse duty cycle to be greater than 30% to minimize the error and a higher frequency to minimize the ripple voltage.

If the voltage across the LED reached below a certain value, there will be no diffusion capacitor, and the LED's voltage will drop logarithmically, causing the large error shown in Figure 7. This value can be estimated from the knees of each curve and depends on the forward current as well, since it depends on how deep the LED is in the conduction region. Figures 8 and 9 show the ripple voltage at 100 kHz with duty cycle of 18% and 40%, respectively. The nonlinearity is clearly shown in Figure 8, where the OFF period was long enough to drive the LED to the weak conduction region while the ripple of Figure 9 is almost linear. It is clear that the ripple is linear for higher duty cycle.

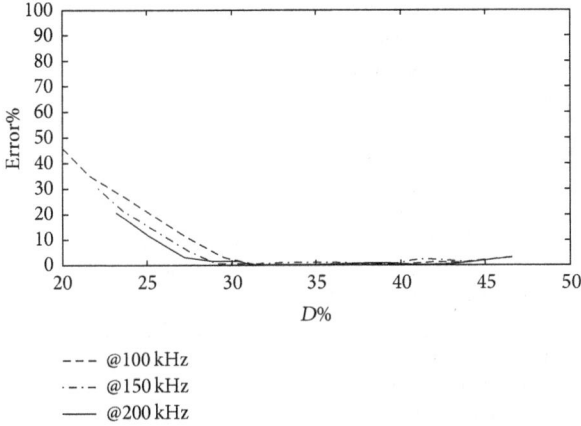

FIGURE 7: The %error in the experimental results with respect to the theory versus the duty cycle.

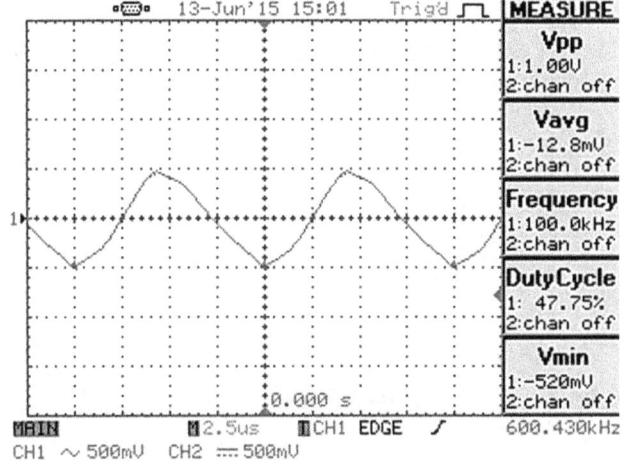

FIGURE 9: Plot of ripple voltage versus time at 100 kHz and $D = 40\%$.

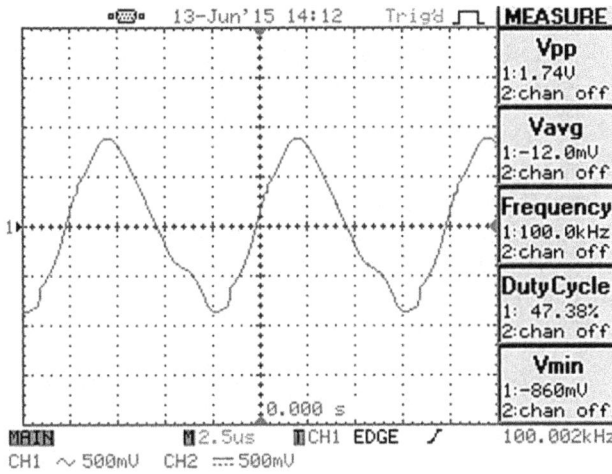

FIGURE 8: Plot of ripple voltage versus time at 100 kHz and $D = 18\%$.

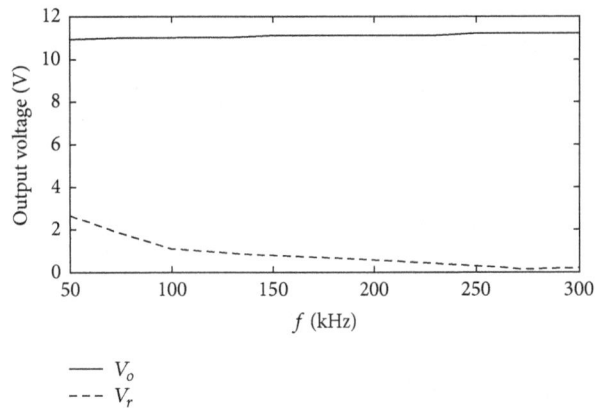

FIGURE 10: The output DC (V_o) and ripple (V_r) voltage versus frequency.

To investigate the changes on the DC output voltages and ripple, the frequency was swept from 50 kHz to 300 kHz at a fixed duty cycle of 40%, and the output was probed. The result is shown in Figure 10. It is clear that the ripple voltage is decreasing as the frequency increases and the DC voltage is almost constant. The minimum ratio of ripple voltage to DC voltage is around 1.4% and it can be decreased further by increasing the frequency.

Efficiency is an important factor in a LED driver. The efficiency was found by measuring the DC output voltage, the output current, the DC input voltage, and the input current for each duty cycle for different frequencies. Experimental results displayed in Figure 11 show that the average efficiency is 85%. The efficiency can be further improved using an inductor with smaller internal resistance and a transistor with smaller ON resistance.

Because of the slight changes in the DC output voltage, the efficiency is barely changing with the change of the frequency, as shown in Figure 12. The average of the efficiency over the frequency range was about 88%. Increasing the frequency further will lead to smaller ripple voltages and smaller components for better integration. However, increasing the

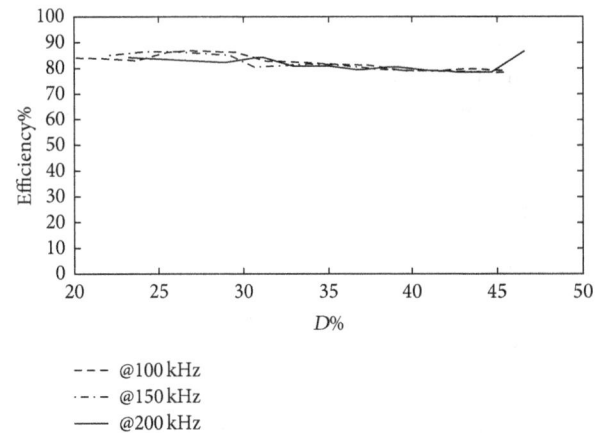

FIGURE 11: The efficiency versus the duty cycle at different frequencies.

switching frequency will reduce the efficiency of the drive because of the switching power loss for light loads [12]. As for LED lighting applications, the LED load needs to draw high current specially when using a capacitor-less drive. This is because it is better to use many parallel LEDs for higher

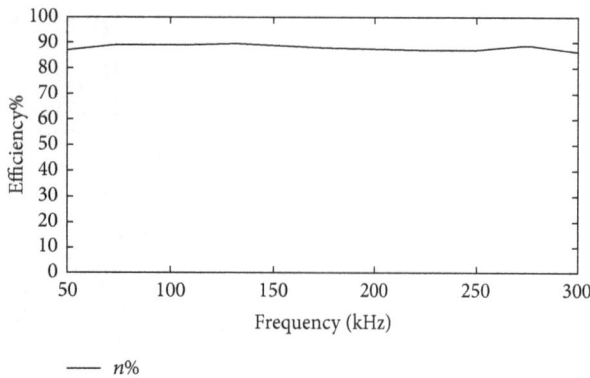

FIGURE 12: The efficiency versus frequency.

summation of LED capacitance, which gives this method one more advantage.

4. Conclusion

A new approach to designing capacitor-less buck DC-DC converter was developed and tested. The proposed single switch circuit is able to reduce the ripple in a compact form and can be extended to any other LED configuration. The design's mathematical model was developed based on experimental verification. The efficiency of the driver is 85% and we expect the lifetime to be much higher than existing drives, as there is no capacitor in the switching path of the driver.

Competing Interests

The authors declare that they have no competing interests.

Acknowledgments

The authors would like to thank King Abdulaziz City for Science and Technology for financial support (Project no. A-T-34-20) and KFUPM for using all facilities to carry out this research.

References

[1] B. Wang, X. Ruan, K. Yao, and M. Xu, "A method of reducing the peak-to-average ratio of LED current for electrolytic capacitor-less AC–DC drivers," *IEEE Transactions on Power Electronics*, vol. 25, no. 3, pp. 592–601, 2010.

[2] S. Wang, X. Ruan, K. Yao, S.-C. Tan, Y. Yang, and Z. Ye, "A flicker-free electrolytic capacitor-less AC-DC LED driver," *IEEE Transactions on Power Electronics*, vol. 27, no. 11, pp. 4540–4548, 2012.

[3] W. Chen and S. Y. R. Hui, "Elimination of an electrolytic capacitor in AC/DC light-emitting diode (LED) driver with high input power factor and constant output current," *IEEE Transactions on Power Electronics*, vol. 27, no. 3, pp. 1598–1607, 2012.

[4] P. S. Almeida, G. M. Soares, D. P. Pinto, and H. A. C. Braga, "Integrated SEPIC buck-boost converter as an off-line LED driver without electrolytic capacitors," in *Proceedings of the 38th Annual Conference on IEEE Industrial Electronics Society (IECON '12)*, pp. 4551–4556, Québec, Canada, October 2012.

[5] H. Ma, J.-S. Lai, Q. Feng, W. Yu, C. Zheng, and Z. Zhao, "A novel valley-fill SEPIC-derived power supply without electrolytic capacitor for LED lighting application," *IEEE Transactions on Power Electronics*, vol. 27, no. 6, pp. 3057–3071, 2012.

[6] H. Ma, W. Yu, C. Zheng, Q. Feng, and B. Chen, "A universal input high-power-factor PFC pre-regulator without electrolytic capacitor for PWM dimming LED lighting application," in *Proceedings of the IEEE Energy Conversion Congress and Exposition (ECCE '11)*, pp. 2288–2295, September 2011.

[7] M. Ryu, J. Kim, J. Baek, and H.-G. Kim, "New multi-channel LEDs driving methods using current transformer in electrolytic capacitor-less AC-DC drivers," in *Proceedings of the 27th Annual IEEE Applied Power Electronics Conference and Exposition (APEC '12)*, pp. 2361–2367, Orlando, Fla, USA, February 2012.

[8] J. C. W. Lam and P. K. Jain, "A high power factor, electrolytic capacitor-less AC-input LED driver topology with high frequency pulsating output current," *IEEE Transactions on Power Electronics*, vol. 30, no. 2, pp. 943–955, 2015.

[9] M. A. Al-Absi, Z. J. Khalifa, and A. E. Hussein, "A new capacitor-less LED drive," in *Proceedings of the 13th International Multi-Conference on Systems, Signals and Devices (SSD '16)*, pp. 354–357, IEEE, Leipzig, Germany, March 2016.

[10] W. Robert and M. Dragan, *Fundamentals of Power Electronics*, vol. 2nd, Kluwer Academic, New York, NY, USA, 2001.

[11] F. Robert, *Semiconductor Device Fundamentals*, Addison-Wesley, New York, NY, USA, 1996.

[12] L. Solymar, D. Walsh, and A. Syms, *Electrical Properties of Materials*, 9th edition, 2014.

Reconfigurable Mixed Mode Universal Filter

Neelofer Afzal and Devesh Singh

Department of Electronics and Communication Engineering, Jamia Millia Islamia University, New Delhi 110025, India

Correspondence should be addressed to Devesh Singh; deva_singh11@yahoo.co.in

Academic Editor: Sudhanshu Maheshwari

This paper presents a novel mixed mode universal filter configuration capable of working in voltage and transimpedance mode. The proposed single filter configuration can be reconfigured digitally to realize all the five second order filter functions (types) at single output port. Other salient features of proposed configuration include independently programmable filter parameters, full cascadability, and low sensitivity figure. However, all these features are provided at the cost of quite large number of active elements. It needs three digitally programmable current feedback amplifiers and three digitally programmable current conveyors. Use of six active elements is justified by introducing three additional reduced hardware mixed mode universal filter configurations and its comparison with reported filters.

1. Introduction

Current feedback amplifier (CFA) plays significant role in area of signal processing/generation because of its higher speed, higher slew rate, simpler circuit realization, and most importantly the independence of gain and bandwidth [1–3]. Introduction of digital control/programming in CFA has further boosted its functional flexibilities and versatility [1]. Programmable characteristic of analog block is essential for controlling the undesired parameter variation caused by temperature and process. Analog programming techniques are widely used in a number of applications [3–8] but the limitation on the allowable range of analog tuning voltage makes it inconvenient for low voltage applications. Hence, in these applications, the digital control is more attractive [9]. Digital programming techniques not only yields better accuracy in avoiding parameter race than their analog counterpart [10] but also offers additional advantages such as better noise immunity, power saving option [11], and most importantly the compatibility to modern mixed mode (analog/digital) systems.

Digitally programmable universal filters (DPUF) are versatile and cost effective from IC realization viewpoint. However, to be compatible to IC realization it should fulfill following two conditions. First, it must be reconfigurable to realize different filter functions (types) without any change in configuration. Second, all its parameters should be independently programmable to set desired frequency response. Obviously, the availability of mixed mode operation will further enhance the versatility of such DPUF.

This paper presents a novel reconfigurable voltage/transimpedance mode (VM/TIM) DPUF using three digitally programmable CFAs (DPCFA) and three digitally programmable second generation current conveyors (DPCCII). The proposed UF can be reconfigured digitally to realize all the standard second order filter functions, namely, lowpass (LP), highpass (HP), bandpass (BP), band reject (BR), and allpass (AP) at single output port. The proposed DPUF configuration is (1) fully programmable as all the coefficients of its transfer function are independently controlled, which makes its parameters, namely, pole frequency (ω_0), quality factor (Q), and gain (G) independently programmable (2) fully cascadable by virtue appropriate (low/high for voltage/current) input and (low for voltage) output port impedances (3) less sensitive to nonidealities and parasitic effects. Use of only two grounded capacitor makes the proposed UF suitable from integration [12].

This paper also introduces three additional reduced hardware DPUFs using "2–4" active elements. These DPUFs are designated as *derived* DPUFs because they are obtained by the reduction of active elements of reconfigurable DPUF.

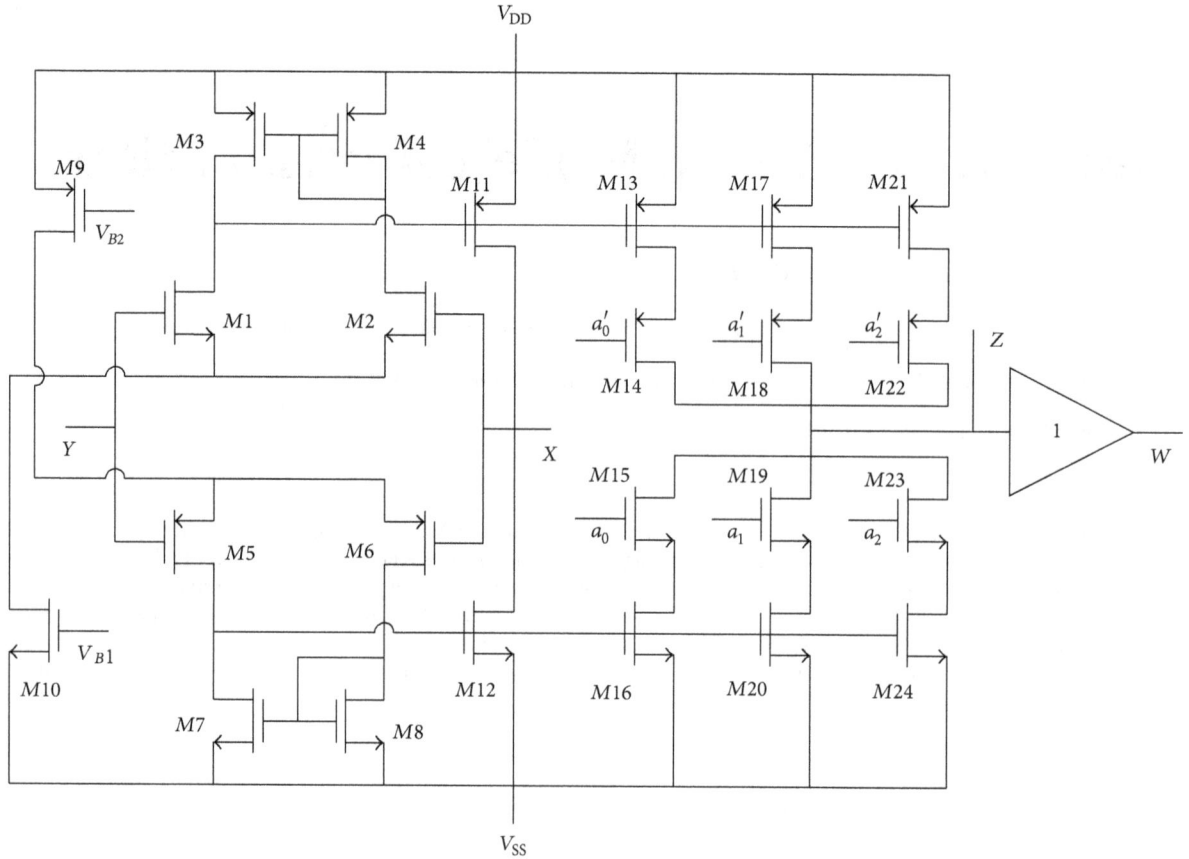

FIGURE 1: CMOS structure of DPCFA with gain K^+ (adopted from [1]).

Obviously, reduction in number of active elements also decreases the features of these derived DPUFs accordingly. But still, these DPUFs (including reconfigurable DPUF) possess more number of features than that of reported DPUFs using equal or more number of active elements. This fact in turn justifies the need of six active elements in reconfigurable filter. One common drawback of second and third derived DPUF is the use of floating capacitors, which is less attractive for integration. However, new integrated circuit technologies are capable of implementing efficient floating capacitor as double poly layer capacitor [13].

This paper is organized as follows. Starting from the introduction, Section 2 presents brief introduction of DPCFA, Section 3 presents the realization of reconfigurable DPUF, Section 4 deals with the derived DPUFs, Section 5 presents the comparison, Section 6 discusses the nonidealities and mismatch effects, Section 7 deals with SPICE simulation, and finally, the paper is concluded in Section 8.

2. Overview of DPCFA/DPCCII

The concept of digital control in DPCFA/DPCCII is based on employing an n-bit current summing network (CSN), which scales up (amplification) or scales down (attenuation) the current gain of conventional CFA/CCII [1]. Figure 1 shows the CMOS structure of 3-array DPCFA. The CSN consist

of transistors M13–M24. Depending on the code-bit values $(a_2 a_1 a_0)$, respective arrays are activated or deactivated to produce port-Z current. DPCFA consists of a voltage follower (VF) between port Z and W. CMOS structure of VF follows the same circuitry as formed by transistors M1–M12. A DPCCII is equivalent to a DPCFA with output VF removed.

Port relation of DPCFA is described by following transfer matrix

$$\begin{bmatrix} I_Y \\ V_X \\ I_Z \\ V_W \end{bmatrix} = \begin{bmatrix} 0 & 0 & 0 & 0 \\ \alpha & 0 & 0 & 0 \\ 0 & \beta K^m & 0 & 0 \\ 0 & 0 & \gamma & 0 \end{bmatrix} \begin{bmatrix} V_Y \\ I_X \\ V_Z \\ V_W \end{bmatrix}, \quad (1)$$

where K denotes the decimal equivalent of applied n-bit codeword ($= a_{n-1}, a_{n-2}, \ldots, a_0$). It is given as

$$K = \sum_{j=0}^{n-1} a_j 2^j, \quad (2)$$

where j denotes the jth bit of applied codeword. Parameter α, β, and γ denotes the nonideal gain transfer ratios. All these gain parameters are unity in ideal condition. Power integer $m = -1$ denotes current attenuation ($K^{-1} = K^- = 1/K$) in range "1 to $1/(2^n - 1)$" while $m = +1$ denotes current amplification ($K^{+1} = K^+ = K$) in range "0 to $(2^n - 1)$". The concept of zero gain (for $K = 0$; that is, all bits zero) is used for programming the generation of various filter functions (types). Figure 1 shows the DPCFA structure with gain K^+. DPCFA structure with gain K^- can be found in [1].

Figure 2 shows the symbolic form of DPCFA and DPCCII. It shows ith DPCFA/DPCCII block with current gain $K_i^\pm$ and applied codeword K.

3. Proposed Reconfigurable DPUF

The proposed VM/TIM DPUF is depicted in Figure 3. It uses three DPCFA, three DPCCII, eight resistors, and two grounded capacitors. The UF offers the following attractive features.

(i) Availability of mixed mode operation.

(ii) Digitally controlled filter functions (type).

(iii) All the filter parameters (ω_0, Q, and G) are independently programmable.

(iv) High (low) impedance input port for voltage (current) signal and low impedance output port enables easy cascading for higher order filter realization without requiring any buffer stage.

In addition to this, use of only two grounded capacitors makes the proposed DPUF suitable for monolithic integration.

The mixed mode operation of proposed reconfigurable DPUF is given by the following output function:

$$V_0 = \frac{N_v(s) + N_i(s)}{D(s)}, \tag{3}$$

where the numerator (N) and denominator (D) functions are given by

$$N_v(s) = V_i \left(s^2 \frac{R_6}{R_7} \{K_4\} \right.$$
$$\left. -s \frac{K_3 R_5}{K_1 C_2 R_4 R_8} \{K_5\} + \frac{K_2 K_3}{C_1 C_2 R_2 R_3} \{K_6\} \right), \tag{4a}$$

$$N_i(s) = s^2 R_6 \{K_4\} I_1 - s \frac{K_3 R_5}{K_1 C_2 R_8} \{K_5\} I_2 + \frac{K_2 K_3}{C_1 C_2 R_2} \{K_6\} I_3, \tag{4b}$$

$$D(s) = s^2 + s \frac{K_3}{K_1 C_2 R_8} + \frac{K_2 K_3}{C_1 C_2 R_1 R_2}. \tag{4c}$$

It is evident from (4a)–(4c) that all the coefficients of numerator and denominator functions are independently programmable by codewords "K_1–K_6". This also justifies the need of minimum six programmable blocks in proposed DPUF. Codeword "K_1–K_3" provides independent programming of ω_0 and Q (discussed below) whereas codewords "K_4–K_6" programming the numerator coefficients, not only governs the generation of various filter functions in both the modes but also provide them the independently programmable gain factors too. Additionally, by setting codewords condition $K_4 > K_6$ ($K_4 < K_6$); high (low) pass notch

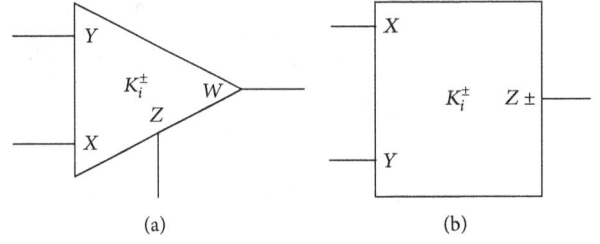

FIGURE 2: Symbol of (a) DPCFA and (b) DPCCII.

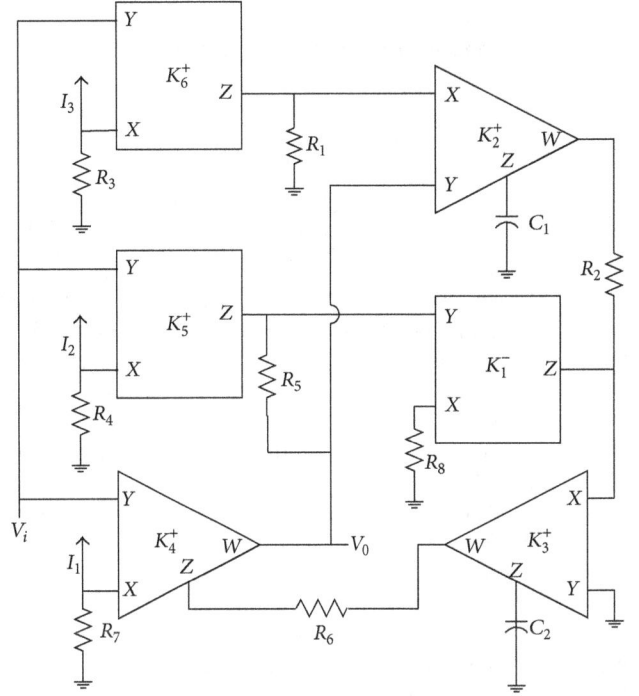

FIGURE 3: Proposed reconfigurable DPUF.

response can be realized. Table 1 summarizes the codeword conditions, realized gain parameters, and component and input (if any) matching conditions for the realization of various filter functions. Only AP and BR response requires component matching constraint.

Setting codeword condition $K_2 = K_3$ modifies (4a)–(4c) as

$$N_v(s) = V_i \left(s^2 \frac{R_6}{R_7} \{K_4\} \right.$$
$$\left. -s \frac{K_2 R_5}{K_1 C_2 R_4 R_8} \{K_5\} + \frac{K_2^2}{C_1 C_2 R_2 R_3} \{K_6\} \right), \tag{5a}$$

$$N_i(s) = s^2 R_6 \{K_4\} I_1 - s \frac{K_2 R_5}{K_1 C_2 R_8} \{K_5\} I_2 + \frac{K_2^2}{C_1 C_2 R_2} \{K_6\} I_3, \tag{5b}$$

$$D(s) = s^2 + s \frac{K_2}{K_1 C_2 R_8} + \frac{K_2^2}{C_1 C_2 R_1 R_2}. \tag{6}$$

TABLE 1: Codeword combination and realized filter functions.

Filter functions	Codeword combination	Gain (G) in different modes VM* (V_0/V_i)	TIM# (V_0/I_i)	Component matching condition	Input matching condition (for TIM only)
HP	$K_5 = K_6 = 0, K_4 \neq 0$	$(K_4 R_6)/R_7$	$K_4 R_6$	NO	NR
BP	$K_4 = K_6 = 0, K_5 \neq 0$	$-(K_5 R_5)/R_4$	$K_5 R_5$	NO	NR
LP	$K_4 = K_5 = 0, K_6 \neq 0$	$(K_6 R_1)/R_3$	$K_6 R_1$	NO	NR
BR	$K_5 = 0, K_4 = K_6 \neq 0$	K_4	$K_4 R_6$	$R_6/R_7 = R_1/R_3 = 1$	$I_1 = I_3 = I_i$
AP	$K_4 = K_5 = K_6 \neq 0$	K_4	$K_4 R_6$	$R_6/R_7 = R_5/R_4 = R_1/R_3 = 1$	$I_1 = I_2 = I_3 = I_i$

*For VM $I_1 = I_2 = I_3 = 0$ is required; #for TIM $V_i = 0$ is required; NR: not required.

Filter parameters ω_0 and Q from (6) are given as

$$\omega_0 = \left[\frac{1}{C_1 C_2 R_1 R_2} \right]^{1/2} [K_2], \qquad (7a)$$

$$Q = R_8 \left[\frac{C_2}{C_1 R_1 R_2} \right]^{1/2} [K_1]. \qquad (7b)$$

It can be seen from (7a)-(7b) that ω_0 and Q of all the responses are independently programmable through codeword $K_2(= K_3)$ and K_1, respectively. Codeword condition $K_2(= K_3)$ makes the parameters ω_0 and Q independently programmable. Thus, independent programming of these two parameters also requires minimum three programmable blocks [14].

One additional advantage offered by reconfigurable and derived (to be described in next section) DPUFs is the *downscale* programming of pole frequency (7a). This is achieved by reversing the gain parameter "K^+" of blocks 2 and 3 by "K^-". In this case pole frequency decreases with increasing codeword (= $a_{n-1}, a_{n-2}, \dots, a_0$). This approach is useful for achieving low frequency operation without requiring *large* component values, which is not favourable from area viewpoint in ICs.

Equations (7a)-(7b) can be rewritten as

$$\omega_0 = \omega_c [K_2], \qquad (8a)$$

$$Q = Q_c [K_1], \qquad (8b)$$

where ω_c and Q_c are defined as component dependent factor of ω_0 and Q, respectively. It is given as

$$\omega_c = \left[\frac{1}{C_1 C_2 R_1 R_2} \right]^{1/2}, \qquad (9a)$$

$$Q_c = R_8 \left[\frac{C_2}{C_1 R_1 R_2} \right]^{1/2}. \qquad (9b)$$

For $C_1 = 2C_2 = C$ and equal resistor values (R), (9a)-(9b) reduces to

$$\omega_c = \frac{1}{\sqrt{2} CR}, \qquad (10a)$$

$$Q_c = \frac{1}{\sqrt{2}}. \qquad (10b)$$

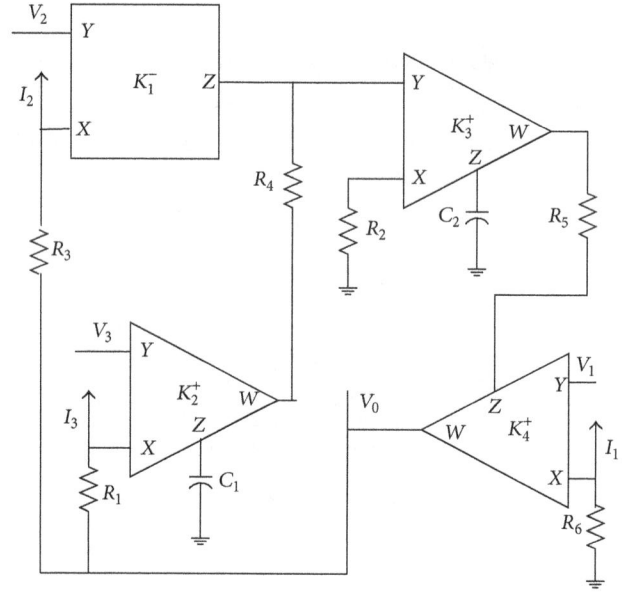

FIGURE 4: First derived DPUF.

Capacitance value $C_1 = 2C_2$ may be set at design level for quality factor of value $1/\sqrt{2}$ as given by (10b). It is required for *maximal flat* LP and HP response. On the other hand, higher Q-values, required for BP and BR are obtained by programming K_1, which can be further increased by adding additional transistor arrays in CSN of block-1.

4. Derived Mixed Mode DPUF

This section introduces three additional mixed mode DPUFs using "2–4" DPCFA/DPCCII. These are obtained by modifying the reconfigurable DPUF. In all the cases output function is same as given by (3). For brevity of discussion, features of all these DPUFs are discussed together in comparison section and also all the equations in this section assume codeword condition $K_2 = K_3$.

4.1. First Derived Mixed Mode DPUF. The first derived DPUF, as depicted in Figure 4 is obtained by deleting the K_5 and K_6 blocks of reconfigurable DPUF. It uses three DPCFA, one DPCCII, six resistors, and two grounded capacitors.

The following set of equations characterize the mixed mode operation of this DPUF:

$$N_v(s) = s^2 \frac{R_5}{R_6}\{K_4\}V_1 + s\frac{K_2 R_4}{K_1 C_2 R_2 R_3}V_2 + \frac{K_2^2}{C_1 C_2 R_1 R_2}V_3, \quad (11a)$$

$$N_i(s) = s^2 R_5\{K_4\}I_1 + s\frac{K_2 R_4}{K_1 C_2 R_2}I_2 + \frac{K_2^2}{C_1 C_2 R_2}I_3, \quad (11b)$$

$$D(s) = s^2 + s\frac{K_2 R_4}{K_1 C_2 R_2 R_3} + \frac{K_2^2}{C_1 C_2 R_1 R_2}, \quad (11c)$$

$$\omega_0 = \left[\frac{1}{C_1 C_2 R_1 R_2}\right]^{1/2}[K_2], \quad (11d)$$

$$Q = \frac{R_3}{R_4}\left[\frac{R_2 C_2}{R_1 C_1}\right]^{1/2}[K_1]. \quad (11e)$$

It can be seen from (11a)–(11e) that this modification retains the programming feature of ω_0 and Q only. Although it provides the gain programming of HP response, it completely misses the programming feature of filter functions (types). Thus, the generation of various filter functions depends on the proper combination of input variables as shown in Table 2. This configuration needs component matching constraints $R_5 = R_6$ for realization of AP and BR response. Additionally, it requires input inversion for AP realization.

4.2. Second Derived Mixed Mode DPUF.

Further deletion of K_4 block results in second derived DPUF (Figure 5). It uses two DPCFA, one DPCCII, four resistors, and two capacitors. This modification leads to one of the capacitor floating. The mixed mode operation of this DPUF is characterized by the following set of equations:

$$N_v(s) = s^2 V_1 + s\frac{K_2 R_4}{K_1 C_2 R_2 R_3}V_2 + \frac{K_2^2}{C_1 C_2 R_1 R_2}V_3, \quad (12a)$$

$$N_i(s) = s\frac{K_2 R_4}{K_1 C_2 R_2}I_2 + \frac{K_2^2}{C_1 C_2 R_2}I_3, \quad (12b)$$

$$D(s) = s^2 + s\frac{K_2 R_4}{K_1 C_2 R_2 R_3} + \frac{K_2^2}{C_1 C_2 R_1 R_2}, \quad (12c)$$

$$\omega_0 = \left[\frac{1}{C_1 C_2 R_1 R_2}\right]^{1/2}[K_2], \quad (12d)$$

$$Q = \frac{R_3}{R_4}\left[\frac{R_2 C_2}{R_1 C_1}\right]^{1/2}[K_1]. \quad (12e)$$

The modification renders with only ω_0 and Q programming. This configuration does not need any component matching constraint but it requires input inversion for AP realization. Moreover, it realizes only BP and LP in TIM.

4.3. Third Derived Mixed Mode DPUF.

Further deletion of K_1 block results in third derived mixed mode UF as shown

TABLE 2: Derived filter functions for various input combinations.

Filter functions	Input combination	
	Voltage	Current
HP	$V_2 = V_3 = 0, V_1 = V_i^\#$	$^@I_2 = I_3 = 0, I_1 = I_i^\#$
LP	$V_1 = V_2 = 0, V_3 = V_i$	$I_1 = I_2 = 0, I_3 = I_i$
BP	$V_1 = V_3 = 0, V_2 = V_i$	$I_1 = I_3 = 0, I_2 = I_i$
BR	$V_2 = 0, V_1 = V_3 = V_i$	$^@I_2 = 0, I_1 = I_3 = I_i$
AP	$V_1 = V_2^* = V_3 = V_i$	$^@I_1 = I_2^* = I_3 = I_i$

*V_2, I_2 are *negative* for first derived DPUF, $^\#V_i$ and I_i show actual applicable input, and $^@$not applicable for second and third derived configurations.

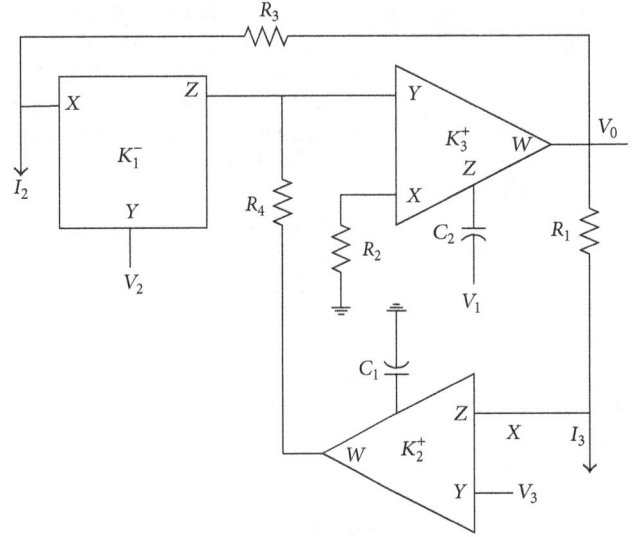

FIGURE 5: Second derived DPUF.

in Figure 6. It uses only one DPCFA, one DPCCII, and minimum number of passive components. Characterizing equations of this DPUF are given as

$$N_v(s) = s^2 V_1 + s\frac{K_2}{C_2 R_2}V_2 + \frac{K_2^2}{C_1 C_2 R_1 R_2}V_3, \quad (13a)$$

$$N_i(s) = s\frac{K_2}{C_2}I_2 + \frac{K_2^2}{C_1 C_2 R_2}I_3, \quad (13b)$$

$$D(s) = s^2 + s\frac{K_2}{C_2 R_2} + \frac{K_2^2}{C_1 C_2 R_1 R_2}, \quad (13c)$$

$$\omega_0 = \left[\frac{1}{C_1 C_2 R_1 R_2}\right]^{1/2}[K_2], \quad (13d)$$

$$Q = \left[\frac{R_2 C_2}{R_1 C_1}\right]^{1/2}. \quad (13e)$$

It is evident from (13a)–(13e) that this DPUF is almost similar to the second derived DPUF. Only the feature it lacks is Q programming.

5. Comparison

The performance of the proposed DPUFs is compared in Table 3 with similar reported filters. For fair comparison, active elements used in various filters are also expressed in terms of equivalent number of followers, that is, total number of current plus voltage followers (CF/VF) (active element and its follower equivalent is given in footnote of Table 3). Comparison is based on the following important features.

(1) Independently programmable ω_0, (2) independently programmable Q, (3) independently programmable gain, (4) programmable filter types, (5) appropriate input port impedance that is, high (or low) for voltage (or current) input, (6) appropriate output port impedance, that is, high (or low) for current (or voltage) output, (7) total number of resistors, (8) total number of capacitors (grounded/floating), (9) number of operating modes, (10) number of filter functions generated in each mode, (11) number of active elements used and (equivalent number of followers in braces), (12) operating frequency (in Hz).

Table 3 clearly indicates the trade-off between the features obtained and the number of active elements used in reconfigurable and derived DPUFs. Obviously, number of features varies proportionately with the number of active elements used.

The potential performance of proposed reconfigurable DPUF (Figure 3) is obvious from Table 3 itself. One important feature, which is not available in any of the reported filter except that of [4] is the programming of filter functions (types). This makes proposed DPUF suitable for integration. Blocks K_4, K_5, and K_6 of proposed DPUF not only govern the generation of different "filter type" but also provide the independent gain programming. It is noteworthy that independently programmable numerator coefficients of (5a) are useful in many ways such as in adaptive filtering applications and in realization of low pass & high pass notch. On the other hand, filter of [4] uses three MOS switches (SW) for programming the filter types and hence, completely lacks the feature of gain programming. Furthermore, it uses analog technique for programming of ω_0 and Q. Obviously, additional active elements will be required to incorporate missing features. It can be seen from Table 3 that all the digitally programmable filters except that of [14, 15] operate either in VM or current mode (CM) only. DPUF of [15] operates in VM and transadmittance mode (TAM) but it realizes three filter functions only and also misses the programmability feature of gain and filter type. Similarly, DPUF of [14] also misses number of features (nos. 3, 4, and 6). Operational transconductance amplifier (OTA) [5–8] and CCII [16, 17] based filter, however, operate in all the four mode but none of them are digitally programmable and also they lack a number of important features (Features-1–4). Apart from this, filters configuration of [6–8, 16, 17] belongs to distributed input topology; thus, they also need current matching constraints as the proposed DPUFs requires. Additionally, filter of [8] requires weighted current/voltage ratios and input (voltage) inversion constraint. OTA based filters [5–8] and that of [16] are considered here because the digital control (as given in [15]) can be easily incorporated in these structures

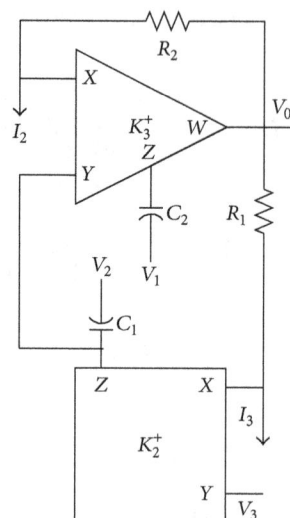

FIGURE 6: Third derived DPUF.

by programming their bias currents. In terms of power consumption, proposed filter shows better performance than that of [7]. Whereas proposed DPUF dissipates 4.88 mW (at 2.6 MHz); filter of [7] consumes quite large power, 30.9 mW at 1 MHz. The proposed configuration, however, lacks in terms of power consumption to that of [5]. Almost at same operating frequency, [5] dissipates power of 1.57 mW only. Power consumption of [15] is also high (6 mW) but it operates at higher frequency of 14 MHz.

All the proposed DPUFs (reconfigurable as well as derived) realize four or all filter functions in VM at appropriate impedance (low) port while all reported VM filters excluding that of [1, 14] realize three or lesser number of functions at inappropriate (high) impedance port. DPUF of [1] not only uses three more followers but also misses the mixed mode operation and programming feature of filter *type*. Similarly, the DPUF of [14] also lacks number of features (3, 4, and 6). However, its component count is low. OTA based filters [4–8] also lacks in providing voltage output at appropriate impedance level.

Comparison of first derived configuration (Figure 4), using four active elements (11 followers), shows that it realizes four filter functions in both the modes but the mixed mode structure of [15] realizes only three functions using five active elements. Compared to the proposed DPUF, filter of [14, 16, 17] uses one and two less followers, respectively, and also they operate in all the four modes but this is due to the use of dual output DPCCIIs. Also, [14] needs an additional VF to take the voltage outputs. It is to be noted that if output DPCFA of all proposed DPUFs are made dual output, that is, by creating additional Z-terminal, all the proposed filters can be operated in all the four modes. DPUF of [9] using one less follower provides the gain programming feature, which is not present in proposed first configuration (except for HP reponse) but it suffers from number of constraints. Filter of [9] uses two filter configurations for programming all the parameters of realized three filter functions. But still, its programming is constrained by the fact that variation of one of the parameters

TABLE 3: Performance comparison of proposed filter.

Reference, year/criteria	1	2	3	4	5	6	7	8	9	10	11	12
[22], 2006	Y	Y	N	N	N	N	2	1/1	CM	3	3DC-CCII {6}	15 K
[23], 2009	Y	N	Y	N	Y	Y	2	2/0	CM	ALL	3DC-CDBA {9}	45.9 K
[18], 2013	Y	Y	N	N	Y	N	4	2/0	VM	2	3DP-CCII {6}, 2CDN	20 M
[19], 2002	Y	Y	N	N	Y	N	3	2/0	VM	2	4DP-CCII {8}	0.28 M
[20], 2013	Y	Y	N	N	Y	N	12	4/0	VM	3	5DC-FDCC {10}	2 M
[9], 2009	Y*	Y*	Y*	N	Y	N	6	2/0	VM	3	5DPCCII {10}	1.56 M
[21], 2011	Y	Y	N	N	Y	N	3	2/0	VM	3	5DC-BOTA	3 M
[5], 2012	Y&	Y&	N	N	Y$^{\varepsilon}$	Y$	3	2/0	ALL	ALL@	(4SO + 2DO) OTA	2.87 M
[6], 2010	Y&	N	N	N	Y	Y$	NA	2/0	ALL	ALL	(3SO + 1TO + 1DO) OTA	1.59 M
[7], 2009	Y&	Y&	N	N	Y	Y$	NA	2/0	ALL	ALL	4OTA, 1DO-OTA	1 M
[8], 2009	Y&	Y&	N	N	Y	Y$	NA	2/0	ALL	ALL	4OTA	1.59 M
[16], 2003	Y&	Y&	N	N	Y$^{\varepsilon}$	Y$	NA	2/0	ALL	<4	4DO-CCCII {8}	12.5 K
[17], 2004	Y&	Y&	N	N	N	Y$	7	2/0	ALL	ALL	(4SO + 1DO) CCII {10}	112 K
[4], 2007	Y&	Y&	N	Y	Y	Y$	NA	2/0	ALL	4	(2SO + 2DO) OTA, 3SW	10 M
[15], 2013	Y	Y	N	N	Y	N	1	2/0	VM/TAM	3/3	5DCCDVCC	14 M
[14], 2014	Y	Y	N	N	Y	N	6	2/0	ALL	ALL	(3SO + 2DO) DPCCII {10}	4 M
[1], 2013	Y	Y	Y	N	Y	Y	7	2/0	VM	ALL	4DPCFA, 2DPCCII {18}	0.4 M
R. DPUF (Figure 3)	**Y**	**Y**	**Y**	**Y**	**Y**	**Y**	**8**	**2/0**	**VM/TIM**	**ALL**	**3DPCFA, DPCCII {15}**	**2.6 M**
1st DPUF (Figure 4)	Y	Y	Y#	N	Y	Y	6	2/0	VM/TIM	4/4	3DPCFA, 1DPCCII {11}	
2nd DPUF (Figure 5)	Y	Y	N	N	Y	Y	4	1/1	VM/TIM	4/2	2DPCFA, 1DPCCII {8}	3–5.5 M
3rd DPUF (Figure 6)	Y	N	N	N	Y	Y	2	0/2	VM/TIM	4/2	1DPCFA, 1DPCCII {5}	

*Conditionally programmable, & not digital control, @ two filter function in TAM, # for HP only, $ for current output only, ε for voltage input only, DC-digitally controlled, SO-single output, DO-dual output, TO-triple output, CDBA-current differencing buffer amplifier- (2CF + 1VF), CCII- (1CF + 1VF), CFA- (1CF + 2VF), {}-equivalent number of followers, BOTA-balanced output transconductance amplifier, and DCCDVCC-digitally current controlled differential voltage current conveyor.

(Gain, ω_0, Q) makes either one or both of the remaining two parameters nonprogrammable, while no such constraint applies over any of the proposed DPUF.

Second derived DPUF (Figure 5) uses one floating capacitor, which is less attractive for integration. However, it needs only three active elements (8 followers) and realizes four filter functions in VM at *low* impedance port. On the other hand, reported VM filters [9, 15, 18–21] using almost equal or more active elements/followers realizes lesser number of filter functions at *high* (not desired) impedance port. Although DPUF of [18] needs six followers and two current division networks (CDN) only, it realizes only two filter functions at high impedance port (not desired). Obviously, it [18] needs additional active elements (more than two followers) to fill the gap. Similarly, CM DPUFs, which are expected to have simpler circuit structure also needs additional circuitry. CM DPUF of [22] using six followers needs additional follower stages to fill the gap of missing cascadability feature and to take the current outputs available in working impedances of filter circuit. Similarly, filter of [23] using nine followers needs additional circuitry for providing weighted replica of input current inputs.

The smallest proposed configuration, that is, third derived configuration (Figure 6) clearly shows the trade-off between number of active elements used and the obtained features. It provides only ω_0 programming and thus, it is suitable for designing the channel select filters. DPUF of [11] provides same feature (ω_0 programming) and realizes three filter functions only but it uses three operational amplifiers, two CDNs, six resistors, and two floating capacitors.

The comparison presented herein verifies the fact that the features of DPUFs vary proportionally with the number of active elements used. Since all the proposed DPUFs show better or comparable performance to that of reported DPUFs using almost equal number of active elements/followers, it implicitly justifies the need of six active elements used in reconfigurable DPUF for obtaining offered programmability features.

6. Nonideal Performance and Mismatch Effects

This section discusses the effect of CFA nonidealities and effect of component and current mismatch over the performance of proposed DPUF.

6.1. Nonideal Analysis. Nonidealities of CFA results from (1) small error in unity transfer gains as described by (1) and (2) CFA parasitics.

The modified output function of UF using (1) is given as

$$V_{0n} = \frac{N_{vn}(s) + N_{in}(s)}{D_n(s)}, \quad (14)$$

where subscript "n" denotes the effect of nonidealities over filter function. Modified numerator and denominator functions are given as

$$N_{vn}(s) = V_i \left(s^2 \alpha_4 \beta_4 \gamma_4 \frac{R_6}{R_7} K_4 \right.$$

$$- s \frac{\alpha_1 \alpha_5 \beta_1 \beta_3 \beta_5 \gamma_3 \gamma_4 R_5 K_3}{C_2 R_4 R_8 K_1} K_5 \qquad (15a)$$

$$\left. + \frac{\alpha_6 \beta_2 \beta_3 \beta_6 \gamma_2 \gamma_3 \gamma_4 K_2 K_3}{C_1 C_2 R_2 R_3} K_6 \right),$$

$$N_{in}(s) = s^2 \beta_4 \gamma_4 R_6 K_4 I_1 - s \frac{\alpha_1 \beta_1 \beta_3 \beta_5 \gamma_3 \gamma_4 R_5 K_3}{C_2 R_8 K_1} K_5 I_2$$

$$+ \frac{\beta_2 \beta_3 \beta_6 \gamma_2 \gamma_3 \gamma_4 K_2 K_3}{C_1 C_2 R_2} K_6 I_3, \qquad (15b)$$

$$D_n(s) = s^2 + s \frac{\alpha_1 \beta_1 \beta_3 \gamma_3 \gamma_4 K_3}{C_2 R_8 K_1} + \frac{\alpha_2 \beta_2 \beta_3 \gamma_2 \gamma_3 \gamma_4 K_2 K_3}{C_1 C_2 R_1 R_2}. \qquad (15c)$$

Modified ω_0 and Q factor using (15c) are given as

$$\omega_0 = \sqrt{\frac{\alpha_2 \beta_2 \beta_3 \gamma_4 \gamma_2 \gamma_3 K_2 K_3}{C_1 C_2 R_1 R_2}}, \qquad (16a)$$

$$Q = \frac{K_1 R_8}{\alpha_1 \beta_1} \sqrt{\frac{\alpha_2 \beta_2 \gamma_2 C_2 K_2}{\beta_3 \gamma_4 \gamma_3 C_1 R_1 R_2 K_3}}. \qquad (16b)$$

Similarly, the gain functions of basic filter functions are given as

$$\text{For VM } G_{HP} = \alpha_4 \beta_4 \gamma_4 \frac{R_6}{R_7} K_4,$$

$$G_{BP} = \frac{\alpha_5 \beta_5 R_5}{R_4} K_5, \qquad G_{LP} = \frac{\alpha_6 \beta_6 R_1}{\alpha_2 R_3} K_6, \qquad (16c)$$

$$\text{For TIM } G_{HP} = \beta_4 \gamma_4 R_6 K_4,$$

$$G_{BP} = \beta_5 R_5 K_5, \qquad G_{LP} = \frac{\beta_6 R_1}{\alpha_2} K_6. \qquad (16d)$$

Thus, the active and passive sensitivities of ω_0, Q, and G may be summarized as

$$0 \leq \left| S_x^{\omega_0} \right| \leq \frac{1}{2}, \qquad (17a)$$

$$0 \leq \left| S_x^{Q,G} \right| \leq 1, \qquad (17b)$$

where "x" denotes various active and passive elements, that is, α_i, β_i, γ_i, resistances, and capacitances. It is evident from (17a)-(17b) that the sensitivity figures are within reasonable limit. Parameter sensitivity of all derived configuration except that of third DPUF also lies in the same range as given by (17a)-(17b). Sensitivity of both parameters (ω_0 and Q) of third derived configuration follows (17a).

Second set of nonidealities include DPCFA parasitics. DPCFA (and DPCCII) have high valued parasitic resistance R_Y (or R_Z) in parallel with low valued parasitic capacitance C_Y (or C_Z) at port Y (or Z) and a low valued series resistance R_X at port X. DPCFA consists of an additional (parasitic) low valued series resistance at port W. To simplify the discussion, parasitic resistances at ports Y and Z are not considered as these are much greater than the external resistance of circuit.

Output function of *reconfigurable DPUF* under the parasitic effect is given as

$$V_{0P} = \left(\frac{s^2 R_{6P} [K_4]}{\{1 + s R_{6P} C_{Z4}\}} \left[\frac{V}{R_{7P}} + \left\{ \frac{R_7}{R_{7P}} \right\} I_1 \right] \right.$$

$$- \frac{s K_3 R_5 [K_5]}{K_1 C_{2P} R_{8P}} \left\{ 1 + \frac{R_{X3} (1 + s R_{2P} C_{Z1})}{R_{2P}} \right\}^{-1}$$

$$\times \{1 + s R_5 (C_{Z5} + C_{Y1})\}^{-1} \left[\frac{V}{R_{4P}} + \left\{ \frac{R_4}{R_{4P}} \right\} I_2 \right]$$

$$+ \frac{K_3 K_2 [K_6]}{C_{1P} C_{2P}} \left\{ 1 + \frac{R_{X2} (1 + s R_1 C_{Z6})}{R_1} \right\}^{-1}$$

$$\times \left\{ \frac{R_{2P} (1 + s R_{X3} C_{Z1})}{1 + s R_{2P} C_{Z1}} + R_{X3} \right\}^{-1}$$

$$\times \left[\frac{V}{R_{3P}} + \left\{ \frac{R_3}{R_{3P}} \right\} I_3 \right] \right)$$

$$\times \left(s^2 \{1 + s R_{W4} C_{Y2}\} + \frac{s K_3}{K_1 C_{2P} R_{8P}} \right.$$

$$\times \left\{ 1 + \frac{R_{X3} (1 + s R_{2P} C_{Z1})}{R_{2P}} \right\}^{-1}$$

$$\times \{1 + s R_5 (C_{Z5} + C_{Y1})\}^{-1}$$

$$+ \frac{K_3 K_2}{C_{1P} C_{2P}} \{R_{2P} (1 + s R_{X3} C_{Z1}) + R_{X3}\}^{-1}$$

$$\left. \times \left\{ \frac{R_1}{1 + s R_1 C_{Z6}} + R_{X2} \right\}^{-1} \right)^{-1}, \qquad (18a)$$

where subscript "P" denotes the modified component values and function under influence of parasitics. For brevity reasons, output function is not segregated in numerator and denominator functions.

Modified component values of (18a)–(18f) are given as

$$\begin{aligned}
R_{2P} &= R_2 + R_{W2}, & R_{3P} &= R_3 + R_{X6}, \\
R_{4P} &= R_4 + R_{X5}, & R_{7P} &= R_7 + R_{X4}, \\
R_{6P} &= R_6 + R_{W3}, & R_{8P} &= R_8 + R_{X1}, \\
C_{1P} &= C_1 + C_{Z2}, & C_{2P} &= C_2 + C_{Z3}.
\end{aligned} \qquad (18b)$$

It can be seen from (18a) that number of parasitics are absorbed in external circuit components. Thus, they do not create any unwanted pole. Expressions enclosed within the braces $\{\}$ of (18a) denotes those parasitic terms, which causes deviation from the ideal response of (5a), (5b) and (6) (braces are used in all other DPUFs to denote the same). Thus, in order to nullify the effect of these parasitic, the following constraint over component values are imposed

$$R_{X2} \ll R_1 \ll \frac{1}{sC_{Z6}}, \qquad R_{X3} \ll R_2 \ll \frac{1}{sC_{Z1}},$$

$$R_3 \gg R_{X6}, \qquad R_4 \gg R_{X5},$$

$$R_5 \gg \frac{1}{s(C_{Z5} + C_{Y1})}, \qquad R_6 \gg \frac{1}{sC_{Z4}}, \qquad (18c)$$

$$R_7 \gg R_{X4}.$$

Since R_X and R_W are of order of few ohms only, lower limit over external resistor values are easily satisfied if it lies in range of few kiloohms ($K\Omega$). On the other hand, parasitic capacitances C_Z and C_Y (of order of few picofarads) offer very high impedance at operational frequencies; thus, upper limit is also satisfied if resistor values lie in few tens of $K\Omega$ (the same argument holds for all *derived* DPUFs also).

It can be seen from (18a) that the equation consist of the following parasitic poles:

$$\frac{1}{R_{X3}C_{Z1}}, \qquad \frac{1}{R_{W4}C_{Y2}}. \qquad (18d)$$

The corresponding pole frequency is too high to affect the circuit operation.

Thus, if conditions of (18c) are met, (18a) reduces to

$$V_{0P} \cong \left(s^2 R_{6P} \left[\frac{V}{R_{7P}} + I_1 \right] [K_4] \right.$$

$$- \frac{sK_3 R_5}{K_1 C_{2P} R_{8P}} \left[\frac{V}{R_{4P}} + I_2 \right] [K_5]$$

$$\left. + \frac{K_3 K_2}{C_{1P} C_{2P} R'_{2P}} \left[\frac{V}{R_{3P}} + I_3 \right] [K_6] \right) \qquad (18e)$$

$$\times \left(s^2 + \frac{sK_3}{K_1 C_{2P} R_{8P}} + \frac{K_3 K_2}{C_{1P} C_{2P} R'_{2P} R_{1P}} \right)^{-1},$$

where $R_{1P} = R_1 + R_{X2}$, $R'_{2P} = R_{2P} + R_{X3}$.

Similarity of (18a) and (5a), (5b), and (6) are obvious. It can be seen from (18b) and (18e) that number of parasitics is absorbed in external components and hence, do not cause any problem.

Modified filter parameters for $K_2 = K_3$ are given by

$$\omega_{0P} = \left[\frac{1}{C_{1P} C_{2P} R_{1P} R'_{2P}} \right]^{1/2} [K_2],$$

$$Q_P = R_{8P} \left[\frac{C_{2p}}{C_{1P} R_{1P} R'_{2p}} \right]^{1/2} [K_1]. \qquad (18f)$$

Similarly, the output function of *first derived DPUF* under the parasitic effect is given as

$$V_{0P} = \left(\frac{s^2 K_4 R_{5P}}{\{1 + sR_{5P} C_{Z4}\}} \left[\frac{V_1}{R_{6P}} + \left\{ \frac{R_6}{R_{6P}} \right\} I_1 \right] \right.$$

$$+ \frac{sK_3 R_{4P}}{K_1 C_{2P} R_{2P} \{1 + sR_{5P} C_{Z4}\} \{1 + sR_{4P} (C_{Z1} + C_{Y3})\}}$$

$$\times \left[\frac{V_2}{R_{3P}} + \left\{ \frac{R_3}{R_{3P}} \right\} I_2 \right]$$

$$+ \frac{K_3 K_2}{C_{1P} C_{2P} R_{2P} \{1 + sR_{5P} C_{Z4}\} \{1 + sR_{4P} (C_{Z1} + C_{Y3})\}}$$

$$\left. \times \left[\frac{V_3}{R_{1P}} + \left\{ \frac{R_1}{R_{1P}} \right\} I_3 \right] \right)$$

$$\times \left(s^2 + \left(sK_3 R_{4P} (K_1 C_{2P} R_{2P} R_{3P} \{1 + sR_{5P} C_{Z4}\} \right. \right.$$

$$\left. \times \{1 + sR_{4P} (C_{Z1} + C_{Y3})\})^{-1} \right)$$

$$+ \left(K_3 K_2 (C_{1P} C_{2P} R_{1P} R_{2P} \{1 + sR_{5P} C_{Z4}\} \right.$$

$$\left. \left. \times \{1 + sR_{4P} (C_{Z1} + C_{Y3})\})^{-1} \right) \right)^{-1}, \qquad (19a)$$

where $R_{1P} = R_1 + R_{X2}$, $R_{2P} = R_2 + R_{X3}$, $R_{3P} = R_3 + R_{X1}$, $R_{4P} = R_4 + R_{W2}$, $R_{5P} = R_5 + R_{W3}$, $R_{6P} = R_6 + R_{X4}$, $C_{1P} = C_1 + C_{Z2}$, and $C_{2P} = C_2 + C_{Z3}$.

Constraints over external component from (19a) are given as

$$R_1 \gg R_{X2}, \qquad R_3 \gg R_{X1},$$

$$R_4 \gg \frac{1}{s(C_{Z1} + C_{Y3})}, \qquad R_5 \gg \frac{1}{sC_{Z4}}, \qquad (19b)$$

$$R_6 \gg R_{X4}.$$

Modified output function satisfying conditions of (19b) is given as

$$V_{0P} \cong \left(s^2 K_4 R_{5P} \left[\frac{V_1}{R_{6P}} + I_1 \right] + \frac{sK_3 R_{4P}}{K_1 C_{2P} R_{2P}} \left[\frac{V_2}{R_{3P}} + I_2 \right] \right.$$

$$\left. + \frac{K_3 K_2}{C_{1P} C_{2P} R_{2P}} \left[\frac{V_3}{R_{1P}} + I_3 \right] \right)$$

$$\times \left(s^2 + \frac{sK_3 R_{4P}}{K_1 C_{2P} R_{2P} R_{3P}} + \frac{K_3 K_2}{C_{1P} C_{2P} R_{1P} R_{2P}} \right)^{-1}. \qquad (19c)$$

Modified filter parameters (for $K_2 = K_3$) from (19c) are given as

$$\omega_{0P} = \left[\frac{1}{C_{1P} C_{2P} R_{1P} R'_{2P}} \right]^{1/2} [K_2],$$

$$Q_P = \frac{R_{3p}}{R_{4p}} \left[\frac{R_{2p} C_{2p}}{R_{1p} C_{1p}} \right]^{1/2} [K_1]. \qquad (19d)$$

TABLE 4: Codeword programming of filter parameters.

| Filter function | Codeword programming | | |
	Pole frequency ($\omega_0 = K_2\omega_c$)	Q-factor ($Q = K_1 Q_c$)	Gain ($G = K_i^* G_c$)
HP $K_5 = K_6 = 0(000)$	$K_2 = K_3 = 1(001), 4(100), 7(111)$ $K_1 = 1(001)[=Q_c]$ $K_4 = 7(111)[=7G_c]$	NA	$K_4 = 1(001), 4(100), 7(111)$ $K_1 = 1(001)[=Q_c]$ $K_2 = K_3 = 4(100)[=4\omega_c]$
LP $K_4 = K_5 = 0(000)$	$K_2 = K_3 = 1(001), 4(100), 7(111)$ $K_1 = 1(001)[=Q_c]$ $K_6 = 7(111)[=7G_c]$	NA	$K_6 = 1(001), 4(100), 7(111)$ $K_1 = 1(001)[=Q_c]$ $K_2 = K_3 = 4(100)[=4\omega_c]$
BP $K_4 = K_6 = 0(000)$	$K_2 = K_3 = 1(001), 4(100), 7(111)$ $K_1 = 7(111)[=7Q_c]$ $K_5 = 7(111)[=7G_c]$	$K_1 = 1(001), 4(100), 7(111)$ $K_2 = K_3 = 4(100)[=4\omega_c]$ $K_5 = 7(111)[=7G_c]$	$K_5 = 1(001), 4(100), 7(111)$ $K_1 = 7(111)[=Q_c]$ $K_2 = K_3 = 4(100)[=4\omega_c]$
BR $K_5 = 0(000)$	$K_2 = K_3 = 1(001), 4(100), 7(111)$ $K_1 = 7(111)[=7Q_c]$ $K_4 = K_6 = 7(111)[=7G_c]$	$K_1 = 1(001), 4(100), 7(111)$ $K_2 = K_3 = 4(100)[=4\omega_c]$ $K_4 = K_6 = 7(111)[=7G_c]$	$K_4 = K_6 = 1(001), 4(100), 7(111)$ $K_1 = 7(111)[=Q_c]$ $K_2 = K_3 = 4(100)[=4\omega_c]$

$^* K_i$ denotes nonzero codeword among K_4, K_5, or K_6.

In order to simplify the further discussion, parasitic resistance R_W is not considered in 2nd and 3rd derived DPUFs as these are of order of few ohms only. Output function of *second derived DPUF* under the parasitic effect is given as

$$V_{0P} = \left(s^2 \left\{ \frac{C_2}{C_{2P}} \right\} V_1 \right.$$

$$+ \frac{sK_3 R_4}{K_1 C_{2P} R_{2P} \left\{ 1 + sR_4 \left(C_{Z1} + C_{Y1} \right) \right\}}$$

$$\times \left[\frac{V_2}{R_{3P}} + \left\{ \frac{R_3}{R_{3P}} \right\} I_2 \right]$$

$$+ \frac{K_3 K_2}{C_{1P} C_{2P} R_{2P} \left\{ 1 + sR_{4P} \left(C_{Z1} + C_{Y3} \right) \right\}}$$

$$\times \left[\frac{V_3}{R_{1P}} + \left\{ \frac{R_1}{R_{1P}} \right\} I_3 \right] \right)$$

$$\times \left(s^2 + \frac{sK_3 R_4}{K_1 C_{2P} R_{2P} R_{3P} \left\{ 1 + sR_4 \left(C_{Z1} + C_{Y1} \right) \right\}} \right.$$

$$+ \left. \frac{K_3 K_2}{C_{1P} C_{2P} R_{1P} R_{2P} \left\{ 1 + sR_4 \left(C_{Z1} + C_{Y3} \right) \right\}} \right)^{-1}, \tag{20a}$$

where $R_{1P} = R_1 + R_{X2}$, $R_{2P} = R_2 + R_{X3}$, $R_{3P} = R_3 + R_{X1}$, $C_{1P} = C_1 + C_{Z2}$, and $C_{2P} = C_2 + C_{Z3}$.

Component constraints from (20a) are given as

$$R_1 \gg R_{X2}, \qquad R_3 \gg R_{X1},$$

$$R_4 \gg \frac{1}{s \left(C_{Z1} + C_{Y3} \right)}, \tag{20b}$$

$$C_2 \gg C_{Z3}.$$

Modified output function satisfying conditions of (20b) is given as

$$V_{0P} = \left(s^2 V_1 + \frac{sK_3 R_4}{K_1 C_{2P} R_{2P}} \left[\frac{V_2}{R_{3P}} + I_2 \right] \right.$$

$$+ \left. \frac{K_3 K_2}{C_{1P} C_{2P} R_{2P}} \left[\frac{V_3}{R_{1P}} + I_3 \right] \right) \tag{20c}$$

$$\times \left(s^2 + \frac{sK_3 R_4}{K_1 C_{2P} R_{2P} R_{3P}} + \frac{K_3 K_2}{C_{1P} C_{2P} R_{1P} R_{2P}} \right)^{-1}.$$

Modified filter parameters (for $K_2 = K_3$) from (20c) are given as

$$\omega_{0P} = \left[\frac{1}{C_{1P} C_{2P} R_{1P} R_{2P}} \right]^{1/2} [K_2],$$

$$Q_P = \frac{R_{3P}}{R_4} \left[\frac{R_{2P} C_{2P}}{R_{1P} C_{1P}} \right]^{1/2} [K_1]. \tag{20d}$$

Output function of *third derived DPUF* under the parasitic effect is given as

$$V_{0P} = \left(s^2 \left\{ \frac{C_2}{C_{2P}} \right\} V_1 + s \frac{K_3 C_1 V_2}{C_{1P} C_{2P} R_{2P}} + \frac{K_3 K_2 V_3}{C_{1P} C_{2P} R_{1P} R_{2P}} \right.$$

$$+ s \frac{K_3 R_2 I_2}{C_{2P} R_{2P}} + \left. \frac{K_3 K_2 R_1 I_3}{C_{1P} C_{2P} R_{1P} R_{2P}} \right)$$

$$\times \left(s^2 + \frac{sK_3}{C_{2P} R_{2P}} + \frac{K_3 K_2}{C_{1P} C_{2P} R_{1P} R_{2P}} \right)^{-1}, \tag{21a}$$

where $R_{1P} = R_1 + R_{X2}$, $R_{2P} = R_2 + R_{X3}$, $C_{1P} = C_1 + C_{Z2} + C_{Y3}$, and $C_{2P} = C_2 + C_{Z3}$.

Component constraints from (21a) are given as

$$C_2 \gg C_{Z3}. \tag{21b}$$

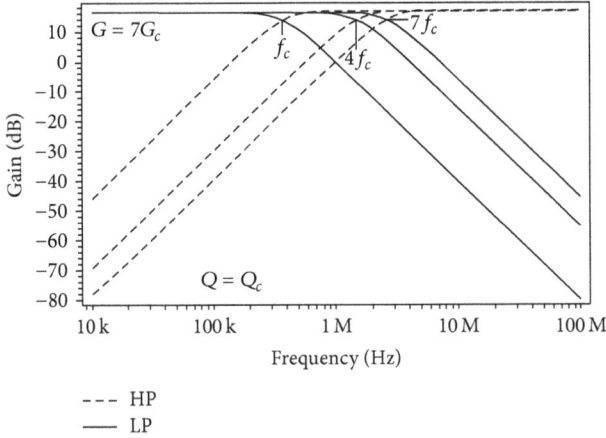

FIGURE 7: Pole frequency programming of HP and LP.

Modified output function satisfying conditions of (21b) is given as

$$V_{OP} = \left(s^2 V_1 + s\frac{K_3 C_1 V_2}{C_{1P} C_{2P} R_{2P}} + \frac{K_3 K_2 V_3}{C_{1P} C_{2P} R_{1P} R_{2P}} \right.$$
$$\left. + s\frac{K_3 R_2 I_2}{C_{2P} R_{2P}} + \frac{K_3 K_2 R_1 I_3}{C_{1P} C_{2P} R_{1P} R_{2P}} \right) \qquad (21c)$$
$$\times \left(s^2 + \frac{sK_3}{C_{2P} R_{2P}} + \frac{K_3 K_2}{C_{1P} C_{2P} R_{1P} R_{2P}} \right)^{-1}.$$

Modified filter parameters (for $K_2 = K_3$) from (21c) are given as

$$\omega_{0P} = \left[\frac{1}{C_{1P} C_{2P} R_{1P} R_{2P}} \right]^{1/2} [K_2],$$
$$\qquad (21d)$$
$$Q_P = \left[\frac{R_{2p} C_{2p}}{R_{1p} C_{1p}} \right]^{1/2}.$$

6.2. Mismatch Effect. It can be seen from Table 1 that the realization of BR and AP response is constrained by component matching conditions. Also, the realization of these responses in TIM requires current matching constraint. Thus, it is necessary to study the effects of component and current mismatch over the filter parameters.

Firstly, the effect of current mismatch is presented. Current matching condition demands for additional active element, for example, multioutput current follower (MOCF) or multioutput CCII (MOCCII). Thus, the effect of current mismatch can be easily predicted by the help of (15b). Modelling the error in copy of current signals I_1, I_2, and I_3 by η_1, η_2, and η_3, respectively, we have $I_1 = \eta_1 I$, $I_2 = \eta_2 I$, and

$I_3 = \eta_3 I$, where "I" denotes the ideal output current of used active element. Under this condition, *only* (15b) modifies as

$$N_{in}(s) = s^2 \beta_4 \gamma_4 R_6 K_4 \eta_1 I$$
$$- s\frac{\alpha_1 \beta_1 \beta_3 \beta_5 \gamma_3 \gamma_4 R_5 K_3}{C_2 R_8 K_1} K_5 \eta_2 I \qquad (22)$$
$$+ \frac{\beta_2 \beta_3 \beta_6 \gamma_2 \gamma_3 \gamma_4 K_2 K_3}{C_1 C_2 R_2} K_6 \eta_3 I.$$

Thus, this modifies the gain parameter (16d) of filter *only*. Current mismatch does not affect ω_0 and Q. The modified gain function is given by

$$G_{HP} = \beta_4 \gamma_4 \eta_1 R_6 K_4, \qquad G_{BP} = \beta_5 \eta_2 R_5 K_5,$$
$$G_{LP} = \frac{\eta_3 \beta_6 R_1}{\alpha_2} K_6. \qquad (23)$$

Gain sensitivity with respect to η_1, η_2, and η_3 follows (17b).

Next, considering the effect of component mismatch, it can be seen from Table 1 that realization of BR and AP requires matching constraint $R_7 = R_6 = R_5 = R_4 = R_3 = R_1$. It is, however, to be noted that any mismatch in resistors R_3, R_4, R_5, R_6, and R_7 does not affect parameters ω_0 and Q ((16a), (16b)); rather it affects the gain parameters ((16c), (16d)) only. For simplicity, error in filter parameters is discussed for only those components, which require matching constraints. Taking the example of VM LP response, its gain parameter from (16c) is given as

$$G_{LP} = \frac{R_1}{R_3}. \qquad (24)$$

Matching condition requires $R_1 = R_3$. Assuming mismatch of ΔR such that $R_3 = R_1 + \Delta R$, gain error is given by

$$\frac{\Delta G_{LP}}{G_{LP}} = \mp \frac{\Delta R}{R_1}. \qquad (25a)$$

Similarly, matching constraint over C_1, C_2 ($C_1 = 2C_2$) introduces error in ω_0 and Q as

$$\frac{\Delta \omega_0}{\omega_0} = \mp \frac{1}{2} \left(\frac{\Delta C_1}{C_1} + \frac{\Delta C_2}{C_2} \right), \qquad (25b)$$
$$\frac{\Delta Q}{Q} = \pm \frac{1}{2} \left(\frac{\Delta C_1}{C_1} - \frac{\Delta C_2}{C_2} \right), \qquad (25c)$$

where $\Delta C_1, \Delta C_2$ is the mismatch in C_1 and C_2, respectively. It is to be noted that mismatch effect is additive in case of ω_0 while it is subtractive in case of Q; that is, the effect of mismatch is lesser over Q-factor.

7. Simulation Results

Performance of proposed UF is verified using $0.25\,\mu m$ TSMC parameters and supply voltage $\pm 0.75\,V$ with same transistor aspect ratios as given in [1]. Taking equal resistor values $R = 2K$ and capacitor values as $C_1 = 0.3\,nF$, $C_2 = 0.15\,nF$, results

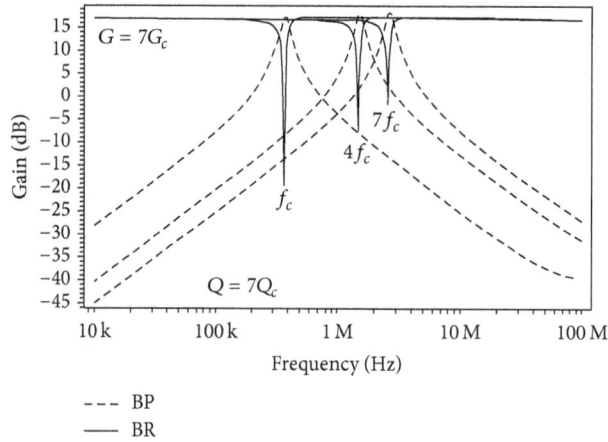

FIGURE 8: Pole frequency programming of BP and BR.

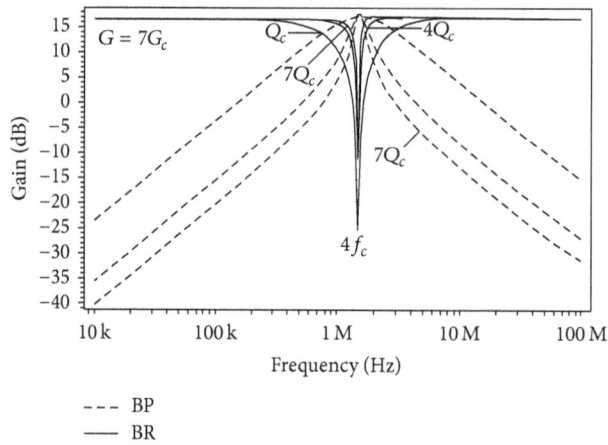

FIGURE 9: Quality factor programming of BP and BR.

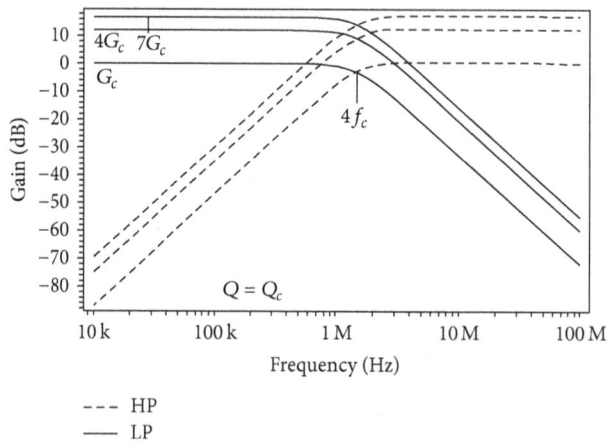

FIGURE 10: Gain programming of HP and LP.

in $f_c = 1/2\sqrt{2}\pi RC = 375.32$ KHz, $Q_C = 1/\sqrt{2}$, and $G_c = 1$. Where, G_c denotes the component dependent factor of gain (G). Simulation results are shown for VM responses.

Pole frequency programming (in range 375 KHz to 2.7 MHz) as described by (7a) is performed by varying $K_2(= K_3)$. Figures 7 and 8 show the ω_0 programming

for various responses for different values of gain and Q-factor. Parameter values are depicted in simulation results. Codeword programming of all the parameters are mentioned in Table 4. Table 4 shows the applied codeword along with its decimal equivalent and corresponding parameter values (in square brackets).

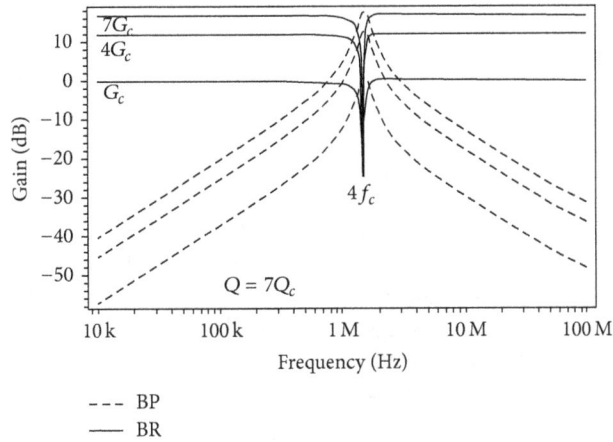

FIGURE 11: Gain programming of BP and BR.

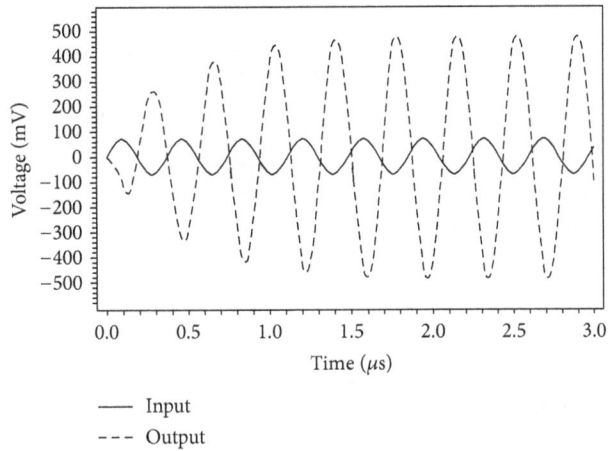

FIGURE 12: Time response of BP response.

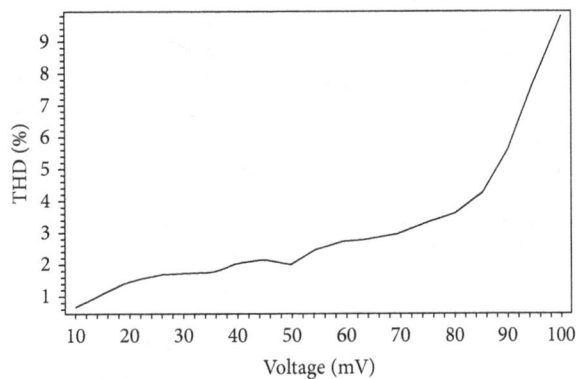

FIGURE 13: THD variation with respect to input sine wave signal.

Quality factor programming through K_1 as described by (7b) is shown in Figure 9 for BP and BR response at frequency $4f_c$ and gain $7G_c$. Quality factor values are chosen depending on the type of response.

Similarly, gain is programmed (Figures 10-11) using suitable codeword as indicated in Table 4. Power dissipation of

proposed reconfigurable DPUF is 4.88 mW. It is obtained for BP response when all the codewords are set to 7(111).

Figure 12 shows the time domain behaviour of BP response with gain 16.4 dB; that is, for $K_5 = 7$. It is obtained by applying 70 mV peak sine wave of frequency 2.7 MHz. Figure 13 shows the variation of the THD with respect to the applied sinusoidal input voltage. The THD values of the circuit remain below 3% for input signals up to 70 mV peak.

8. Conclusion

This paper presents a novel reconfigurable mixed mode DPUF. The proposed VM/TIM DPUF offers several attractive features: generation of all the standard filter function, reconfigurable filter type, independently programmable filter parameters (ω_0, Q, gain), full cascadability, and low sensitivity figures. In addition to this, use of only two grounded capacitors makes the UF suitable for monolithic implementation. Circuit constraints are discussed by presenting parasitic study and mismatch analysis.

Conflict of Interests

The authors declare that there is no conflict of interests regarding the publication of this paper.

Acknowledgments

The authors would like to thank the Department of Electronics and Communication Engineering, Jamia Millia Islamia University, for providing valuable support. Authors are also thankful to the anonymous reviewers for their valuable suggestions.

References

[1] D. Singh and N. Afzal, "Digitally programmable High-Q voltage-mode universal filter," *Radioengineering*, vol. 22, no. 4, pp. 995–1006, 2013.

[2] R. Senani, D. R. Bhaskar, A. K. Singh, and K. V. Singh, *Current Feedback Operational Amplifiers and Their Applications*, Analog Circuits and Signal Processing, Springer, Berlin, Germany, 2013.

[3] M. Siripruchyanun, C. Chanapromma, P. Silapan, and W. Jaikla, "BiCMOS current controlled current feedback amplifier (CC-CFA) and its applications," *WSEAS Transactions on Electronics*, vol. 5, no. 6, pp. 203–219, 2008.

[4] C. M. Chang, J. H. Lo, L. D. Jeng, and S. H. Tu, "Analytical synthesis of digitally programmable versatile-mode high-order OTA-equal C universal filter structures with the minimum number of components," *International Journal of Circuits System and Signal Processing*, vol. 1, no. 2, pp. 101–104, 2007.

[5] M. Kumngern and S. Junnapiya, "Mixed-mode universal filter using OTAs," in *Proceedings of the IEEE International Conference on Cyber Technology in Automation, Control and Intelligent Systems*, Bangkok, Thailand, May 2012.

[6] C.-N. Lee, "Multiple-mode OTA-C universal biquad filters," *Circuits, Systems, and Signal Processing*, vol. 29, no. 2, pp. 263–274, 2010.

[7] H.-P. Chen, Y.-Z. Liao, and W.-T. Lee, "Tunable mixed-mode OTA-C universal filter," *Analog Integrated Circuits & Signal Processing*, vol. 58, no. 2, pp. 135–141, 2009.

[8] C.-N. Lee and C.-M. Chang, "High-order mixed-mode OTA-C universal filter," *International Journal of Electronics and Communications*, vol. 63, no. 6, pp. 517–521, 2009.

[9] T. M. Hassan and S. A. Mahmoud, "Fully programmable universal filter with independent gain-ω_0-Q control based on

[10] D. Biolek, R. Senani, V. Biolkova, and Z. Kolka, "Active elements for analog signal processing: classification, review, and new proposals," *Radioengineering*, vol. 17, no. 4, pp. 15–32, 2008.

[11] K.-P. Pun, C.-S. Choy, C.-F. Chan, and J. E. da Franca, "Current-division-based digital frequency tuning for active RC filters," *Analog Integrated Circuits and Signal Processing*, vol. 45, no. 1, pp. 61–69, 2005.

[12] M. Bhushan and R. Newcomb, "Grounding of capacitor in integrated circuits," *Electronic Letters*, vol. 3, no. 4, pp. 148–149, 1967.

[13] R. J. Baker, H. W. Li, and D. E. Boyce, *CMOS Circuit Design, Layout, and Simulation*, IEEE Press, New York, NY, USA, 1998.

[14] D. Singh and N. Afzal, "Digitally programmable current conveyor based mixed mode universal filter," *International Journal of Electronics Letters*, 2014.

[15] P. Beg, S. Maheshwari, and M. A. Siddiqi, "Digitally controlled fully differential voltage- and transadmittance-mode biquadratic filter," *IET Circuits, Devices and Systems*, vol. 7, no. 4, pp. 193–203, 2013.

[16] M. T. Abuelma'atti, "A novel mixed-mode current-controlled current-conveyor-based filter," *Active and Passive Electronic Components*, vol. 26, no. 3, pp. 185–191, 2003.

[17] M. T. Abuelma'atti and A. Bentrcia, "A novel mixed-mode CCII-based filter," *Active and Passive Electronic Components*, vol. 27, no. 4, pp. 197–205, 2004.

[18] H. Alzaher, N. Tasadduq, O. Al-Ees, and F. Al-Ammari, "A complementary metal-oxide semiconductor digitally programmable current conveyor," *International Journal of Circuit Theory and Applications*, vol. 41, no. 1, pp. 69–81, 2013.

[19] A. A. A. El-Adawy, A. M. Soliman, and H. O. Elwan, "Low voltage digitally controlled CMOS current conveyor," *International Journal of Electronics and Communication*, vol. 56, no. 3, pp. 137–144, 2002.

[20] S. A. Mahmoud and E. A. Soliman, "Digitally programmable second generation current conveyor-based FPAA," *International Journal of Circuit Theory and Applications*, vol. 41, no. 10, pp. 1074–1084, 2013.

[21] P. Beg, I. A. Khan, S. Maheshwari, and M. A. Siddiqi, "Digitally programmable fully differential filter," *Radioengineering*, vol. 20, no. 4, pp. 917–925, 2011.

[22] I. A. Khan, M. R. Khan, and N. Afzal, "Digitally programmable multifunctional current mode filter using CCIIs," *Journal of Active and Passive Electronic Devices*, vol. 1, no. 4, pp. 213–220, 2006.

[23] W. Tangsrirat, D. Prasertsom, and W. Surakampontorn, "Low-voltage digitally controlled current differencing buffered amplifier and its application," *International Journal of Electronics and Communications*, vol. 63, no. 4, pp. 249–258, 2009.

At top right of references column (ref 9 continuation):
new digitally programmable CMOS CCII," *Journal of Circuits, Systems and Computers*, vol. 18, no. 5, pp. 875–897, 2009.

12

Comparative Simulation Analysis of Process Parameter Variations in 20 nm Triangular FinFET

Satyam Shukla,[1,2] Sandeep Singh Gill,[1] Navneet Kaur,[1] H. S. Jatana,[2] and Varun Nehru[2]

[1]*Department of Electronic and Communication Engineering, Guru Nanak Dev Engineering College, Ludhiana 141006, India*
[2]*Semi-Conductor Laboratory, Department of Space, Government of India, Mohali 160071, India*

Correspondence should be addressed to Navneet Kaur; navneetkaur@gndec.ac.in

Academic Editor: Mingxiang Wang

Technology scaling below 22 nm has brought several detrimental effects such as increased short channel effects (SCEs) and leakage currents. In deep submicron technology further scaling in gate length and oxide thickness can be achieved by changing the device structure of MOSFET. For 10–30 nm channel length multigate MOSFETs have been considered as most promising devices and FinFETs are the leading multigate MOSFET devices. Process parameters can be varied to obtain the desired performance of the FinFET device. In this paper, evaluation of on-off current ratio (I_{on}/I_{off}), subthreshold swing (SS) and Drain Induced Barrier Lowering (DIBL) for different process parameters, that is, doping concentration ($10^{15}/cm^3$ to $10^{18}/cm^3$), oxide thickness (0.5 nm and 1 nm), and fin height (10 nm to 40 nm), has been presented for 20 nm triangular FinFET device. Density gradient model used in design simulation incorporates the considerable quantum effects and provides more practical environment for device simulation. Simulation result shows that fin shape has great impact on FinFET performance and triangular fin shape leads to reduction in leakage current and SCEs. Comparative analysis of simulation results has been investigated to observe the impact of process parameters on the performance of designed FinFET.

1. Introduction

To continue with the pace of Moore's law, reduction in transistor dimensions causes very significant short channel effects in device. Methods like (i) variable threshold CMOS, (ii) multithreshold CMOS, (iii) transistor stacking, and (iv) power gating are available to reduce the leakage current to some extent but are not suitable for technologies below 22 nm. FinFETs are considered as most promising device to reduce SCEs and leakage. FinFET is chosen to replace conventional planar CMOS devices below 22 nm [1, 2]. FinFET is a multigate transistor, in which gate is wrapped around the silicon fin channel. Better electrical control is provided by the wrap-around gate structure and thus leakage current and short channel effects are reduced.

FinFET has several advantages compared to planar devices such as well suppressed short channel effects, reduced subthreshold swing (~70 mV/dec), and small threshold voltage roll-off [3]. Rectangular cross section fins are commonly used for design and analysis of FinFET but they are rarely found in industry. In industries, cross section of FinFET is nonuniform and is similar to trapezoidal shape [4]. In a rectangular or trapezoidal FinFET top fin width can be decreased up to minimum possible value to get triangular FinFET keeping all other parameters the same as those of rectangular or trapezoidal FinFET. Hence, shape of fin is approximately triangle in triangular FinFET.

Triangular fin cross section and 3D schematic representation of triangular FinFET are shown in Figure 1. In this paper, evaluation of on-off current ratio (I_{on}/I_{off}), subthreshold swing (SS), and Drain Induced Barrier Lowering (DIBL) for various process parameters, that is, doping concentration (N_{ch}), oxide thickness (T_{ox}), and fin height (H_{fin}), has been presented for 20 nm triangular FinFET device.

In deep submicron technology, quantum effects become significant. Hence, density gradient model is used in design simulation which incorporates the considerable quantum

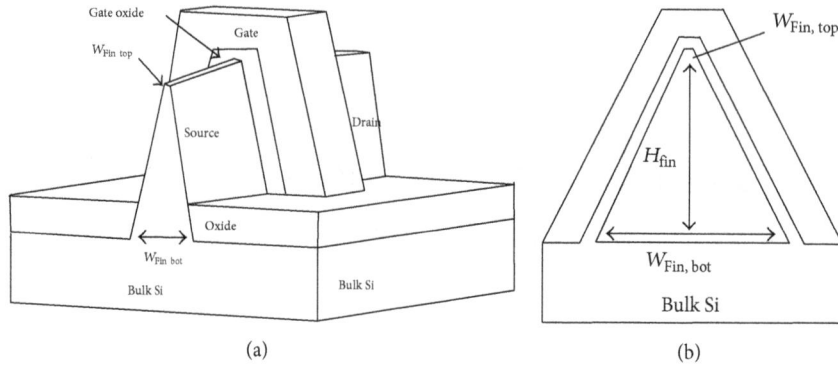

(a)   (b)

FIGURE 1: Triangular FinFET. (a) 3D schematic representation. (b) Triangular fin cross section of FinFET.

effects. By considering the quantum effects, more practical environment is provided for device simulation.

This paper has been formulated as follows. Section 2 presents the earlier work on FinFETs. Subsequent section explains the device design and simulation setup. Results have been discussed in Section 4. Conclusion of the work has been presented in Section 5.

2. Literature Review

In deep submicron technology further downscaling in gate length and oxide thickness can be achieved by varying the device structure of MOSFET. For 10–30 nm channel length double-gate MOSFET (DG-MOSFET) has been considered as most promising device. FinFET process parameters such as T_{ox}, H_{fin}, N_{ch}, fin width (W_{fin}), and gate length (L_{gate}) highly affect the performance of the device. These process parameters can be varied to achieve the desired performance of the FinFET device such as high on-off current ratio, low DIBL, and low SS. The fin thickness should be kept less than 1/3 of channel length to reduce SCEs [5]. Due to reduction in SCEs and leakage performance is improved in bulk FinFETs compared to planar CMOS. Reduction in fin width leads to decrement in leakage due to SCEs. By optimizing input process parameters for a 22 nm triangular FinFET leakage current can be reduced up to 70% as compared to rectangular fin shape with same base fin width [6]. To overcome the gate oxide leakage current in CMOS devices high-k gate stack is used. With high-k gate stack, gate-to channel capacitive coupling can also be improved without any reduction in the gate oxide layer. Variation in work function leads to variation in threshold voltage and it is identified as the main hurdle in scaling of CMOS technology. Below 22 nm technology, due to nonrectangular fin shape, for quantitative estimation of the work-function variation, the work-function values of metallic gate are randomized. Dependence of threshold voltage (V_{th}) on work function can be reduced by 30% for FinFET devices [7]. To characterize Trigate FinFET devices, no complete analytical model is published; in most of the literature, experimental or simulation results are presented.

TABLE 1: Geometry of designed FinFET.

SN	FinFET parameters	Value
(1)	Top fin width (nm)	1
(2)	Bottom fin width (nm)	15
(3)	Oxide thickness (nm)	1
(4)	Fin height (nm)	10 to 40
(5)	Doping concentration (/cm^3)	10^{15} to 10^{18}

Developing compact model for FinFETs is a very challenging task due to 3D structure and ultrasmall dimensions [3].

Due to presence of the wrapped gate over three sides of semiconductor channel in FinFET devices, the electrostatic control of the gate is improved and several problems of planar transistors are solved. Compact models are one of the important components in circuit simulators which establish a link between the device technology and circuit designers. For different doping concentrations rectangular FinFETs can be accurately modelled [4]. Compared to rectangular fin shape triangular fin cross section reduces the SCEs to a greater extent [2]. The body of bulk FinFET should be lightly doped or undoped, to achieve similar on-state performance in Silicon on Insulator (SOI) and bulk FinFET [8]. In FinFETs, immunity to SCE decreases with increase in H_{fin}. It leads to good subthreshold slope and more significant DIBL. Sidewall angle which can be achieved from a particular process limits the fin height [9].

3. Device Design and Simulation Setup

3.1. Device Design Parameters and Material Composition. FinFET of triangular fin shape with 20 nm channel length is designed on Cogenda's GDS2Mesh 3D construction Technology Computer Aided Design (TCAD) tool [10]. The geometrical dimensions used in design are listed in Table 1. Si is used for substrate and fin, and high-k dielectric Hafnium Oxide is used as gate oxide. Device design is shown in Figure 2.

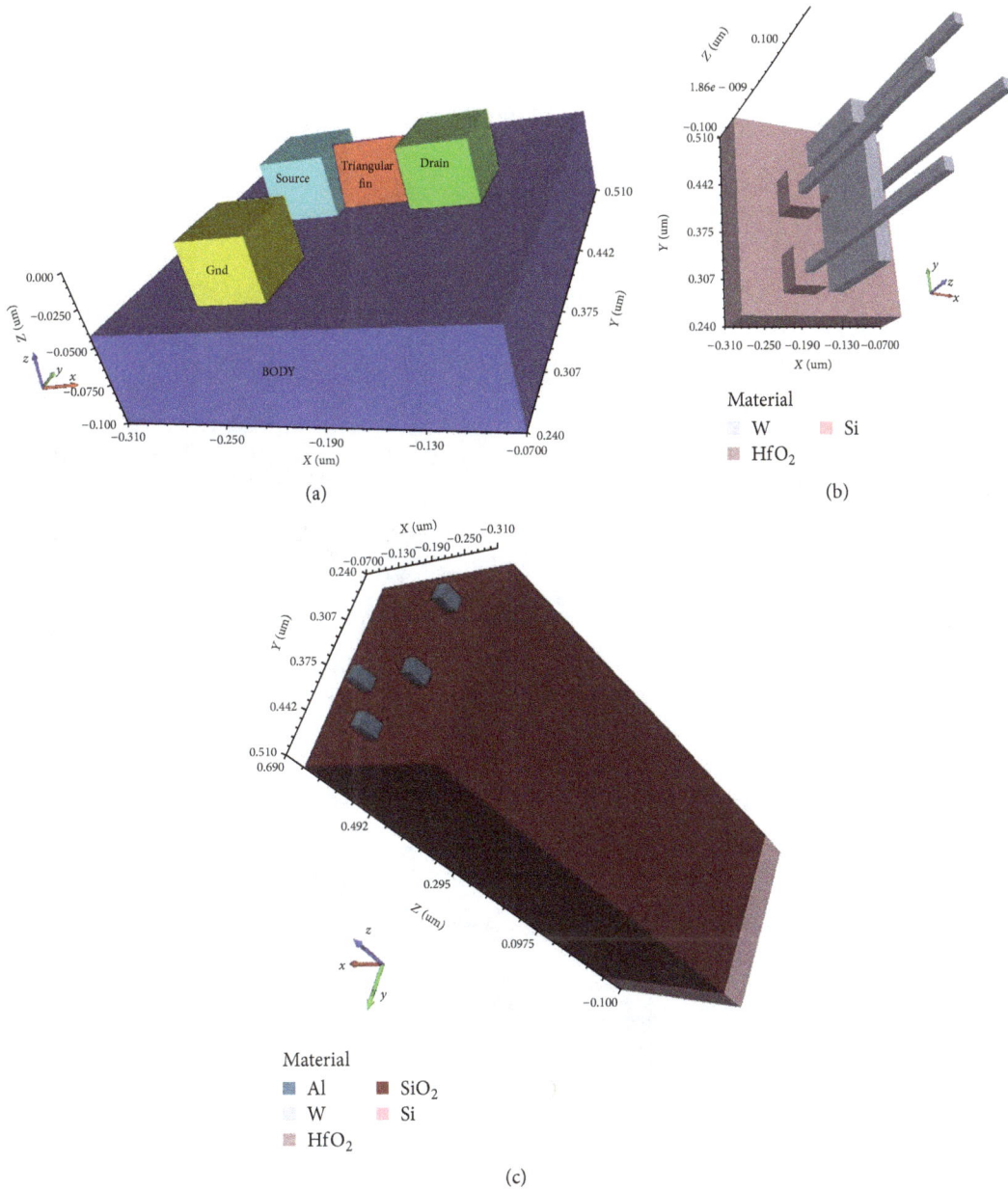

FIGURE 2: Bird's eye view of 3D triangular FinFET device. (a) Active region with triangular fin. (b) Internal view with active region material. (c) Complete device 3D structure.

3.2. Simulation Setup. In this work, triangular FinFET has been simulated at 20 nm gate length for varying H_{fin} and doping using Cogenda's Visual TCAD [11]. In nanoscale devices, quantum effect becomes significant due to very small device size. For such kind of device classical physics model is not suitable for analysis. Hence, quantum physics model is used while simulating the device. The TCAD simulations include density gradient quantum correction model for quantum effects consideration.

An additional quantum potential is included in density gradient (DG) model for calculating the driving force of electrons and holes [10, 11]. The electron and hole quantum correction equation included in simulation is shown as follows [12]:

$$\Lambda_n = -\frac{\hbar^2 \gamma n}{6qm_n^*} \frac{\nabla^2 \sqrt{n}}{\sqrt{n}}$$

$$\Lambda_p = \frac{\hbar^2 \gamma p}{6qm_p^*} \frac{\nabla^2 \sqrt{p}}{\sqrt{p}},$$

(1)

where $\hbar$ is reduced Planck's constant (i.e., $h/2\pi$), m_n^* and m_p^* are the electron effective mass and hole effective mass, respectively, n and p are the electron concentration in

(a)

(b)

(c)

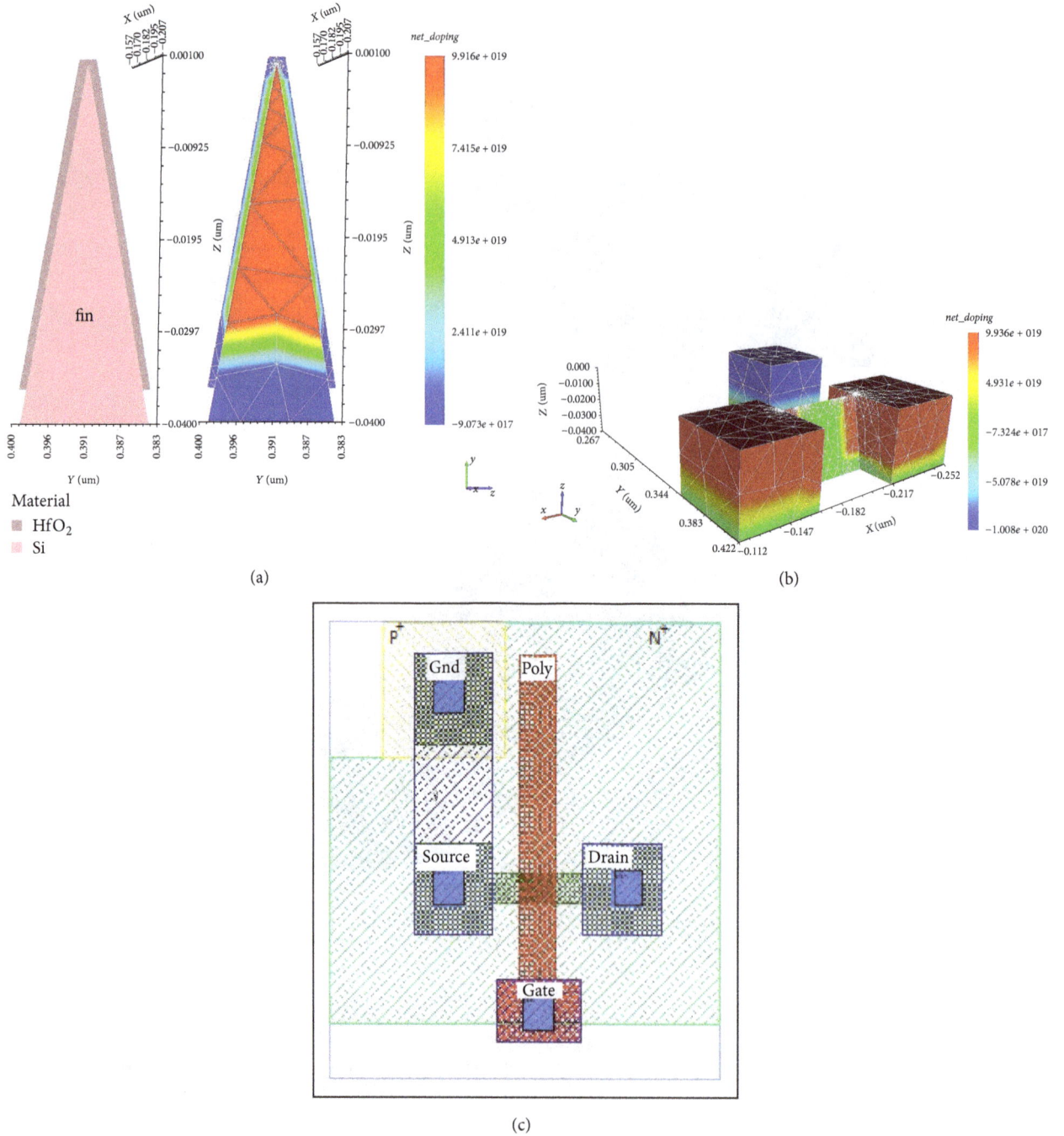

FIGURE 3: Triangular FinFET. (a) Triangular fin material and net doping in triangular fin. (b) Net doping in active region. (c) Mask of designed FinFET.

conduction band and hole concentration in valence band, respectively, and q is electron charge.

For Fermi-Dirac statistics, the electron and hole concentration at Ohmic boundaries must be adjusted as per [12]

$$n = N_c F_{1/2} \left(\frac{E_F - E_{c,\text{eff}} - \Lambda_n}{K_b T} \right)$$

$$p = N_v F_{1/2} \left(\frac{E_{v,\text{eff}} - \Lambda_p - E_F}{K_b T} \right),$$

(2)

where K_b is Boltzmann's constant, T is temperature, N_c is effective density of states (DOS) for electrons in conduction band, N_v is effective density of states (DOS) for holes in valence band, and E_F, $E_{c,\text{eff}}$, and $E_{v,\text{eff}}$ represent fermi level, conduction band edge, and valence band edge, respectively.

Consideration of quantum effects provides more practical environment to design simulation. Net doping profile for fin and active region is shown in Figure 3. Fin height is varied from 10 nm to 40 nm and doping is varied from $10^{15}/\text{cm}^3$ to

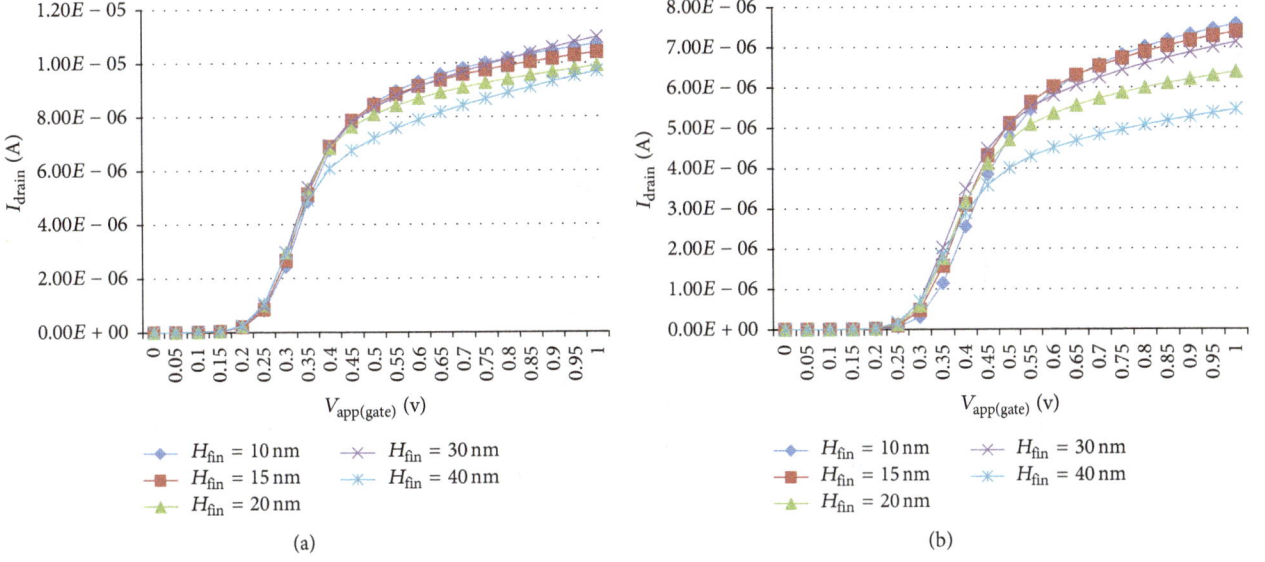

FIGURE 4: Transfer characteristics of triangular FinFET. (a) For low doping level ($10^{15}/cm^3$). (b) For high doping level ($10^{18}/cm^3$).

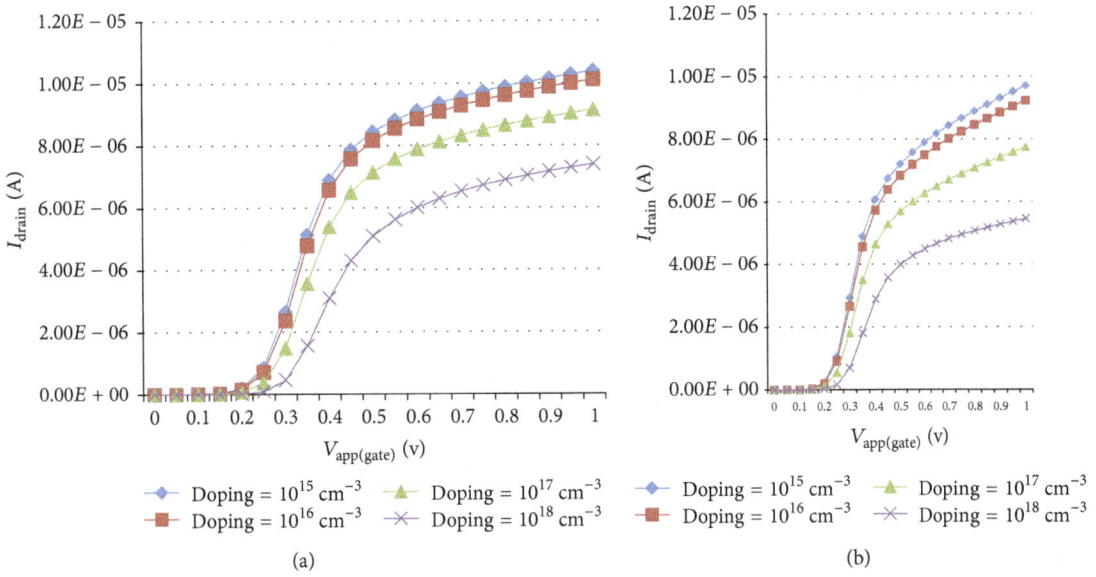

FIGURE 5: Transfer characteristics of triangular FinFET. (a) For 15 nm fin height. (b) For 40 nm fin height.

$10^{18}/cm^3$ for oxide thickness of 1 nm. The impact of doping and fin dimensions like fin height (H_{fin}) on output parameters of FinFET is investigated using TCAD tool. Evaluation of fin height and fin doping concentration corresponding to better device performance is presented.

4. Results and Discussion

Simulation results show that drain current of the triangular FinFET at drain voltage 50 mV decreases with H_{fin} at constant doping concentration of $10^{15}/cm^3$ and $10^{18}/cm^3$ as shown in Figure 4. Variation of drain current with doping is evaluated for different values of fin height. Drain current decreases

with increase in doping level for constant H_{fin} of 15 nm and 40 nm as shown in Figure 5. High channel doping causes more impurity scattering in the crystal and the consequence is that carrier mobility gets reduced, thus resulting in low drain current whereas, in case of low doping concentration, on current is more due to lesser impurity scattering as shown in Figure 5. Drain voltage was kept at 50 mV for measurement of on-off current ratio. On current was studied at gate voltage (V_g) = 1 V and off current was measured at V_g = 0 V.

The variation of on-off current ratio with varying H_{fin} for different doping levels is shown in Figure 6. To maximize this ratio, the doping level should be high (i.e., $10^{18}/cm^3$) because off current is also less in case of high channel doping similar to on current and leads to better on/off current ratio.

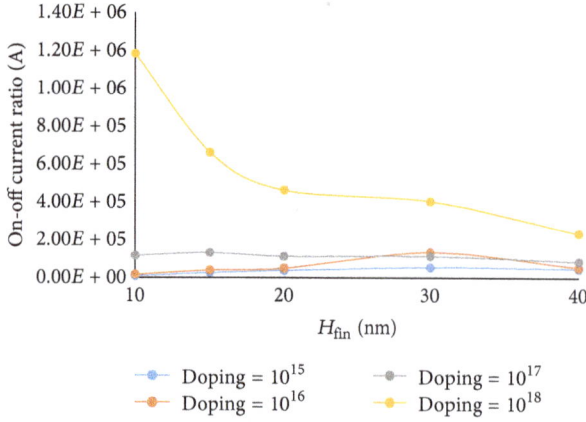

FIGURE 6: Variation of on-off current ratio with fin height for different doping concentration at 1 nm oxide thickness.

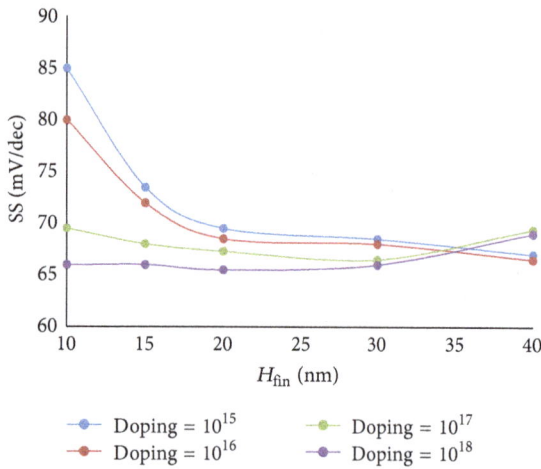

FIGURE 7: Variation of SS with fin height for different doping concentration at 1 nm oxide thickness.

TABLE 2: Variation of W_{eff} and I_{DIBL} with H_{fin}.

SN	H_{fin} (nm)	W_{eff} (nm)	I_{DIBL} (μA)
(1)	10	30.042	0.1502083
(2)	15	40.042	0.2002083
(3)	20	50.042	0.2502083
(4)	30	70.042	0.3502083
(5)	40	90.042	0.4502083

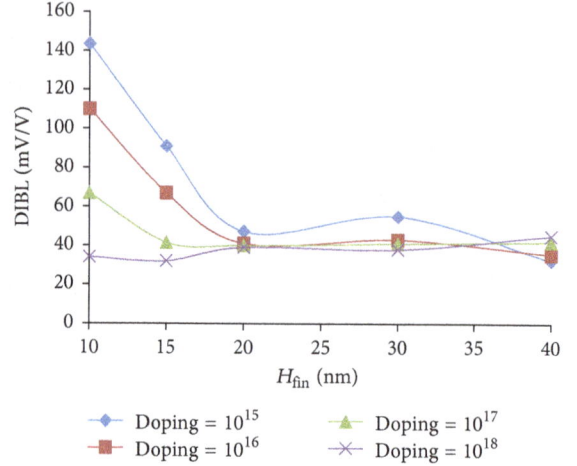

FIGURE 8: Variation of DIBL with fin height for different doping concentration at 1 nm oxide thickness.

Change in gate voltage for a decade change in drain current is known as subthreshold swing (SS). It is calculated using (3) keeping drain voltage at 50 mV.

$$SS = \frac{dV_g}{d \log (I_d)}. \tag{3}$$

For low doping level, SS reduces with H_{fin}. For high doping levels (10^{17}/cm^3 and 10^{18}/cm^3) SS decreases for H_{fin} range of 10 nm to 30 nm and it increases with H_{fin} after 30 nm. Comparison of SS for all the doping levels shows that high doping results in decreased SS as shown in Figure 7.

For DIBL calculations, simulations (at drain voltage 20 mV and 1 V) are performed for each combination of input process parameters. The horizontal displacement of the experimental transfer characteristics for $V_d = 20$ mV and 1 V at constant drain current is defined as DIBL [13].

The constant drain current at which the horizontal displacement of transfer characteristics is observed for the calculation of DIBL is given as [13]

$$I_{\text{DIBL}} = 10^{-7} \times \left(\frac{W_{\text{eff}}}{L} \right) A. \tag{4}$$

W_{eff} is effective channel width and for rectangular FinFET it is defined as

$$W_{\text{eff}} = 2H_{\text{fin}} + W_{\text{fin}}. \tag{5}$$

W_{fin} for rectangular FinFET is constant throughout the fin but for triangular FinFET it varies from $W_{\text{fin,bot}}$ to $W_{\text{fin,top}}$. In [14], the equivalent fin width for trapezoidal FinFET with $W_{\text{fin,top}} = 5$ nm and $W_{\text{fin,bot}} = 15$ nm is given at its orthocenter. The same idea can be extended to find the equivalent fin width for the triangular FinFET at its orthocenter and can be given as

$$W_{\text{fin}} = W_{\text{fin,top}} + \frac{\alpha}{1 + \alpha} \left(W_{\text{fin,bot}} - W_{\text{fin,top}} \right), \tag{6}$$

where

$$\alpha = \frac{2W_{\text{fin,bot}} + W_{\text{fin,top}}}{2W_{\text{fin,top}} + W_{\text{fin,bot}}}. \tag{7}$$

For the designed triangular device $W_{\text{fin,top}} = 1$ nm, $W_{\text{fin,bot}} = 15$ nm and hence from (7), $\alpha = 1.82353$. Since H_{fin} is varied from 10 nm to 40 nm, the value of W_{eff} and I_{DIBL} is listed in Table 2.

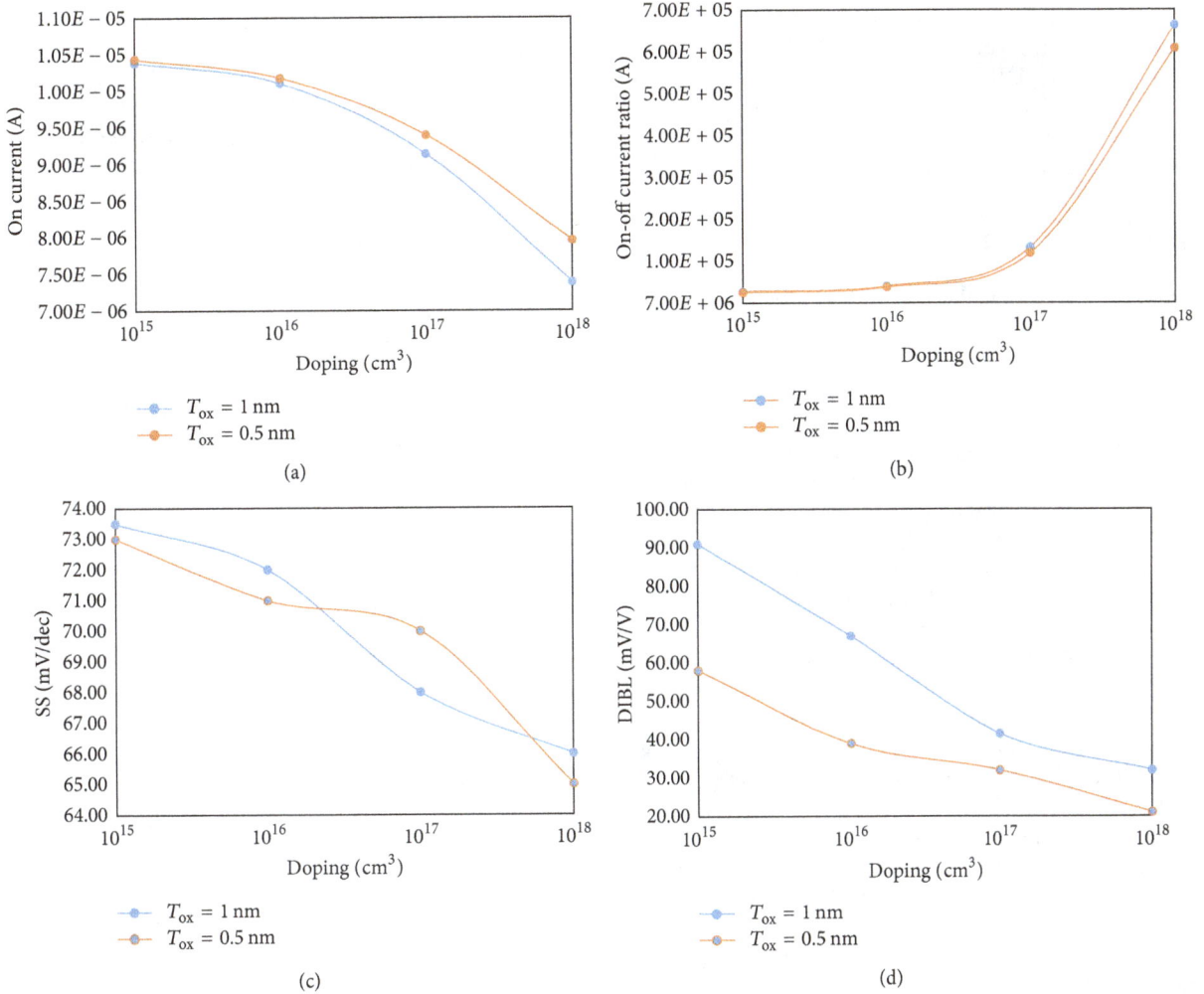

FIGURE 9: Variation of output parameters of FinFET with respect to different doping concentration for T_{ox} = 0.5 nm and 1 nm and H_{fin} = 15 nm. (a) On current. (b) On-off current ratio. (c) SS. (d) DIBL.

DIBL value is improved for high channel doping; this is due to the reason that high doping in the channel lessens the impact of drain potential onto it which further shields channel from drain terminal. In [13], for 25 nm gate length device it is shown that for nonrectangular FinFET device if $W_{fin,top}/W_{fin,bot}$ is less than 0.2 then the range of SS is 74 mV/dec to 76 mV/dec and DIBL is 80 mV/V to 90 mV/V. In this work, simulation results show that, with the gate length reduced to 20 nm, $W_{fin,top}$ = 1 nm, and $W_{fin,bot}$ =15 nm (i.e., $W_{fin,top}/W_{fin,bot}$ = 0.0667), SS can be achieved as low as 65.5 mV/dec and DIBL can be minimized to 32 mV/V as shown in Figures 7 and 8.

On current of device is decreasing with increase in doping concentration; however increasing doping concentration leads to increment in on-off current ratio. SS and DIBL decrease with high doping concentration for 15 nm fin height. Dependence of on current, on-off current ratio, SS, and DIBL on doping concentration for different gate oxide thickness is shown in Figure 9. It is noticed that mobility gets decreased for high doping and thus results in lesser on current and also, leakage current is decreased thus leading to improved on-off

current ratio. Better values of SS and DIBL are obtained with doping of 10^{18}/cm^3 as compared to 10^{15}/cm^3. Oxide thickness of 0.5 nm shows better result for DIBL. This is because of the fact that gate has strong control over three sides of channel for T_{ox} = 0.5 nm compared to that for T_{ox} = 1 nm which reduces the influence of drain over channel. Thus, threshold voltage is less affected by variations in applied drain voltage and results in lower DIBL. Also, the use of HfO$_2$ as gate dielectric can scale the oxide thickness down simultaneously without increasing gate tunneling current.

5. Conclusion

Simulation results show that the variation of process parameters of FinFET has considerable impact on performance parameters of the FinFET. Due to reduced top fin width, gate control over channel is improved and it leads to better performance. On-off current ratio sharply decreases with decrease in doping concentration and increase in fin height. On-off current ratio is improved for high doping (10^{18}/cm^3) and minimum fin height (10 nm). SS is better for maximum doping concentration with 20 nm fin height. DIBL is best for maximum doping concentration with 15 nm fin height. For 1 nm gate oxide, DIBL reduces with doping. For 0.5 nm gate oxide the results are better than that of 1 nm gate oxide provided that fin height is less than 15 nm. For fin height more than 15 nm, performance at 1 nm oxide thickness is better.

Conflicts of Interest

The authors declare that there are no conflicts of interest regarding the publication of this paper.

References

[1] C. Auth, C. Allen, A. Blattner et al., "A 22nm high performance and low-power CMOS technology featuring fully-depleted tri-gate transistors, self-aligned contacts and high density MIM capacitors," in *Proceedings of the Symposium on VLSI Technology (VLSIT '12)*, pp. 131–132, Honolulu, Hawaii, USA, June 2012.

[2] J. Mohseni and J. D. Meindl, "Scaling limits of rectangular and trapezoidal channel FinFET," in *Proceedings of the IEEE Green Technologies Conference*, pp. 204–210, Denver, Colo, USA, April 2013.

[3] B. D. Gaynor and S. Hassoun, "Fin shape impact on FinFET leakage with application to multithreshold and ultralow-leakage FinFET design," *IEEE Transactions on Electron Devices*, vol. 61, no. 8, pp. 2738–2744, 2014.

[4] H. Nam and C. Shin, "Impact of current flow shape in tapered (versus rectangular) FinFET on threshold voltage variation induced by work-function variation," *IEEE Transactions on Electron Devices*, vol. 61, no. 6, pp. 2007–2011, 2014.

[5] J. P. Duarte, N. Paydavosi, S. Venugopalam, A. Sachid, and C. Hu, "Unified FinFET compact model: modelling trapezoidal triple-gate FinFETs," in *Proceedings of the International Conference on Simulation of Semiconductor Processes and Devices (SISPAD '13)*, pp. 135–138, Glasgow, UK, September 2013.

[6] K. Wu, W.-W. Ding, and M.-H. Chiang, "Performance advantage and energy saving of triangular-shaped FinFETs," in *Proceedings of the 18th International Conference on Simulation of Semiconductor Processes and Devices (SISPAD '13)*, pp. 143–146, Scotland, UK, September 2013.

[7] M. Poljak, V. Jovanovic, and T. Suligoj, "SOI vs. Bulk FinFET: body doping and corner effects influence on device characteristics," in *Proceedings of the 14th IEEE Mediterranean Electrotechnical Conference*, pp. 425–430, 2008.

[8] X. Wu, P. C. H. Chan, and M. Chan, "Impacts of nonrectangular fin cross section on the electrical characteristics of FinFET," *IEEE Transactions on Electron Devices*, vol. 52, no. 1, pp. 63–68, 2005.

[9] G. Pei, J. Kedzierski, P. Oldiges, M. Ieong, and E. C.-C. Kan, "FinFET design considerations based on 3-D simulation and analytical modeling," *IEEE Transactions on Electron Devices*, vol. 49, no. 8, pp. 1411–1419, 2002.

[10] N. Fasarakis, T. A. Karatsori, A. Tsormpatzoglou et al., "Compact modeling of nanoscale trapezoidal finFETs," *IEEE Transactions on Electron Devices*, vol. 61, no. 2, pp. 324–332, 2014.

[11] http://www.cogenda.com/article/Gds2Mesh.

[12] "Visual TCAD Brochure," http://www.cogenda.com/article/downloads.

[13] 3D FinFET simulation with Density Gradient (DG) quantum correction model, http://www.cogenda.com/article/examples#FinFET-dg.

[14] N. Thapa, L. Maurya, and R. Mehra, "Performance advancement of High-K dielectric MOSFET," *International Journal of Innovations and Advancement in Computer Science*, vol. 3, pp. 98–103, 2014.

13

On the Memristances, Parameters, and Analysis of the Fractional Order Memristor

Rawid Banchuin (ORCID)

Faculty of Engineering and Graduated School of Information Technology, Siam University, Bangkok, Thailand

Correspondence should be addressed to Rawid Banchuin; rawid_b@yahoo.com

Academic Editor: Stephan Gift

In this work, the analytical expressions of memristances, related parameters, and time domain behavioral analysis of the fractional order memristor have been proposed. Both DC with arbitrary delay and many AC waveforms including arbitrary phase sinusoidal and cosinusoidal waveform along with arbitrary periodic waveform have been taken into account. Unlike the previous works, the formerly ignored dimensional consistency has been taken into account and the analytical modelling of the boundary effect has been performed. Moreover, both transient and asymptotic behaviors of the fractional order memristor excited by AC waveform have been distinguished and analyzed. The effect of phase of AC waveform has also been studied. The influence of the fractional order to the areas of voltage-current hysteresis loop and memristance-current lissajous curve has also been clearly discussed and the usage of fractional order memristor in the memristor based circuit has also been demonstrated.

1. Introduction

Recently, a state-of-the-art electrical circuit element, namely, fractional order memristor, is often cited. This circuit element can be obtained from the generalization of the 4th electrical circuit element, namely, memristor, that has been theoretically found by Leon Chua since 1971 [1], by using the concept of fractional calculus which have been adopted in various disciplines, e.g., biomedical engineering [2, 3], control system [4–6], and electronic engineering [7–9]. For decades after Chua proposed his original work, the memristor has been practically realized by a research group in Hewlett Packard (HP) labs [10] in 2008. As a result, the mathematical modelling and analysis attempts of the memristor have been proposed (e.g., [11–15]).

For the fractional order memristor on the other hand, there also exists such modelling and analysis attempts [16–21]. Some of them generalize the memristor by applying concept of the fractional calculus to the voltage-current relationship [16, 17] and termed such generalized memristor as the fracmemristor [17]. On the other hand, others do so by applying the fractional calculus to the memristor's state equation where the often cited HP memristor has been adopted as the basis [18–21]. However, only the analytical

expression of the area of voltage-current hysteresis loop has been proposed in [20] and those of the memristances proposed in [18, 19, 21] are in terms of the input voltage despite the fact that fractional order memristor of interested is the generalization of the HP memristor which is actually of a charge/current controlled type. Moreover, these previous works also neglected the dimensional consistency [22, 23] related issues and the boundary effect, which is an important characteristic of the HP memristor [10], has not been analytically modelled.

By this motivation, we generalize the HP memristor in the fractional order domain by also concerning the formerly ignored dimensional consistency and formulate the analytical expression of memristance in term of the input current where boundary effect has also been modelled. We also derive the expressions of those related parameters of the fractional memristor excited by various exciting waveforms including DC with arbitrary delay and sinusoidal and cosinusoidal with arbitrary phase and arbitrary periodic which are the AC waveforms. With these expressions, parameters, and numerical simulations with MATHEMATICA, the behaviors of the fractional order memristor have been thoroughly explored. Unlike [18–21], both transient and asymptotic behaviors of the fractional order memristor excited by AC waveform have been distinguished and analyzed. The effect of phase of AC waveform has also been studied. Moreover, the influence of the fractional order to the areas of voltage-current hysteresis loop and memristance-current lissajous curve has been clearly discussed and the usage of fractional order memristor in the memristor based circuit has also been demonstrated.

In the following section, the overview of memristor will be briefly given followed by the memristor's generalization and derivation of our expressions in Section 3 where the behavioral analysis of the fractional order memristor will also be given. The DC waveform will be firstly treated followed by the AC waveforms where the sinusoidal waveform has been emphasized as it is the most fundamental. This is because the memristances and parameters of the memristor excited by the cosinusoidal and arbitrary periodic waveform can be obtained by using those due to the sinusoidal waveform as the basis as will be shown in Section 3 as well. The usage of fractional order memristor in the memristor based circuit will be presented in Section 4 and the conclusion will be finally drawn in Section 5.

2. The Overview of Memristor

Memristor is a nonlinear electrical circuit element. This circuit element relates the instantaneous flux, $\phi(t)$, and charge, $q(t)$, through the following relationship:

$$M(t) = \frac{d\phi(t)}{dq(t)} \tag{1}$$

where $M(t)$ denotes the memristance.

According to [10], $M(t)$ of the HP memristor can be given in terms of the minimum and maximum values of $M(t)$ denoted by M_{on} and M_{off} and the state variable, $x(t)$, as

$$M(t) = M_{on}x(t) + (1 - x(t))M_{off} \tag{2}$$

where $x(t)$, which is dimensionless, can be given in terms of the memristor's current, $i(t)$, by

$$\frac{dx}{dt} = ki(t) \tag{3}$$

Note that $k = \mu M_{on}/D^2$, where μ and D, respectively, stand for the ion mobility and semiconductor film of thickness. Therefore, the dimension of k is $(Asec)^{-1}$.

As can be seen from (3) and also according to [10], $x(t)$ can be simply given as follows:

$$x(t) = kq(t) \tag{4}$$

Therefore, it can be seen that the HP memristor is charge controlled. Since $q(t)$ is a time integration of $i(t)$, it can be stated that the HP memristor is of a current controlled type. Note also that $0 \leq x(t) \leq 1$; thus, $M_{on} \leq M(t) \leq M_{off}$ as long as the memristor is unsaturated. Otherwise, $x(t)$ will be bounded at either 0 or 1 so $M(t)$ will be equal to either M_{on} or M_{off} according to the boundary effect of the device. Traditionally, such boundary effect can be mathematically modelled by multiplying the RHS of (3) with the window function [24].

3. The Fractional Order Domain Generalization of the Memristor and the Memristances, Related Parameters, and Analysis of the Fractional Order Memristor

By generalizing the memristor in the fractional order domain with the fractional calculus, the fractional order memristor can be obtained. Similarly to [18–21], we perform such generalization by applying the fractional calculus to the memristor's state equation, i.e., (3). In these previous works, $d^\alpha x/dt^\alpha = ki(t)$, where α stands for the order of the fractional order memristor which can be arbitrary real value and has been obtained from such generalization. However, as $x(t)$ is dimensionless; the dimension of the LHS of this previous generalized equation is given by $sec^{-\alpha}$ where that of the RHS is sec^{-1} which means that a dimensional inconsistency has always existed.

Therefore, the fractional time component [22], σ, which has the dimension of sec, has been introduced for handling this issue. As a result, unlike [18–21], the following generalized state equation has been used instead.

$$\sigma^{\alpha-1}\frac{d^\alpha x}{dt^\alpha} = ki(t) \tag{5}$$

Similarly to that of the RHS, the dimension of the LHS of (5) is sec^{-1}; thus the dimensional inconsistency issue has been resolved. Note also that (5) is reduced to (3) when $\alpha = 1$ despite the presence of σ as $\sigma^{\alpha-1}$ become 1 with such value of α.

Unlike [18, 19, 21], we derive $M(t)$ of the fractional order memristor as a function $i(t)$ as it has been assumed that the of fractional order memristor is a generalization of the HP memristor which is of a current controlled type as aforementioned. Therefore we directly determine $x(t)$ from (5) by using the Riemann-Liouville fractional order integral [25] as follows:

$$x(t) = x(0) + \frac{k\sigma^{1-\alpha}}{\Gamma(\alpha)}\int_0^t (t-\tau)^{\alpha-1}i(\tau)d\tau \tag{6}$$

where $x(0)$ and $\Gamma()$ denote the initial value of $x(t)$ and the Gamma function [26], respectively.

Since it can be seen from (2) that

$$M(t) = M_{off} - M_d x(t) \tag{7}$$

where $M_d = M_{off} - M_{on}$, the initial memristance value, i.e., $M(0)$, can be immediately given by

$$M(0) = M_{off} - M_d x(0) \tag{8}$$

Thus by substituting (6) into (7) and keeping (8) in mind, $M(t)$ of the fractional order memristor can be obtained as follows:

$$M(t) = M(0) - \frac{k\sigma^{1-\alpha}M_d}{\Gamma(\alpha)}\int_0^t (t-\tau)^{\alpha-1}i(\tau)d\tau \tag{9}$$

which shows that $M(t)$ is current-controlled.

If we let $\alpha = 1$, (9) will be reduced to

$$M(t) = M(0) - kM_d \int_0^t i(\tau)\,d\tau \qquad (10)$$

Since the integer order integration of $i(t)$ gives $q(t)$, we obtain

$$M(t) = M(0) - kM_d q(t) \qquad (11)$$

By using (8), (11) can be simplified under the assumption that $x(0) = 0$ and $M_{off} \gg M_{on}$ as follows:

$$M(t) = M_{off}(1 - kq(t)) \qquad (12)$$

which is similar to the original simplified model of the HP memristor [10]. Such correspondence cannot be found in [18, 19, 21] as the integer order integration of the memristor's voltage, $v(t)$, yields $\phi(t)$.

For traditionally including the boundary effect, the window function must be introduced to the state equation as mentioned above. In [18], the linear window function given by $f(x(t)) = 1$ has been adopted for simplicity as the usage of more accurate yet more complicated window function; e.g., those Joglekar, Biolek, and Prodomakis [24] can be mathematically cumbersome. Unfortunately, using such linear window function is mathematically equivalent to multiplying the RHS of the state equation by 1. As a result, no modification has been made on the state equation; thus the boundary effect modelling has not been performed. Moreover, neither the usage of window function nor alternative boundary effect analytical modelling has been made in both [19] and [21].

In order to model the boundary effect in a simplified manner, we apply two mathematical operators, i.e., $\max[x, y]$ and $\min[x, y]$, which, respectively, selects the maximum value and minimum value among x and y, to (9). As a result, our expression of $M(t)$ due to arbitrary exciting waveform can be finally given as follows:

$$M(t)$$
$$= \min\left[\max\left[M(0) - \frac{KM_d}{\Gamma(\alpha)}\int_0^t (t-\tau)^{\alpha-1}i(\tau)\,d\tau, M_{on}\right], \qquad (13)$$
$$M_{off}\right]$$

where $K = k\sigma^{1-\alpha}$; thus the dimension of K is $A^{-1}\sec^{-\alpha}$.

Since the dimension of fractional integral of $i(t)$ is $A\sec^{\alpha}$, that of $(KM_d/\Gamma(\alpha))\int_0^t (t-\tau)^{\alpha-1}i(\tau)d\tau$ is given by Ω which is physically measurable, similarly to those of M_d and $M(0)$. Therefore, $M(0)$ and $(KM_d/\Gamma(\alpha))\int_0^t (t-\tau)^{\alpha-1}i(\tau)d\tau$, which are at the RHS of (13), can be physically combined as they have the same dimensions and the dimension of $M(t)$, which is the LHS of such equation, has also been found to be such physically measurable Ω; thus it can be seen that our expression of $M(t)$ has dimensional consistency. Moreover, due to the operation of nested $\max[x, y]$ and $\min[x, y]$, $M(t)$ will be equal to $M(0) - (KM_d/\Gamma(\alpha))\int_0^t (t-\tau)^{\alpha-1}i(\tau)d\tau$ if and only if $M(0) - (KM_d/\Gamma(\alpha))\int_0^t (t-\tau)^{\alpha-1}i(\tau)d\tau$ lies within

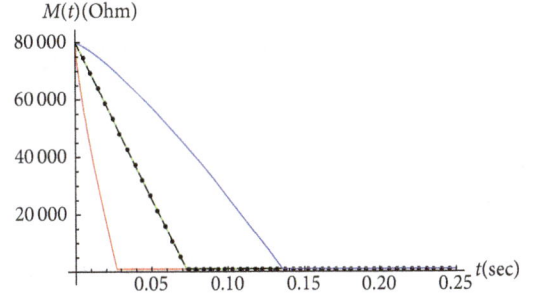

FIGURE 1: $M(t)$ of the fractional order memristor excited by a 110 μA DC waveform: $\alpha = 0.75$ (red), $\alpha = 1$ (green), $\alpha = 1.25$ (blue), and HP memristor (black dots).

$[M_{on}, M_{off}]$ which means that the fractional order memristor remains unsaturated. Otherwise, $M(t)$ will be equal to either M_{on} or M_{off} if $M(0) - (KM_d/\Gamma(\alpha))\int_0^t (t-\tau)^{\alpha-1}i(\tau)d\tau$ is lower than M_{on} or higher than M_{off} which in turn means that the device become saturated at either its on-state or off-state. Therefore it can be seen that the boundary effect has been modelled without any necessity to use the window function and (13) along with its related results is valid to the saturated fractional order memristor. In the following subsections, $M(t)$'s due to due to various exciting waveforms and the behavioral analysis of fractional order memristor will be presented.

3.1. DC Waveform. Mathematically, the DC waveform with arbitrary delay (t_d), which is more generic than the undelay waveform assumed in the previous works [18, 19], can be defined as $i(t) = I_{DC}u(t - t_d)$, where I_{DC} and $u(t)$ denote the magnitude of the waveform and the unit step function. Therefore, the resulting $M(t)$ can be straightforwardly obtained by using (13) as

$$M(t)$$
$$= \min\left[\max\left[M(0) - \frac{KM_d I_{DC}(t-t_d)^{\alpha}}{\Gamma(\alpha+1)}, M_{on}\right], \qquad (14)$$
$$M_{off}\right]$$

By using (14) with $t_d = 0$ sec, $K = 100000$ $A^{-1}\sec^{-\alpha}$, $M_{on} = 1$ $k\Omega$, $M_{off} = 100$ $k\Omega$, and $M(0) = 80$ $k\Omega$, $M(t)$'s of the fractional order memristor with various α's excited by the DC waveform can be numerically simulated as depicted in Figures 1 and 2 where $I_{DC} = 110$ μA and $I_{DC} = -110$ μA have been, respectively, assumed and $M(t)$'s of the HP memristor simulated by using its SPICE model [27] have also been included.

From these figures, the strong agreements between our $M(t)$'s obtained by using (14) with $\alpha = 1$ and those of the HP memristor can be observed. Since the HP memristor is of order 1 in the context of fractional order domain, such strong agreements verify our expression. These figures also show that $M(t)$ of the fractional order memristor can be either the increasing or decreasing function of t with the final value of

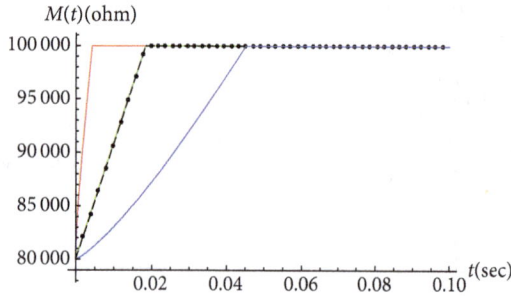

FIGURE 2: $M(t)$ of the fractional order memristor excited by a -110 μA DC waveform: $\alpha = 0.75$ (red), $\alpha = 1$ (green), $\alpha = 1.25$ (blue), and HP memristor (black dots).

M_{on} or M_{off} when the fractional order memristor become saturated if we let $I_{DC} > 0$ or $I_{DC} < 0$. Moreover, the rate of change of $M(t)$ has been found to be inversely proportional to α.

Since it can be seen that the fractional order memristor become saturated at a certain time given by t_{sat}, by using (14), we have

$$M(0) - \frac{KM_d I_{DC}(t_{sat} - t_d)^\alpha}{\Gamma(\alpha + 1)} = \begin{cases} M_{on} & I_{DC} > 0 \\ M_{off} & I_{DC} < 0 \end{cases} \quad (15)$$

As a result, t_{sat} can be immediately given as follows:

$$t_{sat} = \begin{cases} t_d + \left[\dfrac{\Gamma(\alpha+1)(M(0) - M_{on})}{kM_d I_{DC}} \right]^{1/\alpha} & I_{DC} > 0 \\[2ex] t_d + \left[\dfrac{\Gamma(\alpha+1)(M(0) - M_{off})}{kM_d I_{DC}} \right]^{1/\alpha} & I_{DC} < 0 \end{cases} \quad (16)$$

which shows that t_{sat} is directly proportional to the size of the difference between $M(0)$ and $M_{on}(M_{off})$. So, t_{sat} reaches

its maximum value, i.e., $t_{sat(MAX)}$, if and only if $M(0)$ reaches its possible peak value given by either M_{off} when $I_{DC} > 0$ or M_{on} when $I_{DC} < 0$. Thus $t_{sat(MAX)}$ can be found as

$$t_{sat(MAX)} = \begin{cases} t_d + \left[\dfrac{\Gamma(\alpha+1)}{KI_{DC}} \right]^{1/\alpha} & I_{DC} > 0 \\[2ex] t_d + \left[\dfrac{\Gamma(\alpha+1)}{K(-I_{DC})} \right]^{1/\alpha} & I_{DC} < 0 \end{cases} \quad (17)$$

which can be immediately given in a more compact manner as follows:

$$t_{sat(MAX)} = t_d + \left[\frac{\Gamma(\alpha+1)}{K|I_{DC}|} \right]^{1/\alpha} \quad (18)$$

For confirming the aforesaid observation on the relationship between the rate of change of $M(t)$ and α, $t_{sat(MAX)}$'s have been simulated by using (18) under the similar assumptions to those of the simulation of $M(t)$'s shown in Figures 1 and 2 but with varying I_{DC}, as depicted in Figure 3 which shows that $t_{sat(MAX)}$ is directly proportional to α. Due to the definition of $t_{sat(MAX)}$, this confirms such observation. Moreover, it can also be seen from Figure 3 that $t_{sat(MAX)}$ is inversely proportional to $|I_{DC}|$.

3.2. AC Waveforms. Among various AC waveforms, the sinusoidal waveform has been emphasized as it is the foundation of the others. Unlike those previous works, the sinusoidal waveform with arbitrary phase (θ) has been chosen due to its generality. Mathematically, such waveform can be given by $i(t) = I_m \sin(\omega t + \theta)$, where I_m and ω, respectively, denote the peak value and angular frequency of $i(t)$. By using (13), $M(t)$ of the fractional order memristor under the arbitrary phase sinusoidal input can be given by

$$M(t) = \min \left[\max \left[M(0) - \frac{KM_d I_m t^\alpha}{\Gamma(\alpha+1)} \left[\sin(\theta) \, {}_1F_2\left(1; \frac{\alpha}{2} + \frac{1}{2}, \frac{\alpha}{2} + 1; -\frac{1}{4}(\omega t)^2\right) + \frac{\omega t \cos(\theta)}{\alpha+1} \, {}_1F_2\left(1; \frac{\alpha}{2} + 1, \frac{\alpha}{2} + \frac{3}{2}; -\frac{1}{4}(\omega t)^2\right) \right], M_{on} \right], M_{off} \right] \quad (19)$$

where $_1F_2(\ ;\ ;\ )$ denotes a generalized hypergeometric function with $p = 1$ and $q = 2$ [28].

By letting $K = 100000 \, A^{-1} sec^{-\alpha}$, $M_{on} = 1 \, k\Omega$, $M_{off} = 100$ $k\Omega$, and $M(0) = 80 \, k\Omega$ similarly to the previous subsection, $M(t)$'s of the fractional order memristor with various α's excited by the sinusoidal input can be simulated as depicted in Figures 4 and 5, where $I_m = 110\mu A$ and $I_m = -110 \, \mu A$ have been, respectively, assumed and $M(t)$'s of the HP memristor simulated by using its SPICE model have also been included. Moreover, we also let $\omega = 1$ rad/sec and $\theta = 0$ rad. Again,

the strong agreements between our expression based $M(t)$'s with $\alpha = 1$ and those of the HP memristor which verify our expression can be observed. It can be seen that $M(t)$ of the fractional order memristor with $\alpha \leq 1$ is periodic with clipped peaks at M_{on} and M_{off} due to the temporary saturation of the fractional order memristor as the unclipped peaks of $M(t)$ lie outside $[M_{on}, M_{off}]$. On the other hand, $M(t)$ of the device with $\alpha > 1$ is not periodic but time independently equal to either M_{on} or M_{off} when $t \geq T_{sat}$ up to the sign of I_m. This is because the device becomes

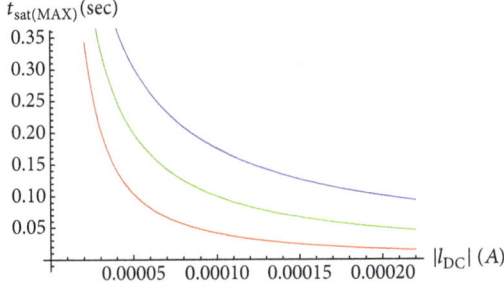

FIGURE 3: $t_{sat(MAX)}$ of the current-controlled fractional order memristor against $|I_{DC}|$: $\alpha = 0.75$ (red), $\alpha = 1$ (green), and $\alpha = 1.25$ (blue).

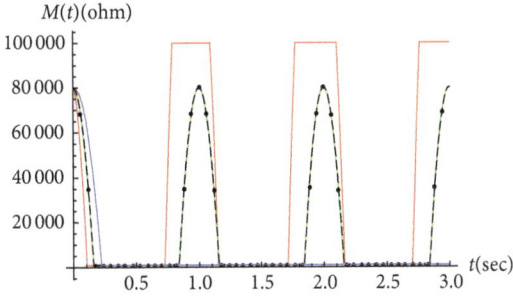

FIGURE 6: $M(t)$ of the boundary effect free fractional order memristor under the sinusoidal input with $I_m = 110\ \mu A$ and $\alpha = 1.25$.

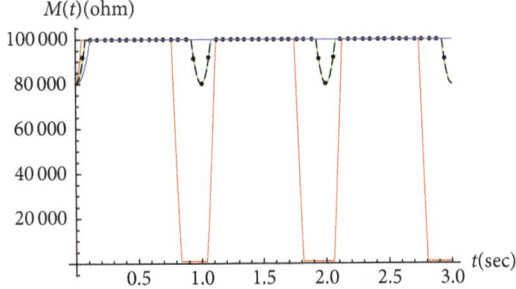

FIGURE 4: $M(t)$ of the fractional order memristor under the sinusoidal input with $I_m = 110\ \mu A$: $\alpha = 0.75$ (red), $\alpha = 1$ (green), $\alpha = 1.25$ (blue), and HP memristor (black dots).

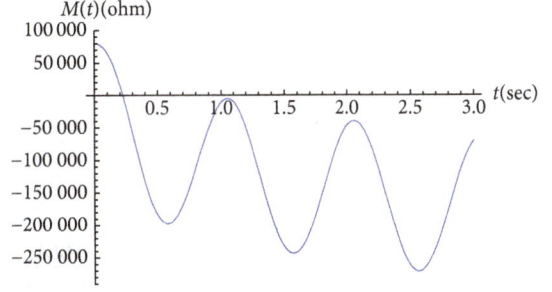

FIGURE 7: $M(t)$ of the boundary effect free fractional order memristor under the sinusoidal input with $I_m = -110\ \mu A$ and $\alpha = 1.25$.

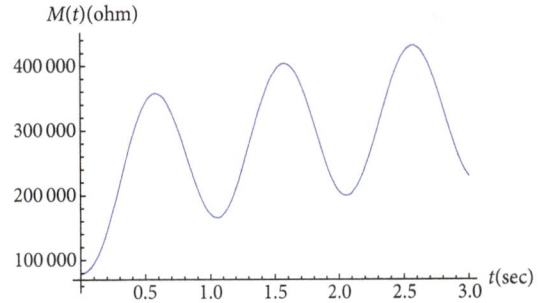

FIGURE 5: $M(t)$ of the fractional order memristor under the sinusoidal input with $I_m = -110\ \mu A$: $\alpha = 0.75$ (red), $\alpha = 1$ (green), $\alpha = 1.25$ (blue), and HP memristor (black dots).

$$\cdot\ {}_1F_2\left(1;\ \frac{\alpha}{2}+1,\ \frac{\alpha}{2}+\frac{3}{2};\ -\frac{1}{4}(\omega t)^2\right)\Bigg]$$

(20)

However, it should be mentioned here that both temporary and permanent saturation do not always occur. Instead, their occurrences are dependent on the conditions on parameters, which makes either the peaks or time proportional term of $M(t)$ be outside $[M_{on}, M_{off}]$ such as those of Figures 4 and 5. Moreover the determination of such time proportional term and T_{sat} will be presented later in this subsection.

Similarly to the memristor, the voltage-current lissajous curve is of interest for the fractional order memristor. Therefore, those curves of the fractional order memristor will be simulated by using (19) as the basis for further studying the behavior of this device under the AC input. Now, we let $0\ \text{sec} < t \leq 19\ \text{sec}$, $\omega = 1\ \text{rad/sec}$, and $\theta = 0\ \text{rad}$; the lissajous curves of the fractional order memristor with various α's can be simulated as depicted in Figures 8–13 where $I_m = 110\ \mu A$, $K = 10000\ \text{A}^{-1}\text{sec}^{-\alpha}$, $M_{on} = 100\ \Omega$, $M_{off} = 16\ \text{k}\Omega$, and $M(0) = 11\ \text{k}\Omega$ have been assumed in Figures 8–10. On the other hand, $I_m = -150\ \mu A$, $K = 10000\ \text{A}^{-1}\text{sec}^{-\alpha}$, $M_{on} = 100\ \Omega$, $M_{off} = 38\ \text{k}\Omega$, and $M(0) = 11.2\ \text{k}\Omega$ have been adopted in Figures 11–13. Based on its SPICE model, the lissajous curves of the HP memristor have also been simulated and compared to those of the fractional order memristor with $\alpha = 1$ as depicted in Figures 9 and 12 where the strong agreements between the fractional order memristor and HP memristor based curves which are both time invariant shaped

permanently saturated after T_{sat} as $M(t)$ contains the time proportional term which starts to lie outside $[M_{on}, M_{off}]$ at $t = T_{sat}$ when $\alpha > 1$. In order to show that the permanent saturation of the fractional order memristor with $\alpha > 1$ excited by sinusoidal input is possible, we simulate $M(t)$ of the device with $\alpha = 1.25$ once again by using (20) which is (19) without the boundary effect. The simulation results have been depicted in Figures 6 and 7 which show that $M(t)$'s eventually lie outside $[M_{on}, M_{off}]$ permanently. This yields the saturation when such effect has been included.

$$M(t) = M(0) - \frac{KM_dI_mt^{\alpha}}{\Gamma(\alpha+1)}\left[\sin(\theta)\right.$$

$$\cdot\ {}_1F_2\left(1;\ \frac{\alpha}{2}+\frac{1}{2},\ \frac{\alpha}{2}+1;\ -\frac{1}{4}(\omega t)^2\right) + \frac{\omega t\cos(\theta)}{\alpha+1}$$

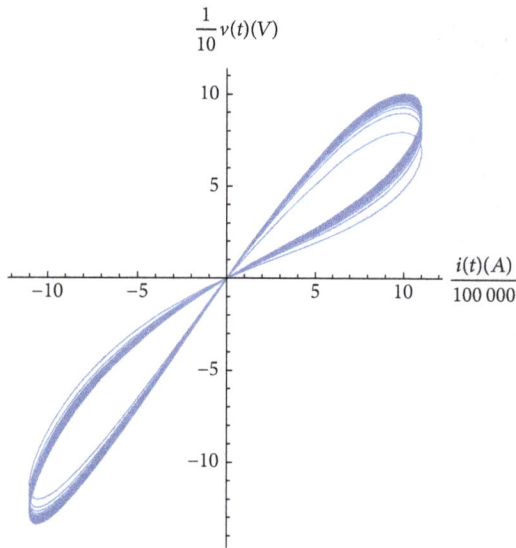

FIGURE 8: $v(t)$-$i(t)$ of the fractional order memristor: $\alpha = 0.75$ and $I_m > 0$.

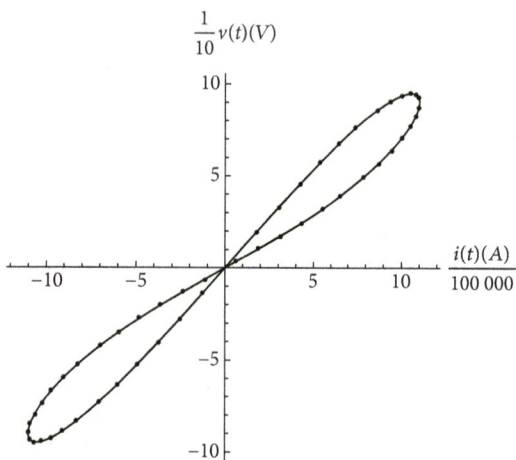

FIGURE 9: $v(t)$-$i(t)$ of the fractional order memristor (green) and HP memristor (black dots): $\alpha = 1$ and $I_m > 0$.

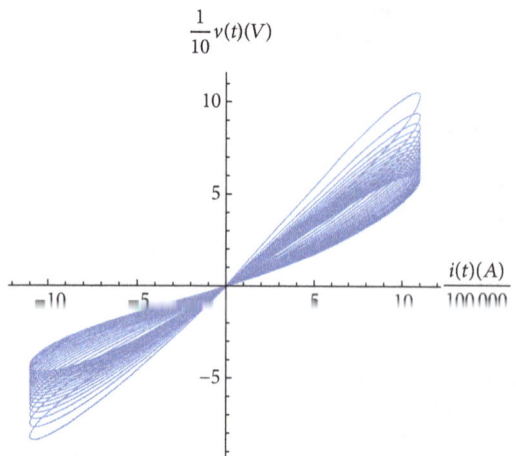

FIGURE 10: $v(t)$-$i(t)$ of the fractional order memristor: $\alpha = 1.25$ and $I_m > 0$.

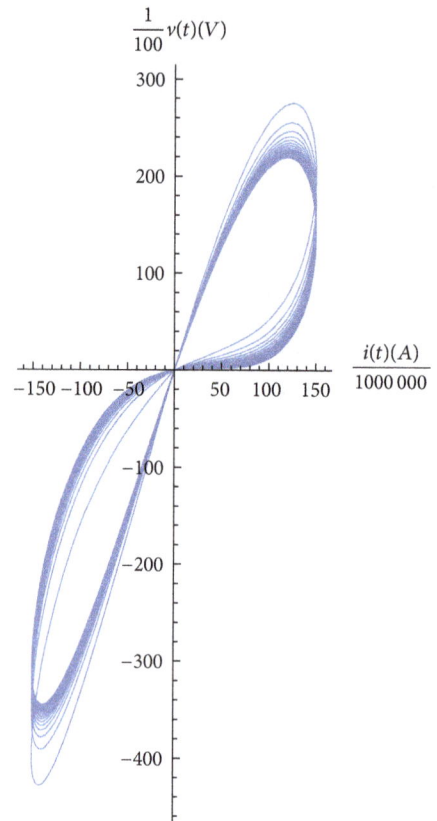

FIGURE 11: $v(t)$-$i(t)$ of the fractional order memristor: $\alpha = 0.75$ and $I_m < 0$.

symmetric pinched hysteresis loops can be observed. Again, this verifies the accuracy of our expression.

From other figures, it can be seen that the lissajous curve of the fractional order memristor with $\alpha \neq 1$ also takes the pinched hysteresis loop shape with pinching point at the origin despite the asymmetricities which means that the fractional order memristor preserves the memristive characteristic [30]. This is unlike those fracmemristor based results previously proposed in [17] which do not display the pinched hysteresis loop at all; therefore such fracmemristor does not employ the memristive characteristic according to [30]. Unlike those of the fractional order memristor with $\alpha = 1$ and HP memristor, the shape of the lissajous curve of the fractional order memristor with $\alpha \neq 1$ keeps changing. This is because $M(t)$ of such fractional order memristor contains the time proportional term. This can be clearly seen from Figures 14–17 which display $M(t)$'s of the fractional order memristor with $\alpha \neq 1$. Unlike the previous ones depicted in Figures 4 and 5, the peaks of these $M(t)$'s with $\alpha < 1$ are unclipped as can be seen from Figures 14 and 16. This is because fractional order memristor is always unsaturated as such peaks lie within $[M_{on}, M_{off}]$. Moreover, the fractional order memristor with $\alpha > 1$ never becomes saturated and can be seen from Figures 15 and 17. This is because the time proportional terms of its $M(t)$'s are always be within $[M_{on}, M_{off}]$. At this point, it can be seen that both temporary and permanent saturation of the fractional order

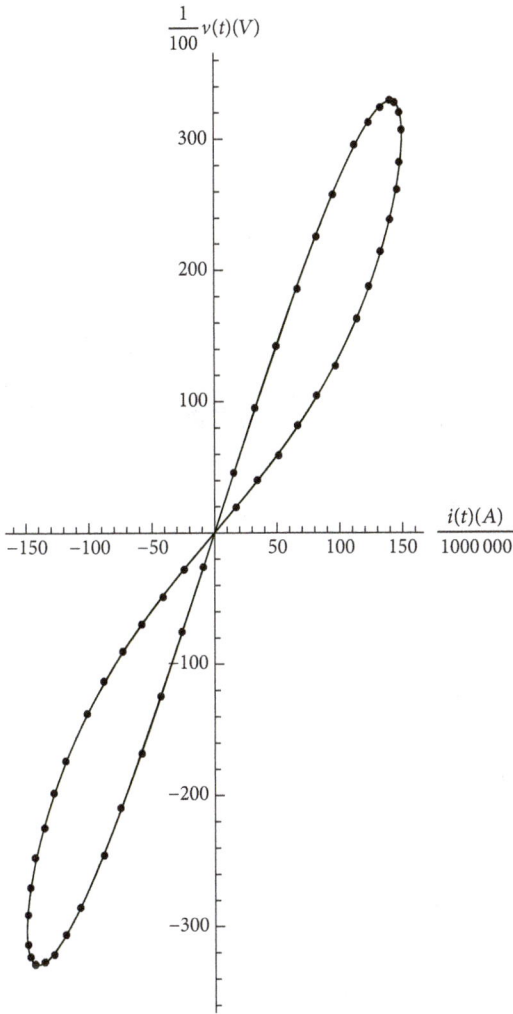

FIGURE 12: $v(t)$-$i(t)$ of the fractional order memristor (green) and HP memristor (black dots): $\alpha = 1$ and $I_m < 0$.

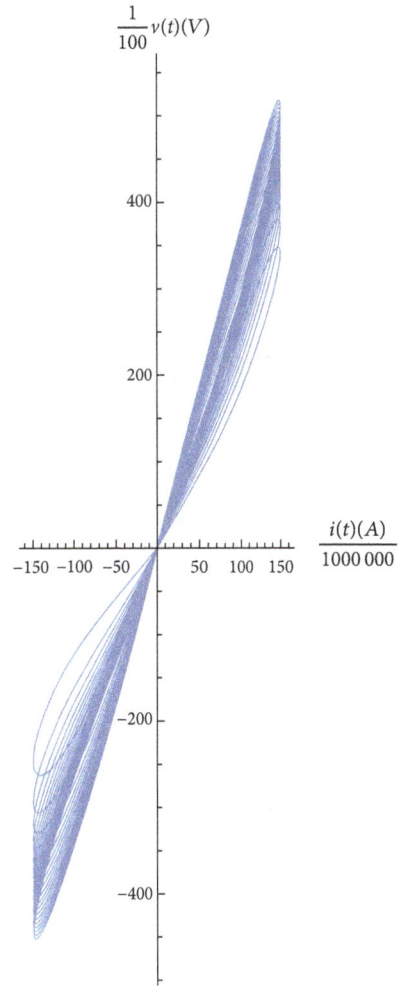

FIGURE 13: $v(t)$-$i(t)$ of the fractional order memristor: $\alpha = 1.25$ and $I_m < 0$.

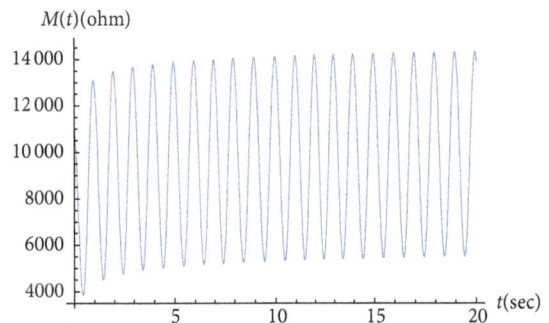

FIGURE 14: $M(t)$ of the fractional order memristor under the sinusoidal input with $I_m > 0$: $\alpha = 0.75$.

memristor under sinusoidal excitation do not always occur but depend on the conditions on parameters and input as aforementioned.

At asymptotic state, it can be seen that such time proportional term of $M(t)$ becomes time independent instead. Therefore, the shape of its lissajous curves of the fractional order memristor with $\alpha \neq 1$ is asymptotically unchanged as can be seen from Figures 18 and 19 which depict the lissajous curves of the fractional order memristor with various α's simulated by assuming that 100 sec < t ≤ 119 sec. Apart from being asymmetric when $\alpha \neq 1$, we have found that the lobe area of the lissajous curve of the fractional order memristor is affected by α similarly to those proposed in [20, 21]. In particular, we have found that the fractional order memristor with lower α yields the lissajous curve with wider lobe area which refers to more memory effect and less linearity. Besides the voltage-current curve, the memristance-current lissajous curve has been found to be also interesting. Therefore, such curves of the fractional order memristor with various α's have also been simulated as depicted in Figures

20 and 21 where the parameters setting similar to those of the voltage-current curve have been assumed. Moreover, we also assume that 100 sec < t ≤ 119 sec. These figures show that the resulting lissajous curves take the elliptical closed loop shape which are unchanged as $M(t)$ enters the asymptotic state at the assumed time interval. The elliptical shaped lissajous curve means that $M(t)$ is periodic as well as

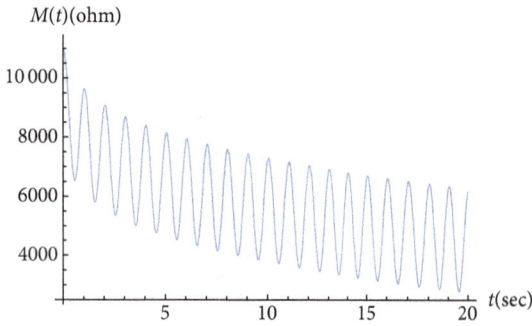

FIGURE 15: $M(t)$ of the fractional order memristor under the sinusoidal input with $I_m > 0$: $\alpha = 1.25$.

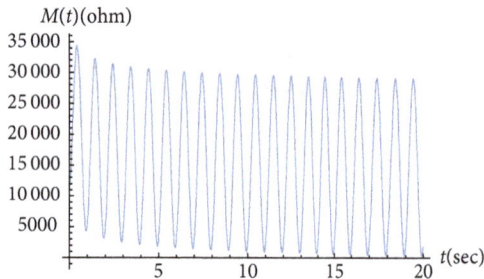

FIGURE 16: $M(t)$ of the fractional order memristor under the sinusoidal input with $I_m < 0$: $\alpha = 0.75$.

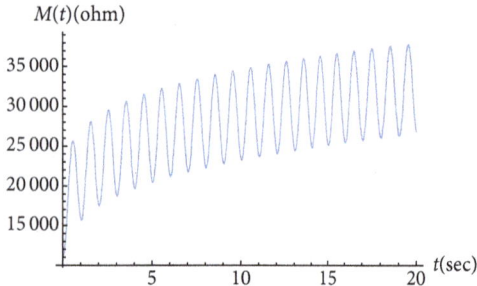

FIGURE 17: $M(t)$ of the fractional order memristor under the sinusoidal input with $I_m < 0$: $\alpha = 1.25$.

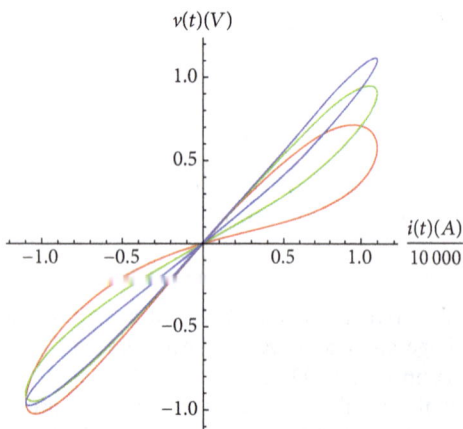

FIGURE 18: $v(t)$-$i(t)$ of the fractional order memristor at asymptotic state ($I_m > 0$): $\alpha = 0.75$ (red), $\alpha = 1$ (green), and $\alpha = 1.25$ (blue).

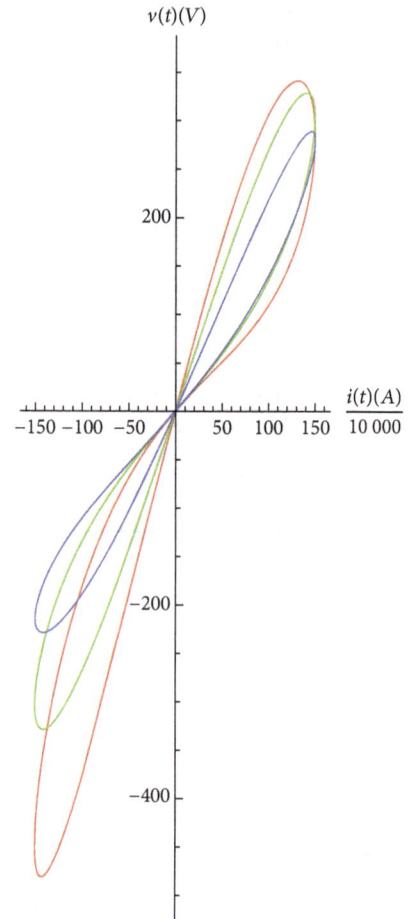

FIGURE 19: $v(t)$-$i(t)$ of the fractional order memristor at asymptotic state ($I_m < 0$): $\alpha = 0.75$ (red), $\alpha = 1$ (green), and $\alpha = 1.25$ (blue).

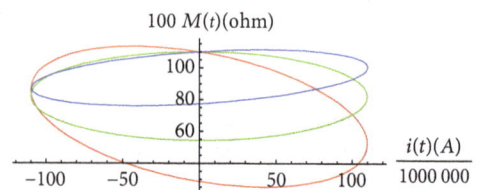

FIGURE 20: $M(t)$-$i(t)$ of the fractional order memristor ($I_m > 0$): $\alpha = 0.75$ (red), $\alpha = 1$ (green), and $\alpha = 1.25$ (blue).

$i(t)$. Similarly to the voltage-current curve, the loop area of the memristance-current lissajous curve is also affected by α. In particular, the fractional order memristor with lower α yields the memristance-current lissajous curve with wider loop area.

At this point, we will analytically show that both upper and lower lobes of the voltage-current lissajous curve have equal sizes of areas which means that the fractional order memristor does not store the energy, and such areas are independent of θ. Here, we let the area of the upper and lower lobes of the lissajous curve be denoted, respectively, by A_U

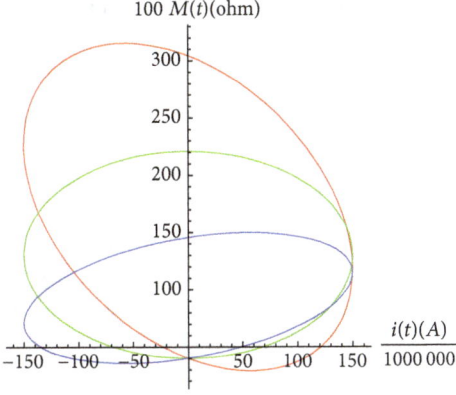

FIGURE 21: $M(t)$-$i(t)$ of the fractional order memristor ($I_m < 0$): $\alpha = 0.75$ (red), $\alpha = 1$ (green), and $\alpha = 1.25$ (blue).

and A_L. Since $v(t) = M_a(t)i(t)$ at asymptotic state where $M_a(t)$ denotes the asymptotic approximation of $M(t)$, A_U and A_L can be given by

$$A_U = \int_{-\theta/\omega}^{(\pi-\theta)/\omega} M_a(t)\,i(t)\,di(t) \tag{21}$$

$$A_L = \int_{(\pi-\theta)/\omega}^{(2\pi-\theta)/\omega} M_a(t)\,i(t)\,di(t) \tag{22}$$

With (13) and the asymptotic approximation of the fractional order integration of sinusoidal function [20], $M_a(t)$ can be found as

$$M_a(t) = \min\left[\max\left[M(0) - \frac{KM_dI_m}{\omega^\alpha}\left[\cos\left(\theta + \frac{1-\alpha}{2}\pi\right) - \cos\left(\omega t + \theta + \frac{1-\alpha}{2}\pi\right)\right], M_{on}\right], M_{off}\right] \tag{23}$$

As $i(t) = I_m\sin(\omega t+\theta)$ and the fractional order memristor is unsaturated for the entire simulation period as can be seen from Figures 18 and 19, A_U and A_L can be finally obtained as follows:

$$A_U = \frac{2}{3}\frac{KM_dI_m^3}{\omega^\alpha}\sin\left(\frac{\alpha\pi}{2}\right) \tag{24}$$

$$A_L = -\frac{2}{3}\frac{KM_dI_m^3}{\omega^\alpha}\sin\left(\frac{\alpha\pi}{2}\right) \tag{25}$$

which show that the upper and lower lobes of the lissajous curve employ equal sizes of areas and such areas are independent of θ. Unlike [20] which considered only A_U, A_L has also been formulated in this work. Moreover, our A_U which has been derived by using a different approach from that used in [20] totally agrees with the previous result that $\theta = 0$ rad has been assumed. This emphasizes the independency from θ of the lobe areas. Moreover, (24) and (25) also show that the fractional order memristor does not store the energy as the summation of A_U and A_L which, respectively, referred to the intake and dissipated energy, is equal to 0.

Besides A_U and A_L, the area within the closed loop of the memristance-current lissajous curve (A_M) can be obtained by also using (23) and

$$A_M = \int_{-\theta/\omega}^{(2\pi-\theta)/\omega} M_a(t)\,di(t) \tag{26}$$

Since it can be seen from Figures 20 and 21 that the device is unsaturated for the whole simulation period and $i(t) = I_m\sin(\omega t + \theta)$, we have

$$A_M = \frac{KM_dI_m^2\pi}{\omega^\alpha}\sin\left(\frac{\alpha\pi}{2}\right) \tag{27}$$

which shows that A_M is affected by α but independent of θ as well as A_U and A_L.

Despite the aforementioned independencies, θ does affect the behavior of the fractional order memristor. For illustration, $M_a(t)$'s due to various θ's have been simulated by using (20) as depicted in Figures 22 and 23 where $\alpha = 1$ and $\omega = 1$ rad/sec has been assumed. It should be mentioned here that $I_m = 110\ \mu A$, $K = 10000\ A^{-1}sec^{-\alpha}$, $M_{on} = 100\ \Omega$, $M_{off} = 16$ kΩ, and $M(0) = 11$ kΩ have been adopted in Figure 20 where $I_m = -150\ \mu A$, $K = 10000\ A^{-1}sec^{-\alpha}$, $M_{on} = 100\ \Omega$, $M_{off} = 38$ kΩ, and $M(0) = 11.2$ kΩ have been assumed in Figure 21. Moreover, $\theta = \pi/4$ rad, $\theta = 3\pi/4$ rad, $\theta = -3\pi/4$ rad, and $\theta = -\pi/4$ rad have been chosen as they are good representatives of those θ's which their coordinates on the Euclidian plane, i.e., $(\cos(\theta), \sin(\theta))$, are located on the portion of unit circle's arc in quadrants 1, 2, 3, and 4 of such plane, respectively. This is because $(\cos(\theta), \sin(\theta))$'s of these chosen θ's are exactly located at the middle points of the portion of unit circle's arc. For example, $(\cos(\theta), \sin(\theta))$ of $\theta = \pi/4$ rad is located at the middle point of the portion of unit circle's arc in quadrant 1, etc.

From both figures, it can be seen that these $M_a(t)$'s contain time independent terms which are formerly the time proportional term of $M(t)$ that become time independent at asymptotic state as aforementioned. Since these time independent terms lie within $[M_{on}, M_{off}]$, $M_a(t)$ takes the shape of sinusoidal waveform. However, the minimum peaks of $M_a(t)$'s due to the input with $I_m < 0$ and $(\cos(\theta), \sin(\theta))$ located on the portion of unit circle's arc in quadrants 2 and 3 of the Euclidian plane have been clipped as can be seen from Figure 23 due to the saturation of the fractional order memristor as these peaks are lower than M_{on}. From Figure 22, it can be seen that the input with $I_m > 0$ and θ with $(\cos(\theta), \sin(\theta))$ located on the portion of unit circle's arc in quadrants 2 and 3 yields $M(t)$ with higher time average. If we have assumed that $M_{on} \leq M_a(t) \leq M_{off}$ is always satisfied, it can be seen from Figure 23 that $M(t)$ with higher time average can be obtained by using the input with $I_m < 0$ and $(\cos(\theta), \sin(\theta))$ located on such portion in quadrants 1 and 4.

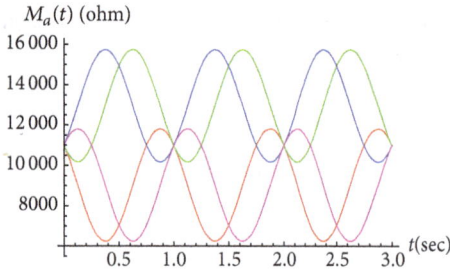

FIGURE 22: $M_a(t)$ of the fractional order memristor under the sinusoidal input with $I_m > 0$: $\theta = 45°$ (red), $\theta = 135°$ (green), $\theta = 225°$ (blue), and $\theta = 315°$ (magenta).

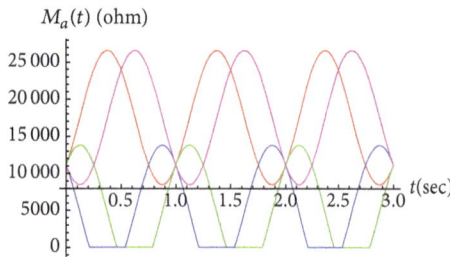

FIGURE 23: $M_a(t)$ of the fractional order memristor under the sinusoidal input with $I_m < 0$: $\theta = 45°$ (red), $\theta = 135°$ (green), $\theta = 225°$ (blue), and $\theta = 315°$ (magenta).

Now, we will derive the analytical expression of time independent term of $M_a(t)$. Let such term be denoted by M_{TI}; it can be given by carefully observing (23) as follows:

$$
M_{TI} = \min\left[\max\left[M(0) - \frac{KM_dI_m}{\omega^\alpha}\left[\cos\left(\theta + \frac{1-\alpha}{2}\pi\right)\right], M_{on}\right], M_{off}\right] \quad (28)
$$

When $\alpha = 1$ as assumed in our simulations of $M_a(t)$, M_{TI} can be reduced to

$$
M_{TI} = \min\left[\max\left[M(0) + \Delta M_{TI}, M_{on}\right], M_{off}\right] \quad (29)
$$

where

$$
\Delta M_{TI} = -K\omega^{-1}M_dI_m\cos(\theta) \quad (30)
$$

Therefore, it can be seen that when $I_m > 0$ ($I_m < 0$), if $(\cos(\theta), \sin(\theta))$ is located on the portion of unit circle's arc in either quadrant 2 or 3 where $\cos(\theta) < 0$, $\Delta M_{TI} > 0$ ($\Delta M_{TI} < 0$) thus $M(0)$ has been increased (decreased). On the other hand, if $(\cos(\theta), \sin(\theta))$ is on the portion of unit circle's arc in either quadrant 1 or 4 where $\cos(\theta) > 0$, $\Delta M_{TI} < 0$ ($\Delta M_{TI} > 0$) as $I_m > 0$ ($I_m < 0$) thus $M(0)$ has been decreased (increased). As a result, the sinusoidal input with $I_m > 0$ ($I_m < 0$) and $(\cos(\theta), \sin(\theta))$ located on the portion of unit circle's arc in either quadrant 2 or 3 (1 or 4) of the Euclidian plane yields higher M_{TI} as graphically observed.

At this point, the influence of α to M_{TI} will be explored. By using (28) with $\theta = 0$ rad and $\omega = 1$ rad/sec, we can simulate M_{TI} as shown in Tables 1 and 2. In Table 1, we assume that $I_m = 110$ μA, $K = 10000$ $A^{-1}sec^{-\alpha}$, $M_{on} = 100$ Ω, $M_{off} = 16$ kΩ, and $M(0) = 11$ kΩ, where $I_m = -150$ μA, $K = 10000$ $A^{-1}sec^{-\alpha}$, $M_{on} = 100$ Ω, $M_{off} = 38$ kΩ, and $M(0) = 11.2$ kΩ have been adopted in Table 2. It can be seen from these tables that M_{TI} is proportional to α when $I_m > 0$ and vice versa when $I_m < 0$.

By subtracting M_{TI} from $M_a(t)$, we obtain the following purely periodic term of $M(t)$, $M_P(t)$

$$
M_P(t)
$$
$$
= \min\left[\max\left[\frac{KM_dI_m}{\omega^\alpha}\cos\left(\omega t + \theta + \frac{1-\alpha}{2}\pi\right), M_{on}\right], \quad (31)
$$
$$
M_{off}\right]
$$

which existed in both initial state and asymptotic state.

Therefore, the time proportional term of $M(t)$, $M_{TP}(t)$ can be formulated by subtracting $M_P(t)$ from $M(t)$ as

$$
M_{TP}(t)
$$
$$
= \min\left[\max\left[M(0) - \frac{KM_dI_mt^\alpha}{\Gamma(\alpha+1)}\left[\sin(\theta)\,{}_1F_2\left(1;\frac{\alpha}{2}+\frac{1}{2},\frac{\alpha}{2}+1;-\frac{1}{4}(\omega t)^2\right) + \frac{\omega t\cos(\theta)}{\alpha+1}\,{}_1F_2\left(1;\frac{\alpha}{2}+1,\frac{\alpha}{2}+\frac{3}{2};-\frac{1}{4}(\omega t)^2\right)\right] - \frac{kM_dI_m}{\omega^\alpha}\cos\left(\omega t + \theta + \frac{1-\alpha}{2}\pi\right), M_{on}\right], M_{off}\right] \quad (32)
$$

As a result, the aforementioned T_{sat} can be determined by solving the following equation:

$$
\left.\begin{array}{ll}M_{on} & I_m > 0 \\ M_{off} & I_m < 0\end{array}\right\} = M(0) - \frac{KM_dI_mT_{sat}^\alpha}{\Gamma(\alpha+1)}\left[\sin(\theta)\right.
$$
$$
\cdot\,{}_1F_2\left(1;\frac{\alpha}{2}+\frac{1}{2},\frac{\alpha}{2}+1;-\frac{1}{4}(\omega T_{sat})^2\right)
$$
$$
+\frac{\omega T_{sat}\cos(\theta)}{\alpha+1}
$$
$$
\cdot\,{}_1F_2\left(1;\frac{\alpha}{2}+1,\frac{\alpha}{2}+\frac{3}{2};-\frac{1}{4}(\omega T_{sat})^2\right)\right]
$$
$$
-\frac{kM_dI_m}{\omega^\alpha}\cos\left(\omega T_{sat}+\theta+\frac{1-\alpha}{2}\pi\right)
$$
$$
\quad (33)
$$

TABLE 1: M_{TI} due to $I_M > 0$.

α	M_{TI} (kΩ)
0.75	6.9296
1.00	8.2175
1.25	9.7365

TABLE 2: M_{TI} due to $I_M < 0$.

α	M_{TI} (kΩ)
0.75	24.4306
1.00	20.2443
1.25	16.4772

At this point, we will derive the expressions of $M(t)$'s due to other AC waveforms by using that due to the sinusoidal waveform, i.e., (19), as the basis. For example, the expression of $M(t)$ due to arbitrary phase cosinusoidal waveform, i.e., $i(t) = I_m \cos(\omega t + \theta)$, can be formulated by using (19) and the relationship between the sinusoidal and cosinusoidal functions is given by $\cos(u) = \sin((\pi/2) - u)$ where u denotes arbitrary angle. As a result, we have

$$M(t) = \min\left[\max\left[M(0) - \frac{KM_d I_m t^\alpha}{\Gamma(\alpha+1)}\left[\sin(\theta)\,_1F_2\left(1;\ \frac{\alpha}{2}+\frac{1}{2},\ \frac{\alpha}{2}+1;\ -\frac{1}{4}\left(\frac{\pi}{2}-\omega t-2\theta\right)^2\right) + \frac{(\pi/2-\omega t-2\theta)\cos(\theta)}{\alpha+1}\,_1F_2\left(1;\ \frac{\alpha}{2}+1,\ \frac{\alpha}{2}+\frac{3}{2};\ -\frac{1}{4}\left(\frac{\pi}{2}-\omega t-2\theta\right)^2\right)\right],\ M_{on}\right],\ M_{off}\right], \quad (34)$$

For arbitrary periodic waveform which has never been considered in those previous works, the resulting expression can also be determined based on (19). This is because such waveform can be given as a series of sinusoidal functions, i.e., $i(t) = \sum_{n=0}^{\infty}[I_{mn}\sin(n\omega t + \theta_n)]$, where I_{mn} and θ_n, respectively, stand for peak value and phase of arbitrary nth term of the series, according to the Fourier theorem. As a result, the expression of $M(t)$ due to arbitrary periodic waveform can be given as follows:

$$M(t) = \min\left[\max\left[M(0) - \frac{KM_d t^\alpha}{\Gamma(\alpha+1)}\sum_{n=0}^{\infty}\left[I_{mn}\left[\sin(\theta_n)\,_1F_2\left(1;\ \frac{\alpha}{2}+\frac{1}{2},\ \frac{\alpha}{2}+1;\ -\frac{1}{4}(n\omega t)^2\right) + \frac{n\omega t\cos(\theta_n)}{\alpha+1}\,_1F_2\left(1;\ \frac{\alpha}{2}+1,\ \frac{\alpha}{2}+\frac{3}{2};\ -\frac{1}{4}(n\omega t)^2\right)\right]\right],\ M_{on}\right],\ M_{off}\right], \quad (35)$$

4. The Usage of Fractional Order Memristor in the Memristor Based Circuit

In this research, the HP memristor based type A Wien oscillator [29] has been chosen as the candidate memristor based circuit. The unique characteristic of such circuit, which is either R_1 or R_2 replaced by the memristive device as depicted in Figure 22, is the fluctuated frequency of oscillation. The smaller range of fluctuation refers to the better chance that the system has sustained oscillation which can be obtained if and only if all poles of the system are fixed in the s-plane [29]. For studying the usage of fractional order memristor, we replace R_1 of the circuit by such fractional order device instead of the HP memristor as traditionally did [29] and analyze the effect of α to the fluctuation in frequency of oscillation which determines the chance that the system has sustained oscillation, as mentioned above. Let the range of such fluctuation be denoted by Δf_{osc}; it can be mathematically defined as given by (36) where f_{up} and f_{low} stand for the upper bound and lower boundary.

$$\Delta f_{osc} = f_{up} - f_{low} \quad (36)$$

According to [29], these boundaries can be obtained by solving (37) and (38) which have been formulated by assuming that $C_1 = C_2 = C$, the memristor is unsaturated, and the frequency of oscillation has been found to lie within the range that the sustained oscillation can be assured. Note also that $I_{mem} = V_{mem}/M(0)$, where V_{mem} denotes the peak value of the voltage dropped across the memristor which can be determined from the oscillator output voltage, $V_{out}(t)$, and also depends on the initial voltages of C_1 and C_2 [29]. Moreover, $V_{out}(t)$ can be obtained from the state space representation and output equation of the fractional memristor based Wien oscillator which are, respectively, given by (39) and (40) where $V_{C1}(t)$ and $V_{C2}(t)$ denote the voltage dropped across C_1 and C_2 and $M(t)$ stands for the memristance of the conventional HP memristor.

$$f_{up} = \sqrt{\frac{4}{16\pi^2 C^2} - \frac{\left(kM_d\,|I_{mem}|\,/2\pi f_{up}\right)^2}{16\pi^2 C^2\left(M(0) - kM_d\,|I_{mem}|\,/2\pi f_{up}\right)R_2}} \quad (37)$$

f_{low}

$$= \sqrt{\frac{4}{16\pi^2 C^2} - \frac{\left(kM_d \left|I_{mem}\right|/2\pi f_{low}\right)^2}{16\pi^2 C^2 \left(M(0) + kM_d \left|I_{mem}\right|/2\pi f_{low}\right) R_2}} \quad (38)$$

$$\left[\begin{array}{c} \dfrac{d}{dt} V_{C_1}(t) \\ \\ \dfrac{d}{dt} V_{C_2}(t) \end{array}\right] \quad (39)$$

$$= \left[\begin{array}{cc} -\dfrac{1}{M(t)C_1} & \dfrac{R_3}{R_4 M(t) C_1} \\ \\ -\dfrac{1}{M(t)C_2} & \dfrac{R_3}{R_4 M(t) C_2} - \dfrac{1}{R_2 C_2} \end{array}\right] \left[\begin{array}{c} V_{C_1}(t) \\ \\ V_{C_2}(t) \end{array}\right]$$

$$V_{out}(t) = \left(\frac{R_3}{R_4} + 1\right) V_{C_2}(t) \quad (40)$$

However, this is not the case when the fractional order memristor has been used as $M(t)$ of such device must be adopted. Therefore $V_{out}(t)$ must be determined based on our derived $M(t)$ instead where f_{up} and f_{low} must be evaluated from (41) and (42). As a result, the corresponding Δf_{osc} will be different from that of the original conventional memristor based circuit and the different chance of obtaining sustained oscillation can be expected.

f_{up}

$$= \sqrt{\frac{4}{16\pi^2 C^2} - \frac{\left(KM_d \left|I_{mem}\right|/\left(2\pi f_{up}\right)^\alpha\right)^2}{16\pi^2 C^2 \left(M(0) - KM_d \left|I_{mem}\right|/\left(2\pi f_{up}\right)^\alpha\right) R_2}} \quad (41)$$

f_{low}

$$= \sqrt{\frac{4}{16\pi^2 C^2} - \frac{\left(KM_d \left|I_{mem}\right|/\left(2\pi f_{low}\right)^\alpha\right)^2}{16\pi^2 C^2 \left(M(0) + KM_d \left|I_{mem}\right|/\left(2\pi f_{low}\right)^\alpha\right) R_2}} \quad (42)$$

Moreover, the condition for ensuring the occurrence of sustained oscillation can be given by () where f_{avr} which stands for the average oscillating frequency is given by ().

$$\frac{R_3}{R_4} = 1 + \frac{1}{R_2}\left[M(0) - \frac{KM_d\left|I_{mem}\right|}{\left(2\pi f_{avr}\right)^\alpha}\right] \quad (43)$$

f_{avr}

$$= \sqrt{\frac{4}{16\pi^2 C^2} - \frac{\left(KM_d \left|I_{mem}\right|/\left(2\pi f_{avr}\right)^\alpha\right)^2}{16\pi^2 C^2 \left(M(0) - KM_d \left|I_{mem}\right|/\left(2\pi f_{avr}\right)^\alpha\right) R_2}} \quad (44)$$

By assuming 0.1 V and -0.95 V as the approximate initial voltages of C_1 and C_2 and also assuming that $R_2 = 5$ kΩ, $C_1 = C_2 = 3.2\mu F$, $K = 100000$ A^{-1}sec$^{-\alpha}$, $M_{on} = 100$ Ω, and $M_{off} = 16$ kΩ, Δf_{osc} of the fractional order memristor based Wien oscillator with $\alpha < 1$ and $\alpha > 1$ can be obtained for various $M(0)$'s by numerically solving (41) and (42). as tabulated in Table 3 where Δf_{osc} of the original HP memristor based circuit which is equivalent to the fractional order memristor with $\alpha = 1$ in the context of this work, determined by solving (37) and (38), has also been included.

TABLE 3: Δf_{osc} (Hz) for various $M(0)$'s.

$M(0)$ (kΩ)	$\alpha = 0.75$	HP ($\alpha = 1$)	$\alpha = 1.25$
4.1	9.16	3.32	1.49
4.4	7.64	2.88	1.32
4.7	6.44	2.52	1.19
5.0	5.49	2.23	1.07
5.3	4.73	1.98	0.97
5.6	4.11	1.78	0.89
5.9	3.59	1.60	0.81

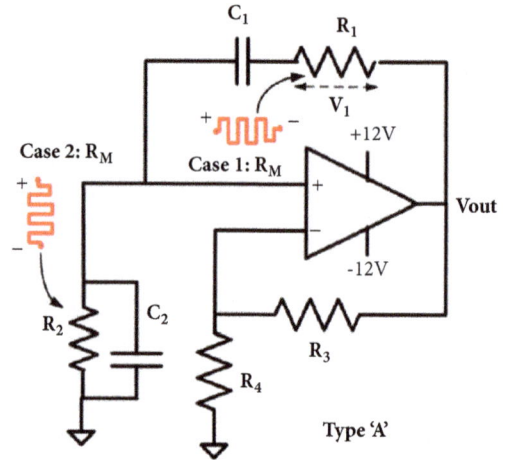

FIGURE 24: The memristor based Wien oscillator [29].

It has been found that Δf_{osc} is inversely proportional to $M(0)$ which is in agreement with [29]. Since 4.1 k$\Omega \leq M(0) \leq 5.9$ kΩ [29] and the probability of obtaining sustained oscillation is inversely proportional to Δf_{osc}, $M(0) = 5.9$ kΩ is recommended as it minimizes Δf_{osc} and thus maximizes such probability. It can also be seen that Δf_{osc} is inversely proportional to α which means that the probability of obtaining sustained oscillation is directly proportional to α. Therefore, the fractional order memristor with larger α is recommended and the fractional order memristor with $\alpha > 1$ should be adopted as it can increase such probability from that of the original circuit which its memristive device has $\alpha = 1$. On the other hand, the fractional order memristor with $\alpha < 1$ should be avoided as it decreases such probability.

Finally, by further assuming that $M(0) = 5$ kΩ, $R_3 = 20.2$ kΩ, and $R_4 = 10$ kΩ, we can simulate $V_{out}(t)$ at asymptotic state and the lissajous patterns of $V_{out}(t)$ and $M(t)$ for various α's as depicted in Figures 23–28 where the units of $100\,V_{out}(t)$, t, and $M(t)$ are V, sec, and Ω, respectively. The comparison of the results with $\alpha = 1$ to the SPICE HP memristor model based counterparts has been made for verification where a strong agreement can be observed. We have found that there exists neither temporary nor permanent saturation of the fractional order memristor. This is because $M(t)$ is always within $[M_{on}, M_{off}]$ in this scenario due to the assumed conditions on parameters and input, as can be seen from the simulated lissajous patterns. Such conditions have been adopted for ensuring the unsaturation which yields the

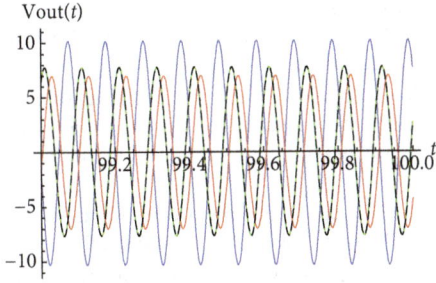

FIGURE 25: $V_{out}(t)$ at asymptotic state of Type A Wien oscillator: fractional memristor with $\alpha = 0.75$ (blue), fractional memristor with $\alpha = 1$ (green), fractional memristor with $\alpha = 1.25$ (red), and HP memristor (black dots).

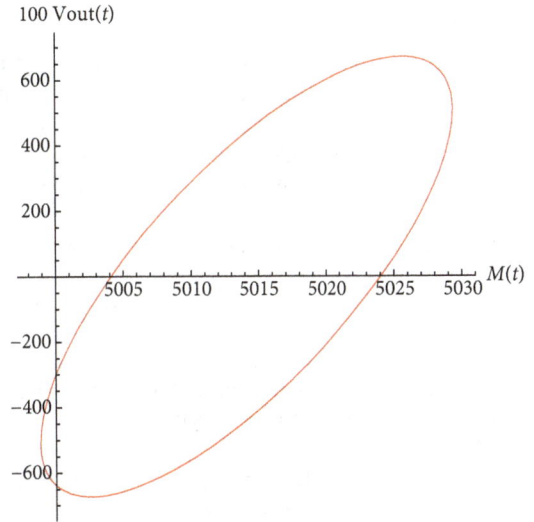

FIGURE 26: $V_{out}(t)$ versus $M(t)$ of Type A Wien oscillator: fractional memristor with $\alpha = 0.75$.

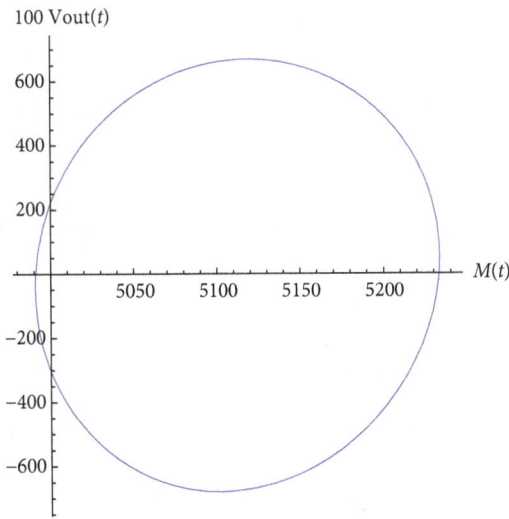

FIGURE 27: $V_{out}(t)$ versus $M(t)$ of Type A Wien oscillator: fractional memristor with $\alpha = 1$ (green line) and HP memristor (orange dashed).

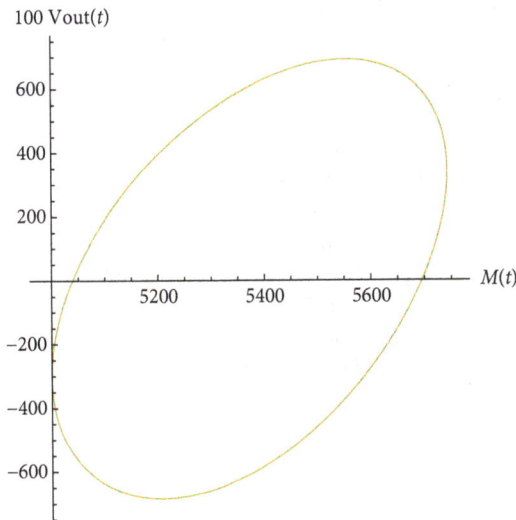

FIGURE 28: $V_{out}(t)$ versus $M(t)$ of Type A Wien oscillator: fractional memristor with $\alpha = 1.25$.

proper operation of the oscillator [29]. From these lissajous patterns, we have also found that $V_{out}(t)$ is of the same frequency as $M(t)$ and the phase difference between $V_{out}(t)$ and $M(t)$, which is less than 90°, is inversely proportional to α.

5. Conclusion

In this work, the HP memristor has been generalized in a fractional order domain by applying the fractional calculus to its state equation. Unlike [18–21], the dimensional consistency has been taken into account. Moreover, the boundary effect has also been modelled. Therefore the analytical expression of $M(t)$ which has been derived as a function of $i(t)$ and its related results are dimensional consistent and valid to such generalized device in its saturation states. By using such expression, $M(t)$'s due to various waveforms and the related parameters have been formulated. With the simulations by using these $M(t)$'s and parameters, the behaviors of the fractional order memristor have been thoroughly explored. Therefore this research gives a precise understanding on the characteristics of such up to date nonlinear electrical circuit element which has been recently applied as the basis of the net grid type fracmemristor [31].

Conflicts of Interest

The author declares that there are no conflicts of interest regarding the publication of this article.

Acknowledgments

The author would like to acknowledge Mahidol University, Thailand, for the online database service which is our primary information resource.

References

[1] L. O. Chua, "Memristor—the missing circuit element," *IEEE Transactions on Circuit Theory*, vol. 18, no. 5, pp. 507–519, 1971.

[2] I. S. Jesus, J. A. Tenreiro MacHado, and J. Boaventure Cunha, "Fractional electrical impedances in botanical elements," *Journal of Vibration and Control*, vol. 14, no. 9-10, pp. 1389–1402, 2008.

[3] C. Tang, F. You, G. Cheng, D. Gao, F. Fu, and X. Dong, "Modeling the frequency dependence of the electrical properties of the live human skull," *Physiological Measurement*, vol. 30, no. 12, pp. 1293–1301, 2009.

[4] A. Charef, "Analogue realisation of fractional-order integrator, differentiator and fractional PIλDµ controller," *IEE Proceedings—Control Theory and Applications*, vol. 153, no. 6, pp. 714–720, 2006.

[5] B. M. Vinagre and V. Feliu, "Optimal fractional controllers for rational order systems: a special case of the Wiener-Hopf spectral factorization method," *Institute of Electrical and Electronics Engineers Transactions on Automatic Control*, vol. 52, no. 12, pp. 2385–2389, 2007.

[6] R. Matusu, "Application of fractional order calculus to control theory," *International Journal of Mathematical Models and Methods in Applied Sciences*, vol. 5, pp. 1162–1169, 2011.

[7] L. Dork, J. Terpk, I. Petr, and F. Dorkov, "Electronic realization of the fractional-order systems," *Acta Montanistica Slovaca*, vol. 12, pp. 231–237, 2007.

[8] B. T. Krishna, K. V. V. S. Reddy, and S. Santha Kumari, "Time domain response calculations of fractance device of order 1/2," *Journal of Active & Passive Electronic Devices*, vol. 3, pp. 355–367, 2008.

[9] A. G. Radwan and A. S. Elwakil, "An expression for the voltage response of a current-excited fractance device based on fractional-order trigonometric identities," *International Journal of Circuit Theory and Applications*, vol. 40, no. 5, pp. 533–538, 2012.

[10] D. B. Strukov, G. S. Snider, D. R. Stewart, and R. S. Williams, "The missing memristor found," *Nature*, vol. 453, pp. 80–83, 2008.

[11] A. G. Radwan, M. A. Zidan, and K. N. Salama, "HP Memristor mathematical model for periodic signals and DC," in *Proceedings of the 2010 53rd IEEE International Midwest Symposium on Circuits and Systems (MWSCAS)*, pp. 861–864, Seattle, Wash, USA, August 2010.

[12] A. G. Radwan, M. A. Zidan, and K. N. Salama, "On the mathematical modeling of Memristors," in *Proceedings of the 2010 International Conference on Microelectronics, ICM'10*, pp. 284–287, Cairo, Egypt, December 2010.

[13] Y. N. Joglekar and S. J. Wolf, "The elusive memristor: properties of basic electrical circuits," *European Journal of Physics*, vol. 30, no. 4, pp. 661–675, 2009.

[14] S. Shin, K. Kim, and S.-M. Kang, "Compact models for memristors based on charge-flux constitutive relationships," *IEEE Transactions on Computer-Aided Design of Integrated Circuits and Systems*, vol. 29, no. 4, pp. 590–598, 2010.

[15] C. Yakopcic, T. M. Taha, G. Subramanyam, R. E. Pino, and S. Rogers, "A memristor device model," *IEEE Electron Device Letters*, vol. 32, no. 10, pp. 1436–1438, 2011.

[16] J. Tenreiro Machado, "Fractional generalization of memristor and higher order elements," *Communications in Nonlinear Science and Numerical Simulation*, vol. 18, no. 2, pp. 264–275, 2013.

[17] Y.-F. Pu and X. Yuan, "Fracmemristor: fractional-order memristor," *IEEE Access*, vol. 4, pp. 1872–1888, 2016.

[18] M. E. Fouda and A. G. Radwan, "On the fractional-order memristor model," *Fractional Calculus and Applied Analysis*, vol. 4, pp. 1–7, 2013.

[19] M. E. Fouda and A. G. Radwan, "Fractional-order Memristor Response Under DC and Periodic Signals," *Circuits, Systems and Signal Processing*, vol. 34, no. 3, pp. 961–970, 2015.

[20] Y. Yu, B. Bao, H. Kang, and M. Shi, "Calculating area of fractional-order memristor pinched hysteresis loop," *The Journal of Engineering*, vol. 2015, no. 11, pp. 325–327, 2015.

[21] M. Shi and S. Hu, "Pinched hysteresis loop characteristics of a fractional-order HP TiO$_2$ memristor," in *Intelligent Computing, Networked Control, and Their Engineering Applications*, vol. 762 of *Communications in Computer and Information Science*, pp. 705–713, Springer, Singapore, Singapore, 2017.

[22] J. F. Gómez-Aguilar, J. J. Rosales-García, J. J Bernal-Alvarado, T. Córdova-Fraga, and R. Guzmán-Cabrera, "Fractional mechanical oscillators," *Revista Mexicana de Física*, vol. 58, pp. 348–352, 2012.

[23] R. Banchuin, "Novel expressions for time domain responses of fractance device," *Cogent Engineering*, vol. 4, no. 1, 2017.

[24] A. G. Radwan and E. Mohammed, "Memristor: models, types, and applications," in *On the Mathematical Modeling of Memristor, Memcapacitor, and Meminductor*, pp. 13–49, Springer, Cham, Switzerland, 2015.

[25] J. Sabatier, O. P. Agrawal, and J. A. Machado, *Advance in Fractional Calculus: Theoretical Developments and Applications in Physics and Engineering*, Springer, Dordrecht, The Netherlands, 2007.

[26] W. H. Beyer, *CRC Handbook of Mathematical Sciences*, CRC Press, Boca Raton, Fla, USA, 1987.

[27] Z. Biolek, D. Biolek, and V. Biolková, "SPICE model of memristor with nonlinear dopant drift," *Radioengineering*, vol. 18, no. 2, pp. 210–214, 2009.

[28] B. Dwork, *Generalized Hypergeometric Functions*, Clarendon Press, Oxford, UK, 1990.

[29] A. Talukdar, A. G. Radwan, and K. N. Salama, "Generalized model for Memristor-based Wien family oscillators," *Microelectronics Journal*, vol. 42, no. 9, pp. 1032–1038, 2011.

[30] L. Chua, "If it's pinched it's a memristor," *Semiconductor Science and Technology*, vol. 29, no. 10, Article ID 104001, 2014.

[31] L. Xu, G. Huang, and Y. Pu, "Numerical Simulation Research of Fracmemristor Circuit Based on HP Memristor," *Journal of Circuits, Systems and Computers*, vol. 27, no. 14, Article ID 1050227, 2010.

14

Design of a SIW Bandpass Filter using Defected Ground Structure with CSRRs

Weiping Li,[1,2] **Zongxi Tang,**[1] **and Xin Cao**[1]

[1]*School of Electronic Engineering, University of Electronic Science and Technology of China, No. 2006, Xiyuan Ave.,*
 West Hi-Tech Zone, Chengdu 611731, China
[2]*School of Information Engineering, East China Jiaotong University, No. 88, Shuanggang Road, Nanchang 330013, China*

Correspondence should be addressed to Weiping Li; lwp8277@126.com

Academic Editor: Guangya Zhou

In this paper, a substrate integrated waveguide (SIW) bandpass filter using defected ground structure (DGS) with complementary split ring resonators (CSRRs) is proposed. By using the unique resonant properties of CSRRs and DGSs, two passbands with a transmission zero in the middle have been achieved. The resonant modes of the two passbands are different and the bandwidth of the second passband is much wider than that of the first one. In order to increase out-of-band rejection, a pair of dumbbell DGSs has been added on each side of the CSRRs. The structure is analyzed using equivalent circuit models and simulated based on EM simulation software. For validation, the proposed filter is fabricated and measured. The measurement results are in good agreement with the simulated ones.

1. Introduction

Substrate integrated waveguide (SIW) was first proposed in 2003 [1]. Based on the TE_{n0} transmission mode, SIW replaces the side metallic walls of the rectangular waveguide with two rows of metallic via holes, which converts conventional waveguides into planar structures [2–4]. Therefore, SIWs not only have the properties of high quality factor and low radiation loss which are similar to metallic waveguides but also have the prominent advantage of compact size due to their planar physical structure. SIWs have been applied to the design of filters, oscillators, power dividers, couplers, and many other microwave components. Complementary split ring resonator (CSRR) was proposed in 2004 as a 3D metamaterial [5] that can exhibit negative permeability near its resonant frequency and therefore can be considered as a composite right/left handed (CRLH) structure. When CSRRs are employed in SIW, passbands based on the evanescent mode below the cutoff frequency of the SIW can be created [6, 7], which can further miniaturize the size of SIW microwave devices. As for the defected ground structure (DGS), it can be regarded as "electromagnetic bandgap" (EBG) structure in the microwave region [8, 9] and the structure is realized by etching periodic or nonperiodic patterns on the metallic layer

in order to create extra transmission zeros. When the DGS is employed in filter design, the out-of-band rejection can be improved without major influence on the insertion loss of the passbands [10].

Based on the three different structures mentioned above, in this paper, a novel dual-band bandpass filter with much compact size and wider passbands is proposed by combining these structures together. Due to the unique resonant properties, two different resonant modes have been achieved to form the passbands and a transmission zero is generated between the two passbands. The corresponding equivalent circuit has been analyzed and the results of full-wave simulation and experimental measurement have been presented.

2. Theory and Analysis

As shown in Figure 1, a SIW consists of three layers, namely, a copper plane, a substrate layer, and a ground plane. Metallic via holes are etched at the edge in parallel position. For a SIW resonant cavity, at TE_{m0n} mode, the resonant frequency can be calculated as

$$f_{m0n} = \frac{c_0}{2\sqrt{\varepsilon_r}}\sqrt{\left(\frac{m}{a}\right)^2 + \left(\frac{n}{l}\right)^2}, \tag{1}$$

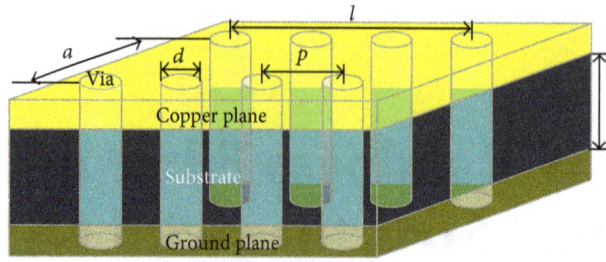

FIGURE 1: The 3D view of an SIW.

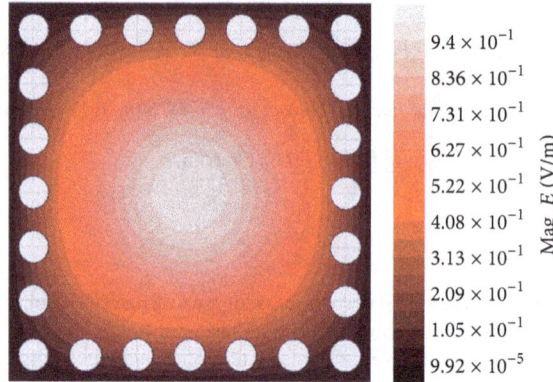

FIGURE 2: The field distribution at TE_{101} resonant mode.

where c_0 is the speed of electromagnetic waves in vacuum and ε_r is the relative dielectric constant. For SIW cavities, TE_{101} is the dominant mode, and its corresponding resonant frequency f_{101} can be given as

$$f_{101} = \frac{c_0}{2\sqrt{\varepsilon_r}} \sqrt{\left(\frac{1}{a}\right)^2 + \left(\frac{1}{l}\right)^2}. \qquad (2)$$

The field distribution at TE_{101} resonant mode is depicted in Figure 2. The electromagnetic energy distributes mainly at the center of the SIW cavity. Therefore, radiation loss can be suppressed and, in this way, the quality factor can be improved. By comparing with conventional microstrip resonators, the insertion loss is reduced and better transmission performance is obtained.

The physical structure of CSRRs is depicted in Figure 3(a). The blue part is metal and the yellow part is the etched CSRRs. They consist of two split resonant rings, and the smaller one is inside the larger one with their openings opposite to each other. While working near the resonant frequency, CSRRs behave like a pair of electric dipoles when they are excited under a vertical axial electric field excitation. Therefore, CSRRs can be regarded as LC parallel circuit, as shown in Figure 3(b), where C_r and L_r represent the self capacitance of the rings and their mutual inductance, respectively; the values of them can be calculated as

$$C_r = (4l_{out} - g_{out}) C_{out} + (4l_{in} - g_{in}) C_{in},$$
$$L_r = (4l_{out} - g_{out}) L_{out} + (4l_{in} - g_{in}) L_{in}, \qquad (3)$$

where C_{out} and L_{out} are the unit characteristic capacitance and inductance of the outside ring, while C_{in} and L_{in} are the unit characteristic capacitance and inductance of the inside ring. Then, the resonant frequency of the CSRRs can be calculated as

$$f_{CSRR} = \frac{1}{2\pi \sqrt{L_r C_r}}. \qquad (4)$$

When CSRRs are employed in SIW, a passband with the evanescent resonant mode below the cutoff frequency of the SIW can be created because CSRRs work as electric dipoles. Therefore, CSRRs can be used to miniaturize the size of the conventional SIWs. In order to increase the out-of-band rejection, defected ground structure (DGS) can be applied in the SIW. The physical structure of the dumbbell DGS is shown in Figure 4(a). The bottom metal layer is etched into "H" shape and this structure can prevent electromagnetic wave from propagating in the transmission line at a certain frequency point, and thus a transmission zero (TZ) can be created (shown in Figure 5). The equivalent circuit of the DGS is shown in Figure 4(b), and if the TZ is designed properly, it can be employed to increase the out-of-band rejection in filters. The circuit elements L_d and C_d can be calculated as

$$C_d = \frac{5f_c}{\pi \left(f_0^2 - f_c^2\right)},$$

$$L_d = \frac{250}{\pi^2 f_0^2 C_d}, \qquad (5)$$

where f_c and f_0 are the cutoff frequency and resonance frequency of the stopband, respectively.

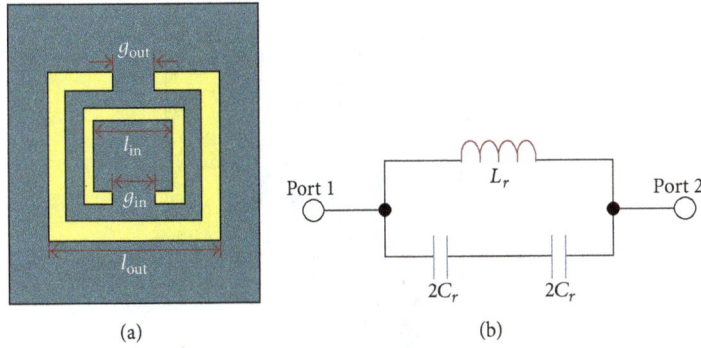

FIGURE 3: The physical structure (a) and the equivalent circuit (b) of CSRRs.

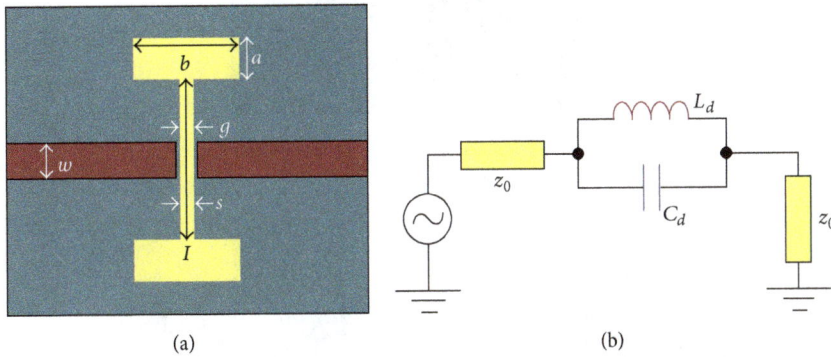

FIGURE 4: The physical structure (a) and the equivalent circuit (b) of dumbbell DGS.

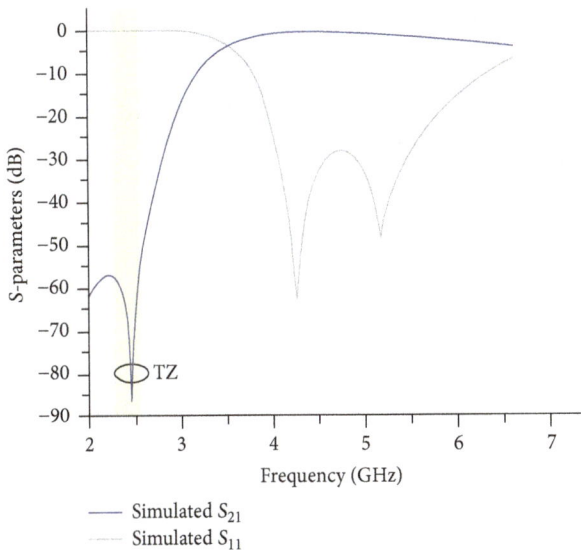

FIGURE 5: The simulated S-parameters of the dumbbell DGS.

In order to implement the proposed filter, firstly, the resonant frequency of the passband produced by the CSRRs based on the evanescent mode below the cutoff frequency of the SIW is calculated and designed. The second step is to add a pair of dumbbell DGSs at each side of the CSRRs to create an extra transmission zero which can increase out-of-band rejection. For CSSRs, they behave like composite right/left-handed resonators, which can generate the evanescent mode of the SIW cavity. This evanescent resonance is the negative resonant mode of the CSSRs, where the electric field, the magnetic field, and the propagation vector satisfy the left-handed rule, which is contrary to the traditional resonant mode where only right-handed rule can be satisfied. H-shaped DGSs are placed near the input and output feed lines. They work as a pair of "gates" which block the transmission of electromagnetic wave at certain frequencies; thus, transmission zeros can be generated consciously. At the frequencies where DGSs act as microwave bandgap structures, electromagnetic energy is radiated to free space through the two h-shaped DGSs. Therefore, the isolation between the passbands can be improved. And in this way, of course, DGSs and CSRRs are not completely independent. The influence between DGSs and CSRRs cannot be neglected due to parasitic effects as well as the weak coupling, and therefore the next step is to tune and make optimizations using EM simulation software. After satisfactory results have been achieved, the final step is fabrication and measurement to verify the correctness of simulation results.

3. Simulation and Measurement

The topology of the proposed filter is shown in Figure 6. The CSRRs are placed in the center of the SIW cavity to create the passbands in the frequency response. Two dumbbell DGSs

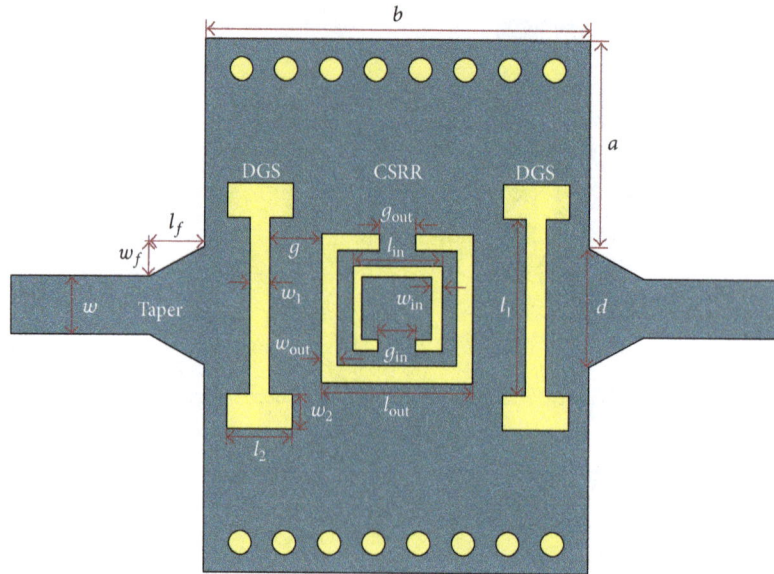

FIGURE 6: The topology of the proposed filter.

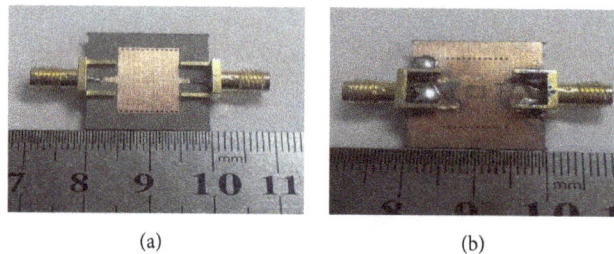

(a)	(b)

FIGURE 7: The top view (a) and the bottom view (b) of the fabricated filter.

are placed at each side of the CSRRs to produce an extra transmission zero which can increase the isolation between these two passbands. The input and output ports are microstrip feed lines with characteristic impedance of 50 ohm. In order to increase the effectiveness of electrical feeding, a pair of tapers function as transition of the transmission line is added between the microstrip line and the SIW.

The filter is fabricated on the Taconic RF-35 substrate with the thickness of 0.508 mm, relative dielectric constant of 3.50, and loss tangent of 0.0018, as shown in Figure 7. The geometric parameters of the filter are tabulated in Table 1. The filter is simulated using the EM full wave simulation software HFSS and measured using the vector network analyzer ZVA40 of the Rohde & Schwarz company. The simulated and measured results are depicted in Figure 8. Two passbands have been created within the frequency range from 2 GHz to 14 GHz. The first passband works at the evanescent mode below the cutoff frequency of the SIW. Its center frequency is 5.4 GHz with the insertion loss of 1.1 dB and the relative 3 dB bandwidth of 17.2%. The second passband works at TE_{101} resonant mode. Its center frequency is 10.1 GHz with the relative 3 dB bandwidth of 55.2%. These two passbands work at different resonant modes, and the bandwidth of the second passband is much wider than that of the first one. A transmission zero is produced at 6.9 GHz

TABLE 1: The geometric parameters of the proposed filter (mm).

Parameter	Value (mm)
a	7.5
l_{in}	2.2
l_{out}	3.0
g_{in}	0.4
l_1	3.5
l_2	2.0
w	1.2
l_f	0.8
g	1.8
b	10.0
w_{in}	0.2
w_{out}	0.2
g_{out}	0.4
w_1	0.1
w_2	0.5
w_f	0.4
d	2.0

with the measured rejection level of 27.2 dB. It can be found that the measured results are in good agreement with the

TABLE 2: Performance comparisons with other recent published SIW filters.

Reference	Center frequencies of the passbands (GHz)	3 dB fractional bandwidth (%)	Insertion loss (dB)	Size
[11]	5.85/6.15	1.3/1.3	2.2/2	$2.11\lambda_0 \times 2.11\lambda_0$
[12]	5.3/8.7	6.8/3.2	1.8/1.94	$0.73\lambda_0 \times 0.73\lambda_0$
[13]	4.8/5.4	3.8/3.9	1.2/1.3	$1.38\lambda_0 \times 0.78\lambda_0$
[14]	9.4/9.98	3.1/3.7	2.24/2.01	$2.67\lambda_0 \times 2.64\lambda_0$
This work	5.4/10.1	17.2/55.2	1.1/1.4	$0.59\lambda_0 \times 0.51\lambda_0$

λ_0 is the guided wavelength at the center frequency of the first passband.

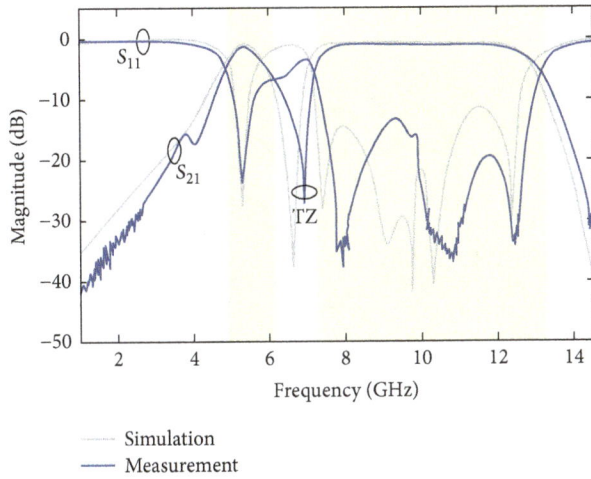

FIGURE 8: The simulated and measured results.

simulated results and the discrepancies are mainly due to the uncertainty of fabrication. The performance comparisons with other recent published SIW filters are given in Table 2. The advantages the proposed filter are quite obvious. Much wider 3 dB fractional bandwidth has been achieved and the size of the filter is very compact, also the insertion loss is acceptable.

4. Conclusion

In this paper, a novel SIW bandpass filter using CSRRs with DGSs has been presented and analyzed. Three different structures are combined together to implement the proposed filter. Two passbands are created by evanescent mode and TE_{101} mode. The rejection level between the passbands is improved due to the transmission zero generated by DGSs. The measured results are in good agreement with the simulation results, and, by comparing with other similar filters, the proposed filter has the advantages of much compact size, lower insertion loss, and wider 3 dB fraction bandwidth, which makes it feasible and applicable in modern microwave communication circuits.

Competing Interests

The authors declare that there is no conflict of interests regarding the publication of this paper.

Acknowledgments

The work is supported by National Natural Science Foundation of China (no. 61563015), Young Foundation of Humanities and Social Sciences of Ministry of Education in China (no. 13YJCZH089), and Young Foundation of Educational Commission of Jiangxi Province of China (no. GJJ14401).

References

[1] D. Deslandes and K. Wu, "Single-substrate integration technique of planar circuits and waveguide filters," *IEEE Transactions on Microwave Theory and Techniques*, vol. 51, no. 2, pp. 593–596, 2003.

[2] X.-P. Chen and K. Wu, "Substrate integrated waveguide filters: design techniques and structure innovations," *IEEE Microwave Magazine*, vol. 15, no. 6, pp. 121–133, 2014.

[3] P. Li, H. Chu, and R. S. Chen, "SIW magic-T with bandpass response," *Electronics Letters*, vol. 51, no. 14, pp. 1078–1080, 2015.

[4] M. Esmaeili and J. Bornemann, "Substrate integrated waveguide dual-stopband filter," *Microwave and Optical Technology Letters*, vol. 56, no. 7, pp. 1561–1563, 2014.

[5] J. D. Baena, J. Bonache, F. Martín et al., "Equivalent-circuit models for split-ring resonators and complementary split-ring resonators coupled to planar transmission lines," *IEEE Transactions on Microwave Theory and Techniques*, vol. 53, no. 4, pp. 1451–1460, 2005.

[6] J. Esteban, C. Camacho-Peñalosa, J. E. Page, T. M. Martín-Guerrero, and E. Márquez-Segura, "Simulation of negative permittivity and negative permeability by means of evanescent waveguide modes—theory and experiment," *IEEE Transactions on Microwave Theory and Techniques*, vol. 53, no. 4, pp. 1506–1513, 2005.

[7] Y. H. Song, G.-M. Yang, and W. Geyi, "Compact UWB bandpass filter with dual notched bands using defected ground structures," *IEEE Microwave and Wireless Components Letters*, vol. 24, no. 4, pp. 230–232, 2014.

[8] S. Biswas, D. Guha, and C. Kumar, "Control of higher harmonics and their radiations in microstrip antennas using compact defected ground structures," *IEEE Transactions on Antennas and Propagation*, vol. 61, no. 6, pp. 3349–3353, 2013.

[9] D. Suhas, C. R. Lakshmi, Z. S. Rao, and D. Kannadassan, "A systematic implementation of elliptic low-pass filters using defected ground structures," *Journal of Electromagnetic Waves and Applications*, vol. 29, no. 15, pp. 2014–2026, 2015.

[10] Y. L. Zhang, W. Hong, K. Wu, J. X. Chen, and H. J. Tang, "Novel substrate integrated waveguide cavity filter with defected ground structure," *IEEE Transactions on Microwave Theory and Techniques*, vol. 53, no. 4, pp. 1280–1286, 2005.

[11] M. Rezaee and A. R. Attari, "A novel dual mode dual band SIW filter," in *Proceedings of the 44th European Microwave Conference (EuMC '14)*, pp. 853–856, Rome, Italy, October 2014.

[12] Y.-D. Wu, G.-H. Li, W. Yang, and T. Mou, "A novel dual-band SIW filter with high selectivity," *Progress in Electromagnetics Research Letters*, vol. 60, pp. 81–88, 2016.

[13] H. Wsx and Y. Wu, "Compact SIW dual-band bandpass filter using novel dual-resonance quasi-SIW-transmission-line-structure resonators," *The Journal of Engineering*, 2016.

[14] G. Zhang, J. Wang, S. Ge, and W. Wu, "A new balanced dual-band SIW bandpass filter with high common-mode suppression," in *Proceedings of the Asia Pacific Microwave Conference (APMC '16)*, New Delhi, India, February 2016.

15

New Pulse Width Modulation Technique to Reduce Losses for Three-Phase Photovoltaic Inverters

Mohannad Jabbar Mnati ⓘ,[1,2] Dimitar V. Bozalakov,[1] and Alex Van den Bossche[1]

[1]*Department of Electrical Energy, Metals, Mechanical Constructions and Systems, Ghent University, Technologiepark Zwijnaarde 913, B-9052 Zwijnaarde, Gent, Belgium*
[2]*Department of Electronic Technology, Institute of Technology Baghdad, Middle Technical University, Al-Za'franiya, 10074 Baghdad, Iraq*

Correspondence should be addressed to Mohannad Jabbar Mnati; mohannad.mnati@ugent.be

Academic Editor: Gerard Ghibaudo

Nowadays, most three-phase, "off the shelf" inverters use electrolytic capacitors at the DC bus to provide short term energy storage. However, this has a direct impact on inverter lifetime and the total cost of the photovoltaic system. This article proposes a novel control strategy called a 120° bus clamped PWM (120BCM). The 120BCM modulates the DC bus and uses a smaller DC bus capacitor value, which is typical for film capacitors. Hence, the inverter lifetime can be increased up to the operational lifetime of the photovoltaic panels. Thus, the total cost of ownership of the PV system will decrease significantly. Furthermore, the proposed 120BCM control strategy modulates only one phase current at a time by using only one leg to perform the modulation. As a result, switching losses are significantly reduced. The full system setup is designed and presented in this paper with some practical results.

1. Introduction

Constantly growing concerns about global warming have forced many countries to change their energy policy towards increasing the share of renewable energy resources and hence reducing CO_2 emissions. Significant amounts of these renewable energy resources are delivered by photovoltaic (PV) systems with small and medium power, interfaced with the distribution grid via power electronic voltage source inverters. In order to bring maximum revenue to the prosumer during the lifetime of the PV system, the power electronic inverter must have as high efficiency as possible and as long a life as possible, at a limited cost. A typical photovoltaic panel has a useful lifetime of 25 years. The inverter lifespan of roughly 10 years is limited, however, primarily due to electrolytic capacitor lifetime. Hence, typically, three inverters will be needed during the exploitation period of a PV system. [1, 2].

The efficiency of a typical power electronic inverter that uses hard switching commutation varies between 95% and 98%, depending on the power electronic switches used (MOSFET, IGBT, SiC, etc.) and power ratings. To increase efficiency, losses in the semiconductor switches and the magnetic components (the output inductive filter) must be reduced. Semiconductor losses can be split into conduction and switching losses. Conduction loss can be reduced by selecting switches with a lower voltage drop across them; nevertheless it is always present. Switching losses, however, can be reduced or even eliminated by using soft-switching techniques such as zero-voltage and zero-current commutation; in [3], the authors achieved peak efficiency of 97% and in [4], 98.4%. Soft switching operates with limited dv/dt and/or di/dt, which have an advantageous effect on electromagnetic compatibility issues, reduces the thermal stresses on the power electronic switches, and combined with the limited dv/dt and/or di/di have a positive impact on the inverter lifetime. Some of the main drawbacks of these techniques are (i) operating at rather fixed loads [3, 5], which is not the case in photovoltaic systems, due to the intermittent nature of solar irradiation; (ii) requiring additional components and control signals [3, 6], which increases the control complexity and physical implementation of the power electronic inverter; and (iii) operating under a variable carrier frequency [6–8], which introduces filtration problems. In [8], a power electronic inverter topology is proposed using hard switching techniques, based on MOSFET and IGBT combination per leg.

The approach uses the MOSFET to perform the current modulation and it is switched with the nominal carrier frequency, while the IGBT is switched with the grid frequency. The advantage of the better free-wheeling IGBT diode is used in this approach, and the authors report decreasing the switching losses by 33%. The advantage of this approach is that no additional components or complex controls are required, which renders the approach extremely attractive. Another possibility for reducing switching losses is to use space vector modulation with 60° bus clamping, as studied in [1, 9]. This is effective at high AC voltage output, where it reduces total switching cycles by one-third. Although the above-mentioned techniques offer increased efficiency, if used in PV applications, they still require a significant amount of filtering capacitance at the DC bus, which limits the lifetime of the power electronic inverter.

An alternative to electrolytic capacitors is film capacitors, which have a lifetime comparable to PV panels, but is not an economically viable solution, due to their small capacitance to price and capacitance to volume ratios. In [9], a control algorithm is proposed that uses two film capacitors of 33 μF connected in a symmetrical half bridge power electronic converter, with dual voltage compensation for stabilizing the DC bus voltage. The proposed approach is tested on a 1 kW inverter, and the results show that the 100 Hz ripple at the DC bus can be completely compensated for in steady state operation, with a minimum capacitance of 33 μF required. The additional control plus the converter react in such a way that capacitance is increased; i.e., the converter includes additional virtual capacitance in the system. Hence, this solution appears particularly suitable for replacing the electrolytic capacitors at the DC side. A disadvantage of the approach is that stabilization is performed via additional hardware, which requires additional space in physical implementation, as well as extra control signals.

In [1], a novel control strategy for a single-phase PV inverter is proposed, which is implemented on a three-phase IGBT module. The inverter consists of an input boost converter (one leg of the module), a film capacitor at the DC bus, and full bridge inverter (leg two and three of the module), the outputs of which are connected to the grid through an LC filter. The control strategy injects the phase current in two stages; i.e., when the grid voltage is lower than the PV voltage, the control strategy drives only the full bridge to modulate the phase current. When the grid voltage is higher than the PV voltage, the control strategy starts controlling the boost converter. By doing so, the boost converter ensures a DC bus voltage margin to the full bridge inverter, which is sufficient for guaranteeing normal current injection into the grid. This control strategy is able to reduce bridge switching losses approximately six times, compared to a state of nonbus clamped control. The other advantage is that the control strategy is implemented on an inverter that uses a film capacitor, which ensures a lifetime comparable with the lifetime of the PV. Moreover, the film capacitor that is used has a value of 22 μF, which is smaller compared to

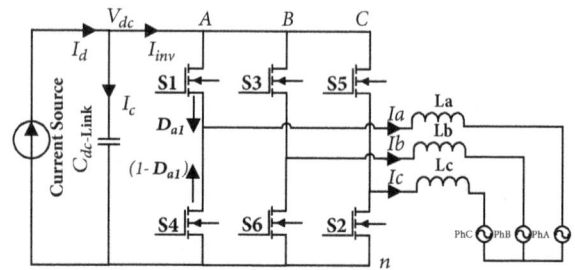

FIGURE 1: Three-phase inverter 120BCM with modulated DC link.

that used in [9] and is suggested as a possible minimum value.

In this paper, a control strategy for a three-phase inverter is proposed that uses a similar approach as [1]. The control is verified by means of simulations in MATLAB/Simulink and practical full system setup on a three-phase, three-leg inverter. The experiments were performed using the experimental setup to verify the performance of the proposed control strategy (a 120° bus clamped PWM (120BCM)). The experimental setup consists of three-phase inverter with L filter and a control board, power board, and voltage and current sensors for DC and AC sides [10, 11]. The control board is composed of dsPIC microcontroller dsPIC33FJ256GP710A and it is the main unit used in the implementation to control the three-phase inverter. It is programmed by MPLAB X IDEV3.15 software compiler and C language.

This paper is organized as follows. Section 2 provides a detailed description of the proposed control strategy and the different time intervals are described. Section 3 provides a detailed simulation of the presented control algorithm, and different modulation strategies are compared using numerical calculations of the switching losses. Section 4 presents the simulation result with details. Section 5 presents the practical of three-phase inverter with results and finally Section 6 presents conclusions of the article.

2. 120° Bus Clamped PWM System Configuration

Natural sampling and centered PWM modulations commutate the semiconductor switches during the entire period of the fundamental current. Thus, if no special measures are taken such as ZVS and ZCS, switching losses are often larger than conduction losses. The proposed PWM technique (120BCM) is based on a partial modulation of the injected phase current; thereby, switching losses can be significantly reduced, compared to the classical PWM technique. In order to evaluate the performance of the 120BCM, the topology presented in Figure 1 is used. This is a classical three-phase inverter topology and is connected to the grid through a star connection. The DC bus capacitor is low value film-based, which, as noted above, will have a beneficial effect on inverter lifetime. Three inductors are used to filter out the current pulsation caused by the modulation frequency. Prior to starting the operational principle, the following assumptions are considered in this paper:

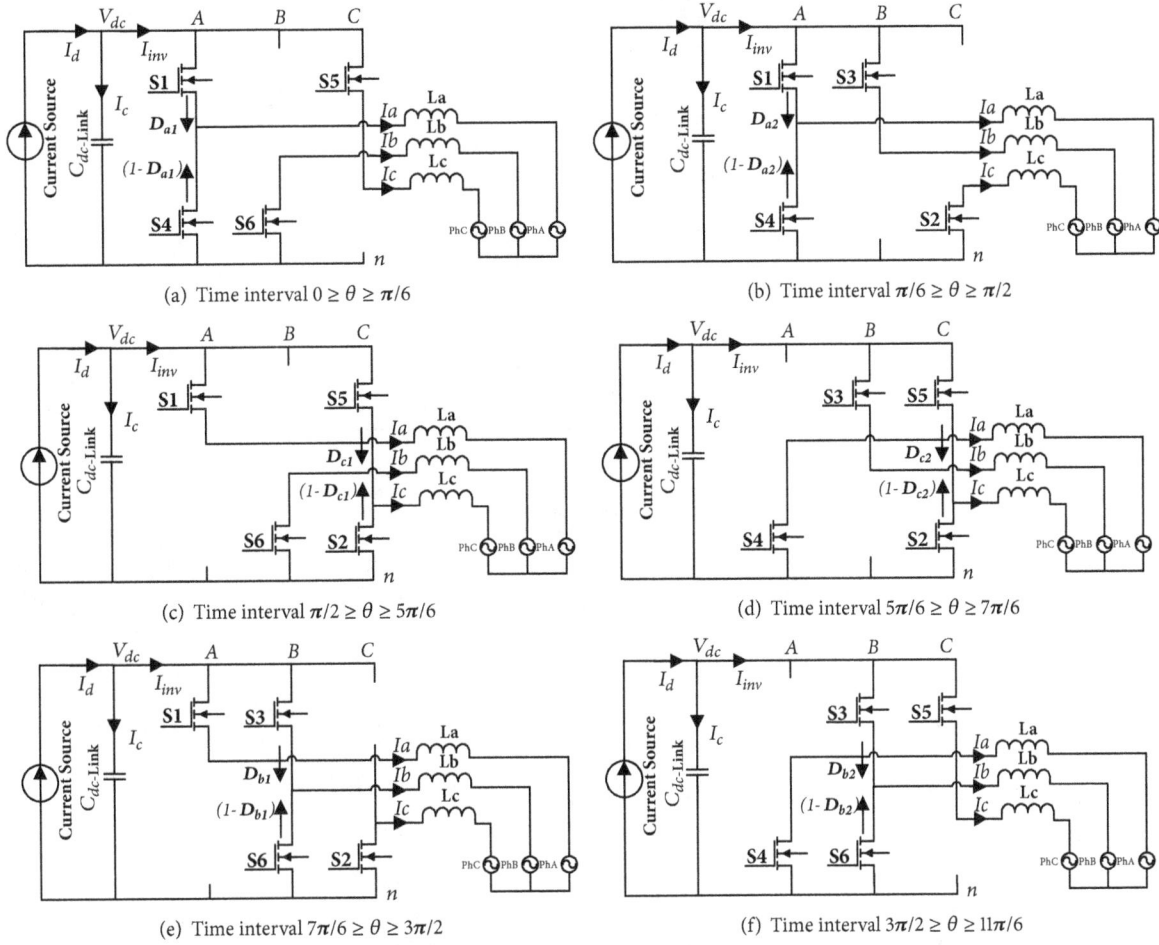

(a) Time interval $0 \geq \theta \geq \pi/6$

(b) Time interval $\pi/6 \geq \theta \geq \pi/2$

(c) Time interval $\pi/2 \geq \theta \geq 5\pi/6$

(d) Time interval $5\pi/6 \geq \theta \geq 7\pi/6$

(e) Time interval $7\pi/6 \geq \theta \geq 3\pi/2$

(f) Time interval $3\pi/2 \geq \theta \geq 11\pi/6$

FIGURE 2: Six switching configurations in 120° BC-PWM control method.

(1) The inverter is supplied with a current source. In practice, this assumption holds if there is a matching circuit (for example, a boost converter) between the renewable source and the inverter that will guarantee maximum power point operation, as well as a current source output characteristic.

(2) It is assumed that the input instantaneous power is constant.

(3) For the sake of maintaining clarity, the matching block between the renewable energy source and the inverter is not considered in this paper.

(4) Only positive-sequence current injection is considered; hence, the instantaneous value of the sum of the three currents is zero.

(5) The inverter operates in a steady state mode.

As noted above, the operational principle of the 120BCM is based on partial modulation of the injected phase current. When only positive-sequence current is being injected, a natural zero crossing of the phase currents occurs every 60 electrical degrees. Considering the above assumptions, the 120BCM operational principle can be explained as follows.

(i) Interval $0° \leq \theta \leq 30°$. This interval begins when the zero crossing of phase current i_a occurs (i_a becomes positive) and the duration of it is 30°. The state of the switches during the first interval is depicted in Figure 2(a), where it can be seen that switches S1, S4, S5, and S6 are being controlled. The modulation for the injected phase currents i_a, i_b, and i_c is performed only by switches S1 and S4, while the other two switches are kept continuously closed. Considering this and the above listed assumptions, the injected currents in this interval can be expressed in

$$D_{a1} = \frac{v_a - v_b}{V_{dc}} = \frac{v_a - v_b}{v_c - v_b} \quad (1)$$

$$i_{b1} = i_{inv} + D_{a1} i_{a1}$$
$$i_{c1} = i_{inv} + \left(1 - D_{a1}\right) i_{a1} \quad (2)$$

where D_{a1} is a duty ratio and i_b and i_c are the modulated phase currents in this interval and i_{inv} is the input inverter current. Since S5 and S6 are closed continuously, the DC bus capacitor is connected to the line-to-line voltage Vbc. However, the line current that flows through it is almost negligible compared to the phase currents, as the DC bus capacitors are rather small (this study uses a DC bus capacitor of 8 μF). Following this

TABLE 1: Mathematical description of the proposed BCPWM technique during the different intervals.

Interval	Equation	Interval	Equation
$(V_{PhB} \geq V_{PhA} \geq V_{PhC})$ $0 \geq \theta \geq 30°$ $330° \geq \theta \geq 360°$ Figure 2(a)	$D_{a1} = \dfrac{V_a - V_b}{V_{dc}} = \dfrac{V_a - V_b}{V_c - V_b}$ $i_{b1} = i_{inv} + D_{a1}i_{a1}$ $i_{c1} = i_{inv} + (1 - D_{a1})i_{a1}$	$(V_{PhC} \geq V_{PhA} \geq V_{PhB})$ $150° \geq \theta \geq 210°$ Figure 2(d)	$D_{a2} = \dfrac{V_a - V_c}{V_{dc}} = \dfrac{V_a - V_c}{V_b - V_c}$ $i_{b2} = i_{inv} + (1 - D_{a2})i_{a2}$ $i_{c2} = i_{inv} + D_{a2}i_{a2}$
$(V_{PhB} \geq V_{PhC} \geq V_{PhA})$ $30° \geq \theta \geq 90°$ Figure 2(b)	$D_{c1} = \dfrac{V_c - V_b}{V_{dc}} = \dfrac{V_c - V_b}{V_a - V_b}$ $i_{a1} = i_{inv} + (1 - D_{c1})i_{c1}$ $i_{b1} = i_{inv} + D_{c1}i_{c1}$	$(V_{PhA} \geq V_{PhC} \geq V_{PhB})$ $210° \geq \theta \geq 270°$ Figure 2(e)	$D_{c2} = \dfrac{V_c - V_a}{V_{dc}} = \dfrac{V_c - V_a}{V_b - V_a}$ $i_{a2} = i_{inv} + D_{c2}i_{c2}$ $i_{b2} = i_{inv} + (1 - D_{c2})i_{c2}$
$(V_{PhC} \geq V_{PhB} \geq V_{PhA})$ $90° \geq \theta \geq 150°$ Figure 2(c)	$D_{b1} = \dfrac{V_b - V_c}{V_{dc}} = \dfrac{V_b - V_c}{V_a - V_c}$ $i_{a1} = i_{inv} + (1 - D_{b1})i_{b1}$ $i_{c1} = i_{inv} + D_{b1}i_{b1}$	$(V_{PhA} \geq V_{PhB} \geq V_{PhC})$ $270° \geq \theta \geq 330°$ Figure 2(f)	$D_{b2} = \dfrac{V_b - V_a}{V_{dc}} = \dfrac{V_b - V_a}{V_c - V_a}$ $i_{a2} = i_{inv} + D_{b2}i_{b2}$ $i_{c2} = i_{inv} + (1 - D_{b2})i_{b2}$

TABLE 2: Switching states of the proposed 120 BCM.

One Cycle (360°)	Leg A S1 and S4	Leg B S3 and S6	Leg C S5 and S2
$0° \leq \theta \leq 30°$ $330° \leq \theta \leq 360°$	PWM	0-1	1-0
$30° \leq \theta \leq 90°$	1-0	0-1	PWM
$90° \leq \theta \leq 150°$	1-0	PWM	0-1
$150° \leq \theta \leq 210°$	PWM	1-0	0-1
$210° \leq \theta \leq 270°$	0-1	1-0	PWM
$270° \leq \theta \leq 330°$	0-1	PWM	1-0

period of 30°, switch S1 is closed (clamps the DC bus) for 120° and modulation is performed by another leg. For this reason, the proposed PWM control strategy is referred to as 120° bus clamped modulation.

(ii) Interval $30° \leq \theta \leq 90°$. This interval begins 30° after the zero crossing of phase current i_a occurs (i_a becomes positive) and its duration is 60°. The states of the switches during the first interval are depicted in Figure 2(c), where it can be seen that switches S1, S6, S5, and S2 are being controlled. The modulation for the injected phase currents i_a, i_b, and i_c is again performed only by switches S5 and S2, while switches S1 and S6 are kept continuously closed. The injected currents in this interval can be expressed in

$$D_{c1} = \frac{V_c - V_b}{V_{dc}} = \frac{V_c - V_b}{V_a - V_b} \tag{3}$$

$$i_{a1} = i_{inv} + (1 - D_{c1})i_{c1}$$
$$i_{b1} = i_{inv} + D_{c1}i_{c1} \tag{4}$$

The operational principle concerning the remaining 4 intervals $90° \leq \theta \leq 150°$, $150° \leq \theta \leq 210°$, $210° \leq \theta \leq 270°$, and $270° \leq \theta \leq 330°$ is similar to the described above. The corresponding equivalent circuits are Figures 2(b), 2(d), 2(e), and 2(f). For more clarity all mathematical equations describing the different intervals are listed in Table 1 and a summary of all switching states is given in Table 2. As can be seen form Figure 2 and Table 2 during all 6 intervals (60° each), the phase currents are always modulated only

by one leg of the power electronic inverter. Thus the total commutations for the entire inverter are decreased 3 times compared to the classical full sine PWM technique. Hence, in general the total switching losses will also decrease. It occurs at a typical current of 0 to 50% of the peak current value. In the simplified assumption that the switching losses would be proportional to voltage and current, the ratio is approximated.

Integral sine −30° to +30° /integral sine −90° to 90° is

$$\frac{(1 - \cos(\pi/6))}{(\cos(0) - \cos(\pi/2))}$$

= 13.4% compared to centred PWM

$$\frac{(1 - \cos(\pi/6))}{(\cos(0) - \cos(\pi/23))} \tag{5}$$

= 26.8% compared to 60° bus clamped PWM

The reality is that switching losses are less than proportional to current, but that the DC link voltage is also typically 15% reduced, so that at least a factor 3-4 reduction occurs in switching loss compared to a 60° bus clamped PWM and 4-8 compared to a centered PWM. The exact value depends on the detailed transistor data and on the way it is controlled.

Note that the duration of the first interval is also 60° and the full length of this interval is $330° \leq \theta \leq 30°$ but for simplicity 0° was chosen for a starting point.

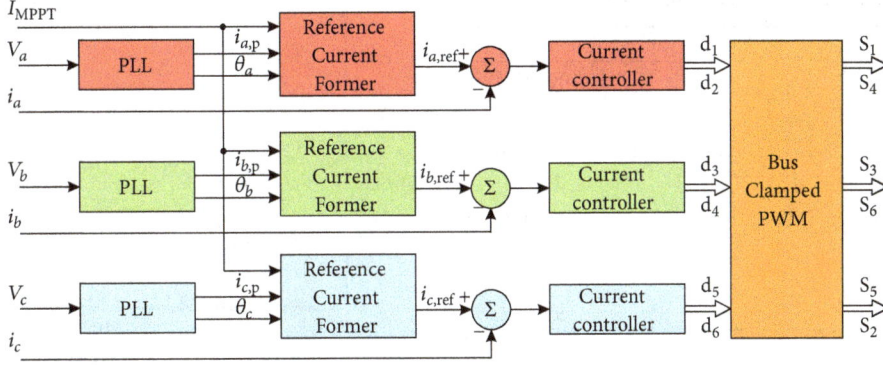

FIGURE 3: Block diagram of the control block of the BCPWM technique.

3. Control Principle of the BCPWM

The designed controller for the power electronic switches equipped with the BCPWM is depicted in Figure 3. The phase angles are extracted from the three-phase voltages using a three-phase locked loop (PLL). The output off the PLL generates three synchronized voltages with the phase voltages and reference amplitude of 1V. The reference current of each phase is created using the PLL signal for the corresponding phase, which is multiplied with the maximum allowed current that can be delivered by the primary source (PV) determined by the MPPT (the MPPT is outside the scope of this work). Then, the reference current is subtracted from the measured phase current, giving an error current signal. This error is passed to a conventional proportional-integral (PI) controller. The output signal of the PI controller is compared with a triangular signal and the modulated signal is processed by a block, where the 120BCM modulation is embedded. In this block, Boolean logic is used to implement the 60° modulations, where the logic equations are derived as shown in (6) to (11).

The "+" means logical OR, multiplication means logical AND, the operators "<" and ">" are used to represent the conditional function "if", $i_{x,p}$ is the instantaneous value of respective reference currents delivered by the PLL, and PWM is the signal generated by the comparison between the PI output and the triangular signal, as shown in Figure 5. In this study, dead time is neglected; however, in real implementation, this should be taken into account.

$$S_{1,PWM} = \left(i_{a,p} > i_{b,p}\right)\left(i_{a,p} > i_{c,p}\right)$$

$$\cdot \left(\left(i_{b,p} \geq i_{a,p}\right)\left(i_{a,p} \geq 0\right) + \left(i_{c,p} \geq i_{a,p}\right)\left(i_{a,p} \leq 0\right)\right.$$

$$\left. + \left(i_{c,p} \geq i_{a,p}\right)\left(i_{a,p} \geq 0\right) + \left(i_{b,p} \geq i_{a,p}\right)\left(i_{a,p} \leq 0\right)\right) \tag{6}$$

$$\cdot PWM$$

$$S_{4,PWM} = \overline{S_{1,PWM}} \tag{7}$$

$$S_{2,PWM} = \left(i_{b,p} > i_{a,p}\right)\left(i_{b,p} > i_{c,p}\right)$$

$$\cdot \left(\left(i_{c,p} \geq i_{b,p}\right)\left(i_{b,p} \geq 0\right) + \left(i_{a,p} \geq i_{b,p}\right)\left(i_{a,p} \leq 0\right)\right.$$

$$\left. + \left(i_{a,p} \geq i_{b,p}\right)\left(i_{b,p} \geq 0\right) + \left(i_{b,p} \geq i_{c,p}\right)\left(i_{c,p} \leq 0\right)\right)$$

$$\cdot PWM$$

$$\tag{8}$$

$$S_{5,PWM} = \overline{S_{2,PWM}} \tag{9}$$

$$S_{3,PWM} = \left(i_{c,p} > i_{b,p}\right)\left(i_{c,p} > i_{a,p}\right)$$

$$\cdot \left(\left(i_{a,p} \geq i_{c,p}\right)\left(i_{c,p} \geq 0\right) + \left(i_{c,p} \geq i_{b,p}\right)\left(i_{c,p} \leq 0\right)\right.$$

$$\left. + \left(i_{b,p} \geq i_{c,p}\right)\left(i_{c,p} \geq 0\right) + \left(i_{c,p} \geq i_{a,p}\right)\left(i_{a,p} \leq 0\right)\right) \tag{10}$$

$$\cdot PWM$$

$$S_{6,PWM} = \overline{S_{3,PWM}} \tag{11}$$

4. Validation of the Proposed 120BCM Technique by Means of Simulation

The performance of the 120BCM is evaluated in a simulation environment using MATLAB-Simulink. The connection diagram for the three-phase inverter equipped with the 120BCM control algorithm is depicted in Figure 1. The three-phase inverter is supplied with a DC current of 13.5A and it is connected to the grid via three differential inductor filters, L_a, L_b, and L_c. The grid is represented by a line impedance of $0.1+j0.0314$, which is a typical impedance value for low voltage distribution grids [12, 13] and an ideal three-phase voltage source. Detailed data concerning the connection diagram is shown in Table 3.

4.1. Simulation Results. The driving pulses used by S1 and S4 generated by the PWM block are depicted in Figure 4, where it can be seen that the driving pulses are present during the first 1/6th and 5/6th of the period. This interval is described in Figures 2(a) and 2(b). The current that flows through switches S1 and S4 is also depicted in Figure 6.

As can be seen, the modulation of the current injected in phase (a) is being modulated only in the first and last 30° interval of the half-wave, while the rest of the current waveform is being modulated by the remaining legs of the inverter. In the transition intervals, a small transient of

TABLE 3: Data used in the Simulink model.

Parameter	Value
Grid Voltage	400V(L-L)
Grid Frequency f_g	50Hz
Line impedance	0.1+j0.0314
Filter Inductor L_a, L_b, L_c	2.5mH
Switching Frequency	25kHz
Film capacitor C_{dc}	8 μF
Inverter rms dc current I_{inv}	13.5A
Total injected active power	6.6 kW

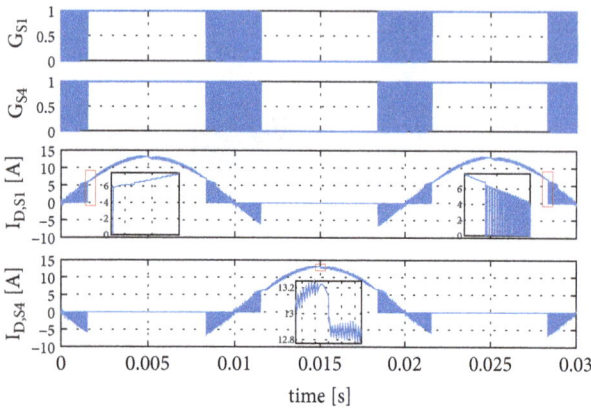

FIGURE 4: The modulation of the current injected in phase (A).

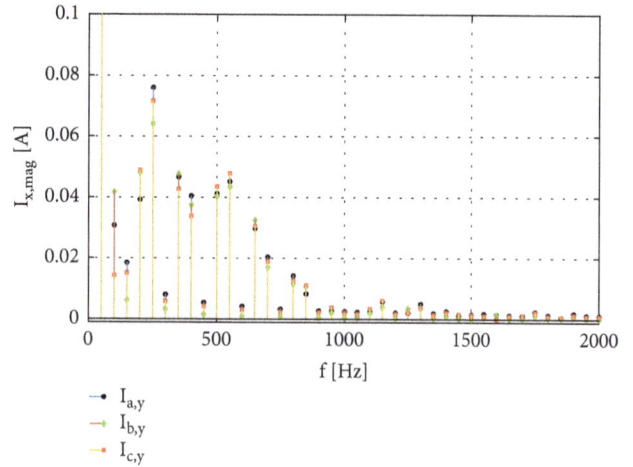

FIGURE 5: Harmonic content of the three-phase currents where y ∈ [1..40]-IEC61000-3-2.

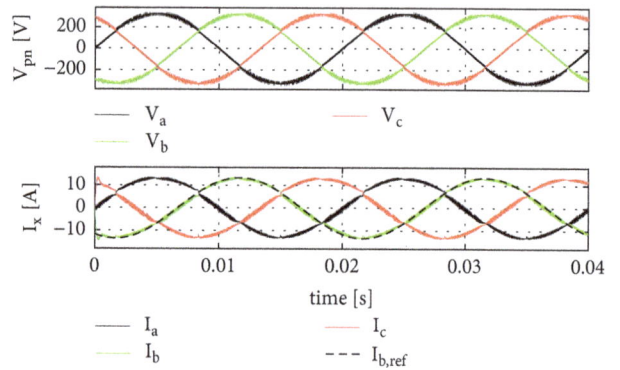

FIGURE 6: Phase voltages at the inverter terminals, injected phase currents, and reference current comparison.

the current occurs, which is also shown in Figure 4. These transients are always present when the 120BCM switches between time intervals, which is the reason for the small transient at the top of the sine wave of the injected current. The other phase currents, i_b and i_c, have identical waveforms, but are 120° phase shifted.

Another important aspect of the grid connected inverters is their total harmonic distortion limit (THD). According to IEC 61000-3-2 [14], the THD of the inductor currents must not exceed 5%. Due the small transients between the different intervals, a spectrum analysis is performed on the phase currents, which is depicted in Figure 5. The harmonics under investigation range from fundamental up to the 41st harmonic, which is the maximum covered by [14].

The fundamental peaks up to 13.6 A, which is its nominal value; however, more attention is paid to higher order harmonics. Figure 5 shows that the higher order harmonics magnitude is relatively low, complying with [14]. Furthermore, despite the partial modulation technique (60 electrical degrees per interval), the THD for each phase does not exceed 2.9%, which complies with [14] and makes the proposed 120BCM extremely suitable for renewable energy applications.

Finally, the obtained simulation results of the phase voltages at the inverter terminals and the injected currents are presented in Figure 6. The results show that despite the current being modulated in split intervals using independent controllers, the wave forms have pure sinusoidal forms with very small distortions. Furthermore, the reference current

$I_{b,ref}$ is depicted on the top of the injected ones by the black dashed line, and it is noted that the PI controller is able to track the reference currents very well, without the need for feed forward controls, which significantly simplifies practical implementation. The other two reference currents are also tracked with the same performance, but are not depicted in Figure 6.

As noted above, the three-phase inverter connection is without a neutral. The simulation results obtained for the DC bus voltage are shown in Figure 7, together with the line-to-line voltages V_{ab}, V_{bc}, and V_{ca}, as well as the inverter voltage, which in this particular modulation technique is a sufficient margin to allow for proper current injection into input current I_{inv}. It is noted that the DC bus voltage has three-phase full bridge rectifier; additionally, it can be seen that the DC bus voltage is slightly higher that the line-to-line waveform, significantly resembling the one obtained after a grid. A standard inverter will need a margin of 10V to 20V between the peak value of the grid line-to-line voltage and the DC bus voltage [13]. Additionally, taking into account the grid voltage fluctuation set by EN50160 [15], which allows for a 10% upper threshold, the DC bus voltage becomes at least

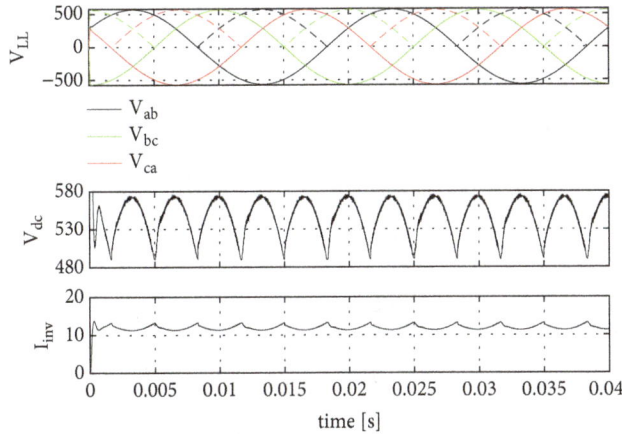

FIGURE 7: Obtained simulation results of the dc bus voltage V_{dc} and inverter dc bus current I_{inv}.

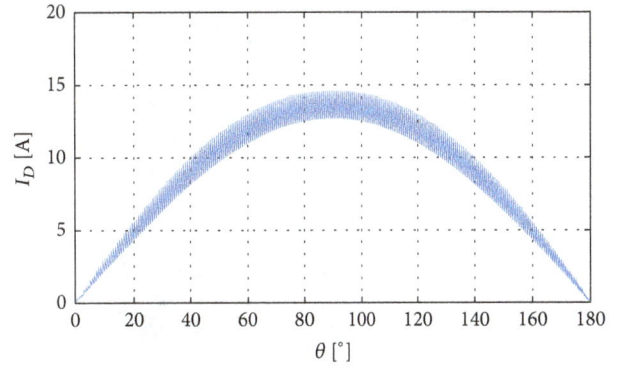

FIGURE 8: Current ripple during one half-sine, used for the loss switching calculation.

620V, and the margin mentioned above the final value will be at least 630 V.

4.2. Switching Loss Calculation in Semiconductor Devices. Switching loss calculation can sometimes be a difficult endeavor, especially when PWM is involved. Since the majority of losses of the inverter are composed of conduction losses, switching losses, and losses in the magnetic material in the magnetic components (output filters), then it is very difficult to segregate and evaluate each loss independently by means of experiment. Furthermore, the assessment of each of the losses becomes even more difficult because of the different modulation intervals that are introduced by the partial modulation of 120BCM and 60PWM. Therefore, a numerical calculation can be used to assess the switching losses so that an easy and fast evaluation can be performed. In [12], the author proposes a cycle-by-cycle calculation for switching losses, and then the individual losses are summarized. This method provides quite accurate results; however, it is designed for central modulation. In [16], a methodology of loss calculation is presented that uses linear approximations to obtain the different parameters, such as turn-on and turn-off energy, needed for the switching loss calculation, with satisfying results. In this article a combination of the cycle-by-cycle approach and linear approximation of the required parameters are used to calculate the losses of the different PWM techniques. Based on these two approaches, the following equation for losses calculation of the half-sine modulated signal for 0° to 180° is obtained:

$$
\begin{aligned}
P_{tot} = \Bigg(& \sum_0^n E_{on} \left(\frac{i_{on}(n)}{I_D} \right)^\alpha \left(\frac{V_{dc}(n)}{V_{cc}} \right)^\beta \\
& + \sum_0^n E_{off} \left(\frac{i_{on}(n)}{I_D} \right)^\alpha \left(\frac{V_{dc}(n)}{V_{cc}} \right)^\beta \Bigg) 2f_g
\end{aligned}
\tag{12}
$$

where E_{on} is the turn-on energy loss, E_{off} is the turn-off energy loss, $i_{on}(n)$ is the instant value of the current at turn-on, $i_{off}(n)$ is the instant value of the current at turn-off, I_D is the drain current and V_{CC} is the voltage at which the energy

is measured, and $V_{dc}(n)$ is the instant value of the DC bus voltage. Since this is the loss calculated for one half period, a transistor, multiplication by two is needed to obtain the full leg period. Suppose that the current flowing through the switches is the one shown in Figure 8 where the current ripple plus the fundamental are used in (12), and also the switching loss has been taken proportional to current (corresponding with the data) and linear approximations are used to calculate the losses as suggested in [17]. The exponent coefficients α and β represent the current and voltage dependency of the variation from the nominal values used in the datasheets. A special case is formed: if $\alpha=1$ and $\beta=1$, then the switching losses are linear with the variation from the nominal values. According to [18] for IGBT switches the current dependency is $\alpha=1$ which is a linear approximation while $\beta=1.2..1.4$. Nevertheless, these coefficients are dependent on the transistor manufacturer and also transistor type. In [19–22], more accurate methodologies for assessing the switching losses are proposed; however, their drawback is the high complexity and also difficulty to being implemented in practice.

It was noted previously that a standard inverter needs some degree of a DC bus voltage margin. However, to better assess the performance of the different modulation control strategies, the DC bus voltage for all three calculations is kept at 560V. The switching losses equation used for the 60° bus clamped method is similar to (12), but the switches clamp to the DC bus voltage for a 60° period, where switching losses are not present. Then (12) can be rewritten as

$$
\begin{aligned}
P_{tot} = \Bigg(& \sum_0^{\pi/3} E_{on} \frac{i_{on}(n)}{I_D} \frac{V_{dc}(n)}{V_{cc}} + \sum_{2\pi/3}^n E_{on} \frac{i_{on}(n)}{I_D} \frac{V_{dc}(n)}{V_{cc}} \\
& + \sum_0^{\pi/3} E_{off} \frac{i_{off}(n)}{I_D} \frac{V_{dc}(n)}{V_{cc}} \\
& + \sum_{2\pi/3}^n E_{off} \frac{i_{off}(n)}{I_D} \frac{V_{dc}(n)}{V_{cc}} \Bigg) 2f_g
\end{aligned}
\tag{13}
$$

The switching losses for 120BCM can be calculated by (13), but the intervals must be changed $\pi/6$ for the first part of the half-sine and $5\pi/6$ for the last part of the half-sine. In this

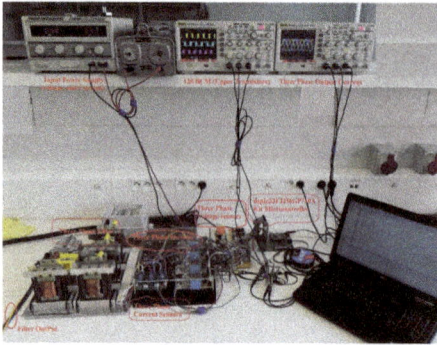

FIGURE 9: Full system designed setup of three-phase inverter under (120BCM).

TABLE 4: Switches data used for the cycle-by-cycle switching loss calculation.

IRG4PH40UDPbF – IGBT		
Parameter	Value	Conditions
E_{tot}	7.04 mJ	@ TJ =150°C, I_D=21 A, VCC=800V tail and diode Qrr included
E_{on}	3.39 mJ	
E_{off}	3.64 mJ	
C3M0065090D		
E_{tot}	0.316 mJ	@ TJ =150°C, ID=20 A, VCC=400V
E_{tot}	0.225 mJ	
E_{tot}	0.091 mJ	

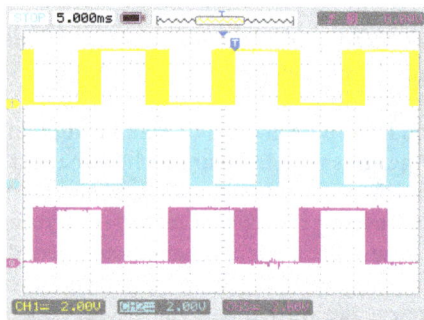

(a)

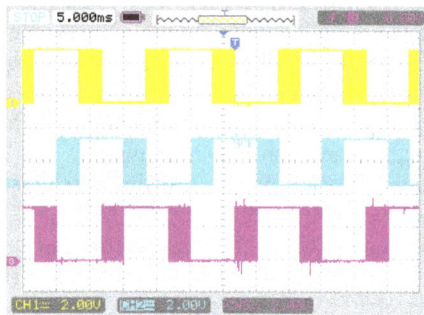

(b)

FIGURE 10: Upper and Lower PWM of three-phase inverter under 120-BCM.

modulation technique, however, the DC bus voltage is not constant, as shown in Figure 9, and there the instant value it is used in (13).

The switching losses are calculated for two different switches: IRG4PH40UD and C3M0065090D, where the first is an IGBT and the second is a SiC type. The energy at turn-on and turn-off is shown in Table 4. The original values of the currents and voltages at which the energies are obtained are also listed in the same table.

The results of the per cycle calculations for the two types of transistors are listed in Table 5. As expected, the central PWM results in significant switching losses per switch, as well as total losses when IGBT are used. By the improved switching performance of the SiC transistors, the losses are

further decreased. When the 60° bus clamped modulation is used, the switching losses are reduced almost by half, which is valid for both types of transistors. This is due to the fact that the transistor clamps the DC bus voltage at high phase currents, and switching loses are no longer present. When the 120° bus clamped method is used, the switching losses in both IGBT and SiC switches are decreased more than eight times, compared to the central PWM. The reason for this significant improvement in losses is as follows. Firstly, the 120BCM operates at a lower DC bus voltage compared to the other two modulation techniques. Secondly, the switches are clamped for a much longer period, which avoids commutating high currents. The third and final reason is that only one leg performs the modulation of all phase currents for a period of 60° at lower current values.

The obtained results may differ for different brands and types of switches. If a less expensive design is required, IGBT can be used while still delivering good performance. If SiC devices are used, switching losses are almost entirely eliminated and the heat sink will be much smaller. This is also the case for the DC link capacitor; therefore, the total converter benefits from the specific control strategy.

5. Experimental Setup and Results

Figure 9 presents the full system design of three-phase inverter; the main components of the system are power semiconductor switching, a gate driver circuit that is described in [23, 24], a three-phase filter, a DC power supply, three-phase voltage measurement [10], current measuring circuits, and finally the dsPIC33FG256GB710A as microcontroller. A 900V SiC MOSFET type C3M0065090D from CREE is chosen for the power switch with gate driver circuit [23, 25].

The dsPIC33FG256GB710A microcontroller with Explore 16 Kit that is presented in Figure 9 is used to generate the pulse width modulation and control the three-phase inverter under 120 BCM. The dsPIC is programmed by C language and MPLAB X IDEV3.15 software compiler for programing and debugging [26–28]. This controller is used due to low power consumption and having 9 pins for comparing PWM and 16 pins for 10-Bit A/D converter. The dsPIC is often used in variety of industrial electronic applications.

Figure 10 shows the upper and lower PWM waveforms under 120 BCM generated by dsPIC microcontroller (for

TABLE 5: Calculation results of the switching losses when different modulation techniques are used.

Modulation type	Central PWM	60° PWM	120BCM
RG4PH40UDPbF-IGBT			
Turn-on losses [W]	25.83	12.96	2.88
Turn-off losses [W]	33.01	17.05	3.96
Total losses [W]	58.84	30.01	6.84
Total inverter losses* [W]	353.07	180.06	41.04
C3M0065090D-SiC			
Turn-on losses [W]	3.59	1.80	0.40
Turn-off losses [W]	1.73	0.89	0.21
Total losses [W]	5.32	2.69	0.61
Total inverter losses *[W]	31.92	16.14	3.66

* **Sum of all switching losses by all six switches.**

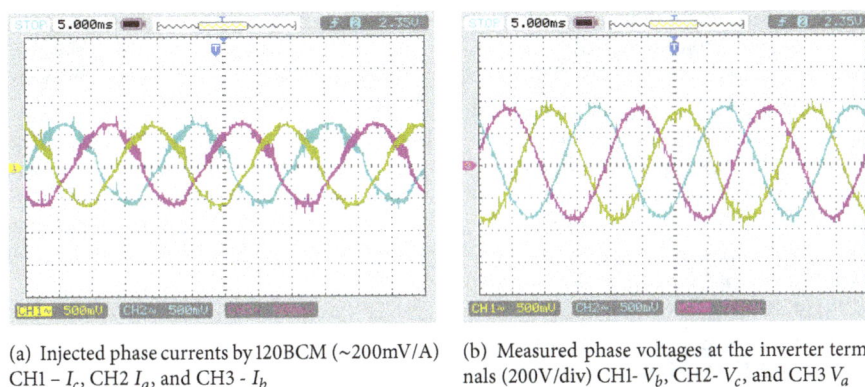

(a) Injected phase currents by 120BCM (~200mV/A) CH1 – I_c, CH2 I_a, and CH3 - I_b

(b) Measured phase voltages at the inverter terminals (200V/div) CH1- V_b, CH2- V_c, and CH3 V_a

FIGURE 11: Three-phase inverter output current and grid voltage.

all test waveforms, PhA-Yelow, PhB-Cyan, and PhC-Pink). The results of the signal detection that performs the partial modulation are depicted in Figure 10. As can be seen, every 60 degrees only one phase is under the PWM and the others are ON/OFF, respectively (upper and lower transistors are opposite under for each phase and PWM).

According to the system setup depicted in Figure 9, the expectable measurement results of three-phase inverter output currents and three-phase grid voltages are shown in the Figure 11. The measurements are taken from the current sensors of the inverter. In this particular application three Hall sensors type CASR 15 NP are used in two-turn configuration which results in an approximate sensitivity of 200 mV/A. The gain of the voltage probe used in this inverter is 1/200 which is enough to attenuate the phase voltages to the acceptable levels of the ADC of the microcontroller.

6. Conclusions

The performance of the proposed 120BCM was extensively verified by means of simulations. The simulation results showed that the 120BCM approach significantly reduces switching losses, compared to state-of-the-art PWM techniques. The advantage in switching losses is in the order of eight times compared to the centered PWM, and roughly four

times compared to the 60° bus clamped PWM, for switching losses proportional to current and frequency.

Moreover, the 120BCM is suitable for all types of transistor technologies: MOSFET, IGBT, SiC, and GaN. However, the loss amount will differ depending on the individual switch characteristics; in general, however, switching losses will be significantly reduced. It was also demonstrated by means of simulation that despite the partial modulation of 60 electrical degrees, the quality of the injected phase currents using independent controllers remains significantly below the acceptable level of 5%. Furthermore, the combination of a small nonelectrolytic capacitor, which ensures a long lifetime for the inverter, and improvements in efficiency makes the proposed bus clamped PWM control strategy extremely suitable for renewable energy applications.

Conflicts of Interest

The authors declare that there are no conflicts of interest regarding the publication of this paper.

Acknowledgments

The first author appreciates the Ministry of Higher Education and Scientific Research/IRAQ and Special Research of Ghent University for the financial support during this work.

References

[1] A. Van Den, J. Bossche, V. Bikorimana, and F. Feradov, "Reduced losses in PV converters by modulation of the DC link voltage," *International Journal of Energy and Power Engineering*, vol. 3, no. 3, pp. 125–131, 2014.

[2] J. M. V. Bikorimana UGent, A. Van ded Bossche, and J. B. Ndikubwimana, *Euro-African business partnership on the forefront: opportunities and challenges with case study of Photovoltaic converter business*, 2015.

[3] C. M. D. O. Stein, H. A. Gründling, H. Pinheiro, J. R. Pinheiro, and H. L. Hey, "Zero-current and zero-voltage soft-transition commutation cell for PWM inverters," *IEEE Transactions on Power Electronics*, vol. 19, no. 2, pp. 396–403, 2004.

[4] J.-G. Cho, J.-W. Baek, G.-H. Rim, and I. Kang, "Novel zero-voltage-transition PWM multiphase converters," *IEEE Transactions on Power Electronics*, vol. 13, no. 1, pp. 152–159, 1998.

[5] T. Mishima, "ZVS-PWM bridgeless active rectifier-applied GaN-HFET zero voltage soft-switching multi-resonant converter for inductive power transfers," in *Proceedings of the 32nd Annual IEEE Applied Power Electronics Conference and Exposition, APEC 2017*, pp. 3751–3758, USA, March 2017.

[6] H. S. Inverter, "A Novel Zero-Voltage-Switching Push – Pull," *IEEE Journal of Emerging and Selected Topics in Power Electronics*, vol. 4, no. 2, pp. 421–434, 2016.

[7] R. H. Ashique and Z. Salam, "A Family of True Zero Voltage Zero Current Switching (ZVZCS) Nonisolated Bidirectional DC-DC Converter with Wide Soft Switching Range," *IEEE Transactions on Industrial Electronics*, vol. 64, no. 7, pp. 5416–5427, 2017.

[8] A. Marinov and V. Valchev, "Power loss reduction in electronic inverters trough IGBT-MOSFET combination," in *Proceedings of the 6th International Conference on Mining Science and Technology, ICMST 2009*, pp. 1539–1543, China, October 2009.

[9] Y. Tang, Z. Qin, F. Blaabjerg, and P. C. Loh, "A Dual Voltage Control Strategy for Single-Phase PWM Converters with Power Decoupling Function," *IEEE Transactions on Power Electronics*, vol. 30, no. 12, pp. 7060–7071, 2015.

[10] M. J. Mnati, R. F. Chisab, and A. Van den Bossche, "A smart distance power electronic measurement using smartphone applications," in *Proceedings of the 2017 19th European Conference on Power Electronics and Applications, EPE 2017 ECCE Europe*, pp. 1–11, Jan, 2017.

[11] M. J. Mnati, A. Van den Bossche, and R. F. Chisab, "A smart voltage and current monitoring system for three phase inverters using an android smartphone application," *Sensors*, vol. 17, no. 4, 2017.

[12] D. Bozalakov, T. L. Vandoorn, B. Meersman, C. Demoulias, and L. Vandevelde, "Voltage dip mitigation capabilities of three-phase damping control strategy," *Electric Power Systems Research*, vol. 121, pp. 192–199, 2015.

[13] M. Farasat, A. Arabali, and A. M. Trzynadlowski, "Flexible-voltage DC-bus operation for reduction of switching losses in all-electric ship power systems," *IEEE Transactions on Power Electronics*, vol. 29, no. 11, pp. 6151–6161, 2014.

[14] G. Narayanan, H. K. Krishnamurthy, D. Zhao, and R. Ayyanar, "Advanced bus-clamping PWM techniques based on space vector approach," *IEEE Transactions on Power Electronics*, vol. 21, no. 4, pp. 974–984, 2006.

[15] H. Markiewicz and A. Klajn, "Voltage Disturbances," *Power Qual. Appl. Guid.*, vol. 5.4.2, pp. 4–11, 2004.

[16] D. V. Bozalakov, T. L. Vandoorn, B. Meersman, G. K. Papagiannis, A. I. Chrysochos, and L. Vandevelde, "Damping-Based Droop Control Strategy Allowing an Increased Penetration of Renewable Energy Resources in Low-Voltage Grids," *IEEE Transactions on Power Delivery*, vol. 31, no. 4, pp. 1447–1455, 2016.

[17] Y. Khersonksy, M. Robinson, and D. Gutierrez, "The Hexfred Ultrafast diode in power switching circuits," *Application Note, International Rectifier*, 1992.

[18] U. Nicolai and A. Wintrich, *Application Note AN 1403*, Semikron, 2014.

[19] A. D. Rajapakse, A. M. Gole, and P. L. Wilson, "Electromagnetic transients simulation models for accurate representation of switching losses and thermal performance in power electronic systems," *IEEE Transactions on Power Delivery*, vol. 20, no. 1, pp. 319–327, 2005.

[20] A. Shahin, A. Payman, J.-P. Martin, S. Pierfederici, and F. Meibody-Tabar, "Approximate novel loss formulae estimation for optimization of power controller of DC/DC converter," in *Proceedings of the 36th Annual Conference of the IEEE Industrial Electronics Society, IECON 2010*, pp. 373–378, USA, November 2010.

[21] H. Liu, J. Ma, H. Ma, and Z. Bai, "Improved switching loss calculation in neutral point clamped inverter via waveforms linearization," in *Proceedings of the 2013 IEEE 22nd International Symposium on Industrial Electronics, ISIE 2013*, pp. 1–6, Taipei, Taiwan, May 2013.

[22] D. Oustad, S. Lefebvre, M. Petit, D. Lhotellier, and M. Ameziani, "Comparison of modeling switching losses of an IGBT based on the datasheet and an experimentation," in *Proceedings of the 18th European Conference on Power Electronics and Applications, EPE 2016 ECCE Europe*, pp. 1–6, Germany, September 2016.

[23] M. J. Mnati, A. Van Den Bossche, and J. M. V. Bikorimana, "Design of half-bridge bootstrap circuit for grid inverter application controled by pic24fj128ga010," in *Proceedings of the 5th IEEE Int. Conf. Renew. ENERGY Res. Appl.*, pp. 85–89, Birmingham, UK, 2016.

[24] M. J. Mnati, A. Hasan, and A. V. Bossche, "An improved sic-mosfet gate driver circuit controlled by dspic33fj256gp710a," in *Proceedings of the 6th European Conference on Renewable Energy Systems*, Istanbul , Turkey, June 2018.

[25] N. E. Mode, "C2M1000170J Silicon Carbide Power MOSFET," no. 1, pp. 1–10, 2015.

[26] G. Purpose and F. Microcontrollers, *dsPIC33FJ256GP710A Family*, Microchip, 2012.

[27] D. Ibrahim, *PIC Microcontroller Projects in C Basic to Advanced*, 2nd Edition , 2014.

[28] Microchip Technology Inc., MPLAB ® X IDE User's Guide. 2012.

16

Operational Simulation of LC Ladder Filter using VDTA

Praveen Kumar,[1] **Neeta Pandey,**[2] **and Sajal Kumar Paul**[1]

[1]*Department of Electronics Engineering, Indian School of Mines, Dhanbad, India*
[2]*Department of Electronics and Communications, Delhi Technological University, Delhi, India*

Correspondence should be addressed to Neeta Pandey; n66pandey@rediffmail.com

Academic Editor: Jiun-Wei Horng

In this paper, a systematic approach for implementing operational simulation of LC ladder filter using voltage differencing transconductance amplifier is presented. The proposed filter structure uses only grounded capacitor and possesses electronic tunability. PSPICE simulation using 180 nm CMOS technology parameter is carried out to verify the functionality of the presented approach. Experimental verification is also performed through commercially available IC LM13700/NS. Simulations and experimental results are found to be in close agreement with theoretical predictions.

1. Introduction

Current mode approach has received a considerable attention in the last few years for analog signal processing applications due to their low power consumption, large dynamic range, higher frequency ranges of operation, better accuracy, higher slew rate, and less complexity. As a result, a large number of current mode active elements such as operational transconductance amplifier (OTA), current conveyor (CC), current controlled conveyor (CCC), current feedback amplifier (CFOA), operational transresistance amplifier (OTRA), differential voltage current conveyor (DVCC), current differencing buffered amplifier (CDBA), current differencing transconductance amplifier (CDTA), and voltage differencing transconductance amplifier (VDTA) are published. A literature review of such analog active block is presented in [1, 2]. The VDTA is a recently proposed analog building block composed of two transconductance amplifiers and may be used to implement different analog processing application such as floating and grounded inductor simulation [3, 4], analog filter [5–10], and oscillators [11–13].

For the active simulation of higher-order LC ladder filter, mainly three methods exist, which are wave active method, topological simulation, and operational simulation. In wave active approach, a wave equivalent is developed for inductor in series branch and then it is configured for other

passive components by making suitable connection [14–21]. Large numbers of active blocks are used in this approach. In the second method, topological simulation or element replacement method, the inductor of LC ladder structure is replaced by appropriate configured active elements [22, 23]. The drawback of this configuration is that a floating capacitor is generally required and this degrades the performance of the derived filter topology in high frequency application. In the third approach, operational simulation or leap-frog method [23–30], simulation is carried out for the operation of ladder rather than its component.

Literature survey reveals the operational simulation of ladder filter using operational amplifier (OA) and current controlled conveyor (CCCII) [24], OTA [25, 26], CC [27], multiple output second generation current controlled conveyor (MO-CCCII) [28], current feedback amplifier (CFA) [29], and CFOA [30]. This paper presents a systematic approach for operational simulation of LC ladder filter using voltage differencing transconductance amplifier (VDTA). The proposed operational simulation of LC ladder using VDTA has the following advantage over existing circuits:

(i) Lesser numbers of active blocks are used as compared to [24, 26, 28–30].

(ii) There is no use of resistors in realization, while [25, 29, 30] use both floating and grounded resistors and [27] uses only grounded resistors.

(iii) Only grounded capacitors are used in proposed implementation, while [25, 29] use floating capacitors too.

(iv) Proposed operational simulation of LC ladder also possesses electronic tunability of cut-off frequency, while [27, 29, 30] do not.

As an example, a fourth-order Butterworth low pass filter is simulated by outlined approach and the workability of the filter is confirmed through PSPICE simulation using 180 nm CMOS technology parameter. The functionality of the ladder filter is also tested experimentally through IC LM13700/NS.

2. VDTA

The voltage differencing transconductance amplifier is consisting of two transconductance amplifiers [5]. Figures 1 and 2 represent the symbolic representation and CMOS implementation of VDTA.

The port relationship of VDTA in matrix form is characterized by the following equation:

$$\begin{bmatrix} I_Z \\ I_{X+} \\ I_{X-} \end{bmatrix} = \begin{bmatrix} g_{mi} & -g_{mi} & 0 \\ 0 & 0 & g_{mo} \\ 0 & 0 & -g_{mo} \end{bmatrix} \begin{bmatrix} V_P \\ V_N \\ V_Z \end{bmatrix}, \qquad (1)$$

where g_{mi} and g_{mo} are the input and output transconductance gain of VDTA. The input transconductance amplifier

converts the input voltage difference $(V_P - V_N)$ into current at Z terminal and the voltage developed at Z terminal is converted into current at $X+$ and $X-$ terminal by output transconductance amplifier. In this paper, VDTA is used as an active analog building block because of

(i) the simple CMOS implementation of VDTA,

(ii) presence of two transconductance amplifiers giving resistorless realization,

(iii) the transconductance gain of VDTA which can vary via bias current, therefore providing the electronic tunability to designed filter.

3. Operational Simulation Using VDTA

The operational simulation method takes a different approach from topological simulation or wave active method, as it simulates the operation of ladder rather than its component [23]. The circuit equations and voltage-current relationship of each element are written using KVL and KCL. Then these equations are represented by block diagrams or signal flow graph. Each block represents some analog operation such as summation, integration, and subtraction. The final circuit is obtained by properly combining these blocks.

To explain the above statement, a fourth-order low pass Butterworth filter of Figure 3 has been taken as a prototype. The transfer function of this prototype filter can be expressed as

$$\frac{V_o}{V_{in}} \qquad\qquad (2)$$

$$= \frac{R_L}{s^4 C_1 C_2 L_1 L_2 R_L + s^3 \left(C_1 L_1 L_2 + C_1 C_2 L_2 R_s R_L\right) + s^2 \left(L_1 C_1 R_L + L_1 C_2 R_L + L_2 C_2 R_L + L_2 C_1 R_s\right) + s\left(R_s R_L C_1 + R_s R_L C_2 + L_1 + L_2\right) + \left(R_s + R_L\right)}.$$

To develop operational simulation in a systematic manner, consider the general ladder of Figure 4, where the series branch elements are labelled by admittance Y_i and the shunt branch elements are labelled by impedance Z_i. The ladder of Figure 4 can be described by the voltage and current equation as in (3a), (3b), (3c), and (3d) as follows:

$$I_1 = Y_1 \left(V_{in} - V_2\right), \qquad (3a)$$

$$V_2 = Z_1 \left(I_1 - I_3\right), \qquad (3b)$$

$$I_3 = Y_2 \left(V_2 - V_o\right), \qquad (3c)$$

$$V_o = Z_2 \left(I_3 - I_5\right); \qquad$$

$$\text{assume } I_5 = 0; \qquad (3d)$$

$$\text{then } V_o = Z_2 I_3,$$

where

$$Y_1 = \frac{1}{R_s + sL_1},$$

$$Y_2 = \frac{1}{sL_2},$$

$$Z_1 = \frac{1}{sC_1},$$

$$Z_2 = \frac{1}{sC_2 + 1/R_L}. \qquad (4)$$

Both voltage and current terms are present in (3a), (3b), (3c), and (3d). This problem can be easily resolved by scaling these equations by a resistor R_V.

$$R_V I_1 = R_V Y_1 \left(V_{in} - V_2\right) \Longrightarrow$$

$$V_{I1} = \frac{R_V}{R_s + sL_1} \left(V_{in} - V_2\right), \qquad (5a)$$

$$V_2 = \frac{Z_1}{R_V} \left(R_V I_1 - R_V I_3\right) \Longrightarrow$$

$$V_2 = \frac{1}{sC_1 R_V} \left(V_{I1} - V_{I3}\right), \qquad (5b)$$

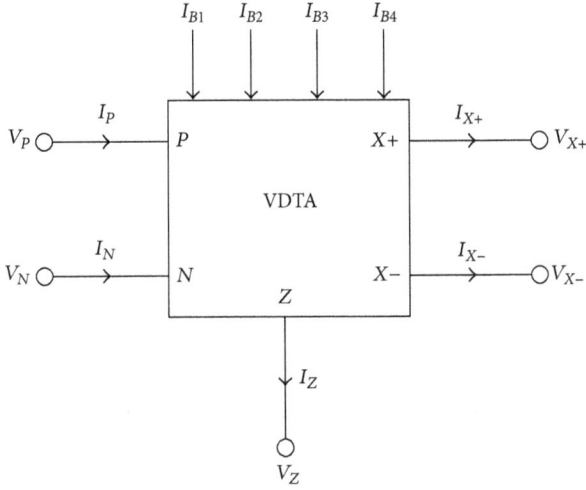

FIGURE 1: Symbolic representation of VDTA.

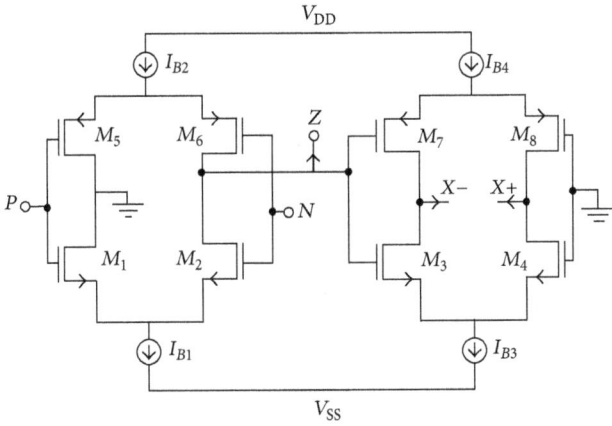

FIGURE 2: CMOS representation of VDTA.

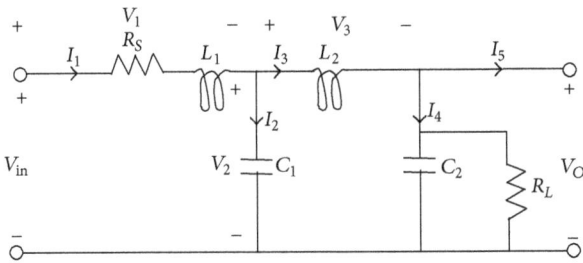

FIGURE 3: Fourth-order Butterworth low pass LC ladder.

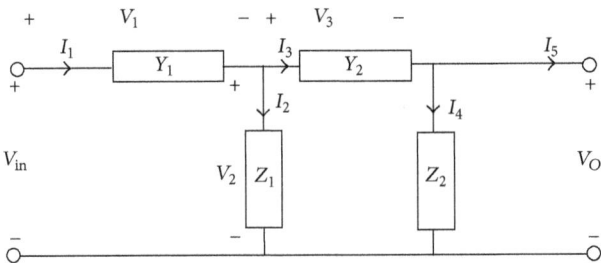

FIGURE 4: The ladder of Figure 3 with admittance in series arm and impedance in shunt arm.

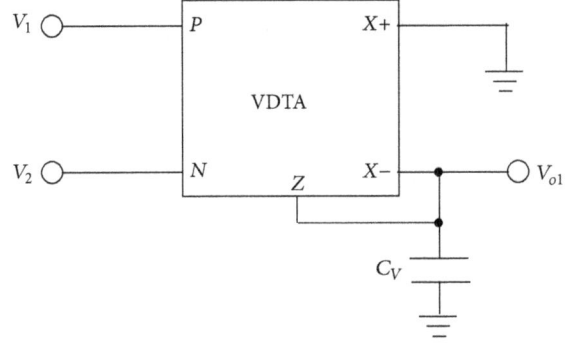

FIGURE 5: Lossy integrator using VDTA.

$$R_V I_3 = R_V Y_2 (V_2 - V_o) \implies$$

$$V_{I3} = \frac{R_V}{sL_2}(V_2 - V_o), \tag{5c}$$

$$V_o = \frac{Z_2}{R_V} R_V I_3 \implies$$

$$V_o = \frac{1}{sC_2 R_V + R_V/R_L} V_{I3}, \tag{5d}$$

where $V_{I1} = R_V I_1$; $V_{I3} = R_V I_3$.

The subscript I with voltages represents the fact that this voltage is derived from a current in the circuit.

Realization of (5a) to (5d) gives the operational simulation of prototype ladder filter of Figure 3. Implementation of (5a) and (5d) requires lossy integrator, while implementation of (5b) and (5c) requires lossless integrator. The lossy and lossless integrator can be easily realized using VDTA as discussed in the following section.

3.1. *Lossy Integration.* The implementation of lossy integration using VDTA is shown in Figure 5. The expression for output voltage of lossy integrator can be written as

$$V_{O1} = \frac{1}{1 + s\tau}(V_1 - V_2), \tag{6a}$$

where

$$\tau = \frac{C_V}{g_m} \quad (\text{with } g_{mi} = g_{mo} = g_m). \tag{6b}$$

3.2. *Lossless Integrator.* Lossless integrator can be implemented using VDTA as shown in Figure 6 and its output voltage expression is

$$V_{o2} = \frac{1}{s\tau}(V_1 - V_2). \tag{7a}$$

Again

$$\tau = \frac{C_V}{g_m} \quad (\text{with } g_{mi} = g_{mo} = g_m). \tag{7b}$$

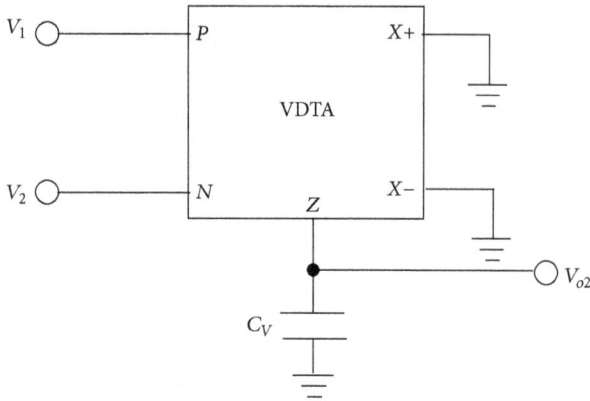

FIGURE 6: Lossless integrator using VDTA.

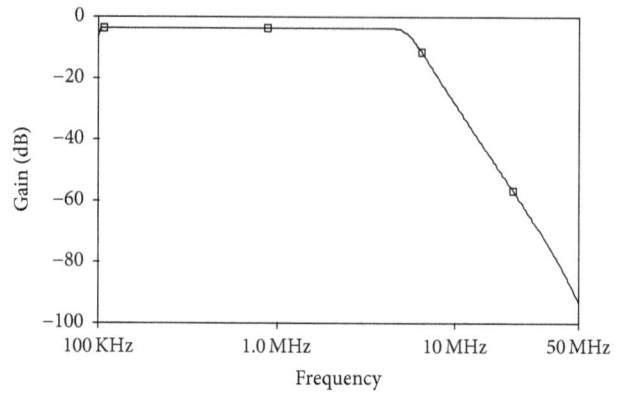

FIGURE 8: Simulated frequency response of 4th-order Butterworth low pass filter.

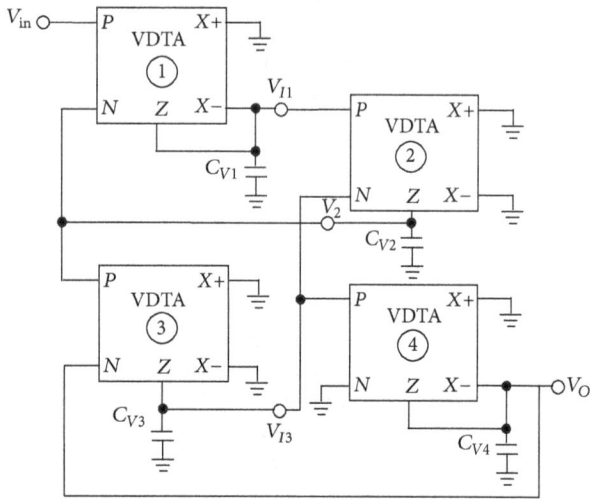

FIGURE 7: VDTA implementation of Figure 3 using operational simulation approach.

3.3. Complete Realization Using VDTA. With the help of lossy and lossless integrator of Figures 5 and 6, the complete realization of prototype 4th-order filter using operational simulation approach is shown in Figure 7.

The value of capacitor used in VDTA 1 and VDTA 4 can be calculated by comparing (6a) and (6b) with (5a) and (5d) as follows.

From (6a) and (6b) and (5a),

$$\frac{R_s}{R_V} = 1 \implies \tag{8}$$

$$R_s = R_V.$$

And $\tau = C_{V1}/g_m = L_1/R_V \implies C_{V1} = L_1 g_m/R_V$

Take the value of scaling resistor

$$R_V = \frac{1}{g_m}. \tag{9}$$

Then

$$C_{V1} = L_1 g_m^2. \tag{10}$$

And from (6a) and (6b) and (5d)

$$\frac{R_V}{R_L} = 1 \implies \tag{11}$$

$$R_L = R_V,$$

$$\tau = \frac{C_{V4}}{g_m} = C_2 R_V \implies \tag{12}$$

$$C_{V4} = C_2.$$

Similarly, the value of capacitor used in VDTA 2 and VDTA 3 can be calculated by comparing (7a) and (7b) with (5b) and (5c) as follows.

From (7a) and (7b) and (5b),

$$\tau = \frac{C_{V2}}{g_m} = C_1 R_V \implies \tag{13}$$

$$C_{V2} = C_1.$$

And from (7a) and (7b) and (5c),

$$\tau = \frac{C_{V3}}{g_m} = \frac{L_2}{R_V} \implies \tag{14}$$

$$C_{V3} = L_2 g_m^2.$$

4. Simulation

The normalized component values of the prototype filter of Figure 3 are $R_s = 1$, $L_1 = .7654$, $C_1 = 1.8485$, $L_2 = 1.8485$, $C_2 = .7654$, and $R_L = 1$. The aspect ratio of various transistor used in CMOS implementation of VDTA is given in Table 1. The values of supply voltage and bias current for VDTA are $V_{DD} = V_{SS} = -0.9$ V and $I_{B1} = I_{B2} = I_{B3} = I_{B4} = 150$ μA ($g_{mi} = g_{mo} = g_m = 627$ μS), respectively.

For cut-off frequency of 5 MHz, the values of capacitor used in Figure 7 can be calculated by (10), (12), (13), and (14) as $C_{V1} = 15.28$ pF, $C_{V2} = 36.9$ pF, $C_{V3} = 36.9$ pF, and $C_{V4} = 15.28$ pF. Figure 8 shows the frequency response of the low pass fourth-order Butterworth filter. The simulated cut-off

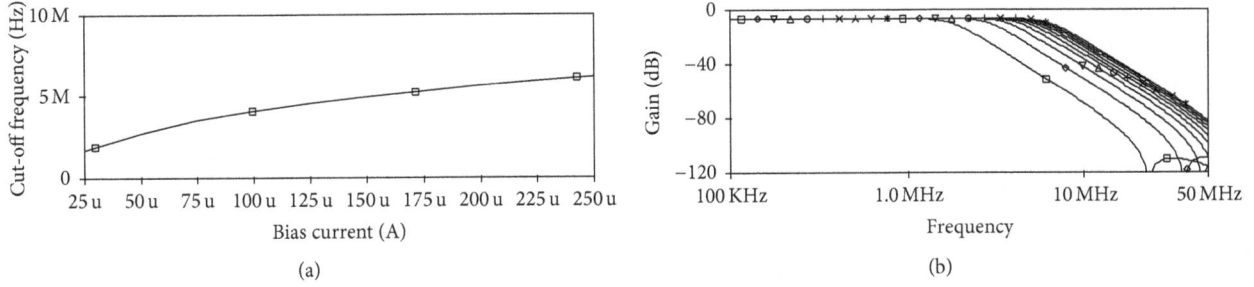

FIGURE 9: Electronic tuning demonstration. (a) Cut-off frequency variation with bias current. (b) Frequency response for various bias currents.

TABLE 1: Aspect ratio of various transistors used in CMOS implementation of VDTA.

Transistors	Aspect ratios (W (μm)$/L$ (μm))
M_1–M_4	3.6/.36
M_5–M_8	16.64/.36

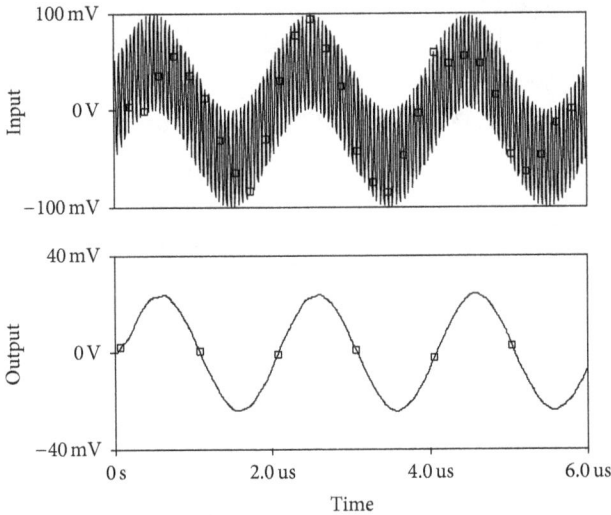

FIGURE 11: Frequency spectrum of input and output signals.

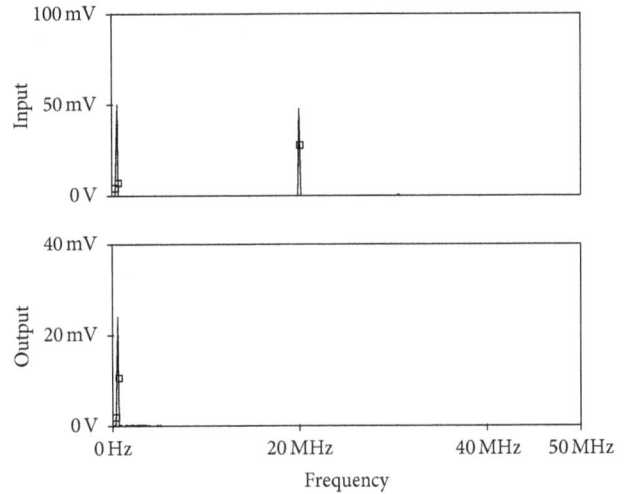

FIGURE 10: Transient response of input and output signals.

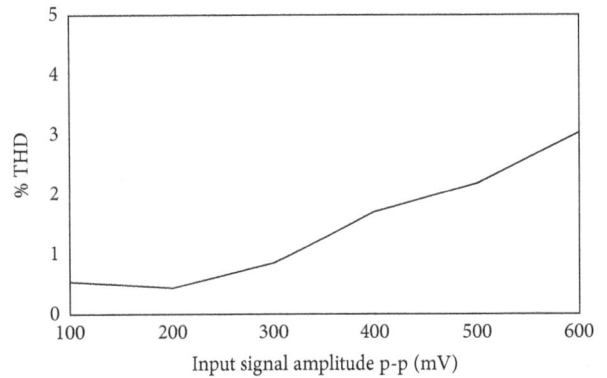

FIGURE 12: % THD variation with p-p input signal amplitude.

frequency is 4.99 MHz, which is very close to the theoretical cut-off frequency of 5 MHz. The electronic tunability of the filter through simulation is demonstrated in Figure 9 by varying bias current from 25 μA to 250 μA. Time domain analysis is studied by applying two signals of frequency 500 KHz and 20 MHz and of magnitude 50 mV at input. The transient response and its spectrum are shown in Figures 10 and 11, respectively. The proposed filter structure is also tested for total harmonic distortion at output and it is found that it is within acceptable limit of 3% up to 600 mV p-p signal of frequency 1 MHz as shown in Figure 12.

Noise analysis is also carried out for the proposed circuit by determining noise at output of the filter through simulation. The output noise variation within pass band frequencies is depicted in Figure 13 which shows that noise is in acceptable limit of nanovolt range. To examine effect of

temperature variation on proposed filter circuit, the circuit is simulated at five different temperatures, 10°C, 25°C, 27°C, 50°C, and 100°C, and the results are depicted in Figure 14. The values of cut-off frequency for these temperatures are listed in Table 2. It is observed that cut-off frequency shifts towards lower frequencies as temperature decreases. This is due to the fact that the transconductance decreases with increases in temperature due to decrease in mobility. This shifting in cut-off frequency can be compensated through bias current

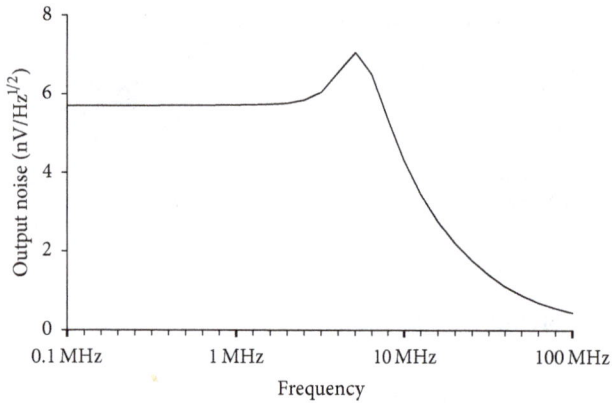

FIGURE 13: Output noise variation of proposed filter with frequency.

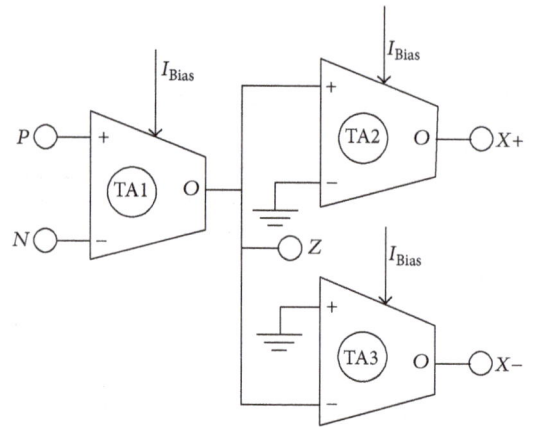

FIGURE 15: VDTA implantation using OTA.

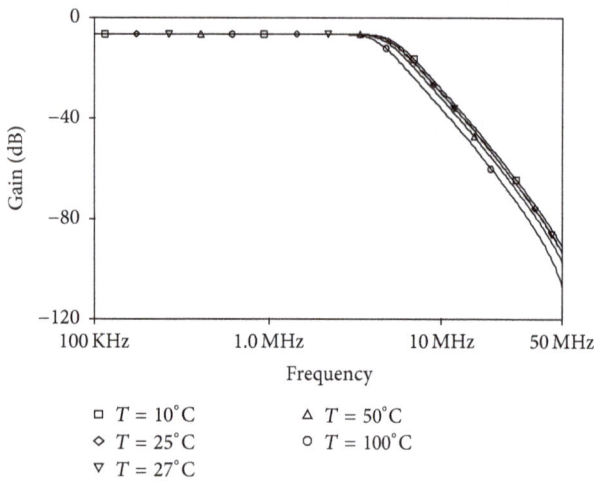

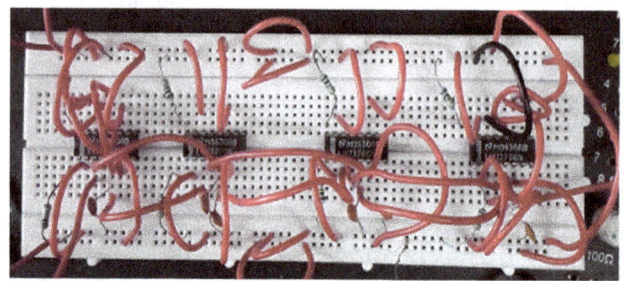

FIGURE 16: Bread-boarded circuit of Figure 7.

FIGURE 14: Demonstration of effect of temperature on proposed filter.

TABLE 3: Key parameters of simulated 4th-order low pass ladder filter.

Bias current	150 μA
VDTA transconductance, g_m	627 μS at bias current of 150 μA
Theoretical cut-off frequency	5 MHz
Simulated cut-off frequency	4.99 MHz
Roll-off rate	80 dB/decade
Total power consumption	2.16 mW
Total output noise voltage	5.7 nV/Hz$^{1/2}$
% THD	<3% for input signal up to 600 mV p-p

TABLE 2: Cut-off frequency at various temperatures.

Temperature	Cut-off frequency
10°C	5.2 MHz
25°C	5 MHz
27°C	4.98 MHz
50°C	4.7 MHz
100°C	4.17 MHz

variation from 104 μA (for f_0 = 4.17 MHz at 100°C) to 164 μA (for f_0 = 5.2 MHz at 10°C).

All the key parameters of the proposed filter structure are summarized in Table 3. The total power dissipated and output noise in simulation of the prototype filter are 2.16 mW and 5.7 × 10^{-9} V/Hz$^{1/2}$, while simulated values of these parameters for the VDTA implementation of the same-order filter using wave active method are 6.48 mW and 1.65 × 10^{-8} V/Hz$^{1/2}$ [20].

Experimental verification is carried out for proposed circuit through commercially available IC LM13700/NS. The

VDTA implementation using IC LM13700/NS is shown in Figure 15. The circuit of Figure 7 is bread-boarded as shown in Figure 16 for experimental testing. Supply voltage of ±15 V is used. The bias current of 1.35 mA is set to obtain the transconductance of 24.89 mA/V. The capacitor values are selected as $C_{v1} = C_{v4}$ = 10 nF and $C_{v2} = C_{v3}$ = 25 nF for cut-off frequency of 303 kHz. The measured magnitude response along with simulated response is depicted in Figure 17. The experimental cut-off frequency is observed to be 292 kHz.

5. Conclusion

The paper presents a systematic methodology for active implementation of operational simulation of LC ladder filter. To explain the outlined approach, a 4th-order Butterworth

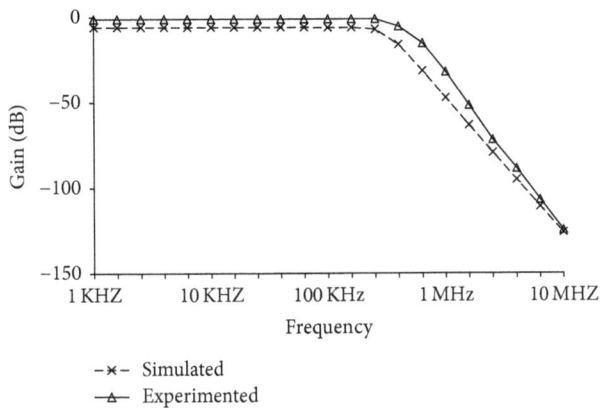

FIGURE 17: Simulated and experimented magnitude response of 4th-order low pass filter.

filter is taken as prototype, and, for active implementation, VDTA is used as an analog building block. The proposed implementation is resistorless and uses only grounded capacitors, which is suitable for IC implementation. The proposed structure also possesses electronic tunability of cut-off frequency. Workability of the proposed implementation is verified through PSPICE simulation using 180 nm TSMC technology parameters. The functionality of proposed LC ladder is also verified experimentally through IC LM13700/NS.

Competing Interests

The authors declare that they have no competing interests.

References

[1] K. K. Abdalla, D. R. Bhaskar, and R. Senani, "A review of the evolution of current-mode circuits and techniques and various modern analog circuit building blocks," *Nature and Science*, vol. 10, no. 10, 2012.

[2] D. Biolek, R. Senani, V. Biolkova, and Z. Kolka, "Active elements for analog signal processing: classification, review, and new proposals," *Radioengineering*, vol. 17, no. 4, pp. 15–32, 2008.

[3] D. Prasad and D. R. Bhaskar, "Grounded and floating inductance simulation circuits using VDTAs," *Circuits and Systems*, vol. 3, no. 4, pp. 342–347, 2012.

[4] W. Tangsrirat and S. Unhavanich, "Voltage differencing transconductance amplifier-based floating simulators with a single grounded capacitor," *Indian Journal of Pure and Applied Physics*, vol. 52, no. 6, pp. 423–428, 2014.

[5] A. Yeşil, F. Kaçar, and H. Kuntman, "New simple CMOS realization of voltage differencing transconductance amplifier and its RF filter application," *Radioengineering*, vol. 20, no. 3, pp. 632–637, 2011.

[6] A. Yeşil and F. Kaçar, "Electronically tunable resistorless mixed mode biquad filters," *Radioengineering*, vol. 22, no. 4, pp. 1016–1025, 2013.

[7] J. Satansup and W. Tangsrirat, "Compact VDTA-based current-mode electronically tunable universal filters using grounded capacitors," *Microelectronics Journal*, vol. 45, no. 6, pp. 613–618, 2014.

[8] D. Prasadl, D. R. Bhaskar, and M. Srivastava, "Universal voltage-mode biquad filter using voltage differencing transconductance amplifier," *Indian Journal of Pure and Applied Physics*, vol. 51, no. 12, pp. 864–868, 2013.

[9] J. Satansup, T. Pukkalanun, and W. Tangsrirat, "Electronically tunable current-mode universal filter using VDTAs and grounded capacitors," in *Proceedings of the International MultiConference of Engineers and Computer Scientists (IMECS '13)*, pp. 647–650, Hong Kong, China, March 2013.

[10] A. Uygur and H. Kuntman, "DTMOS-based 0.4V ultra low-voltage low-power VDTA design and its application to EEG data processing," *Radioengineering*, vol. 22, no. 2, pp. 458–466, 2013.

[11] D. Prasad, M. Srivastava, and D. R. Bhaskar, "Electronically controllable fully-uncoupled explicit current-mode quadrature oscillator using VDTAs and grounded capacitors," *Circuits and Systems*, vol. 4, no. 2, pp. 169–172, 2013.

[12] D. Prasad and D. R. Bhaskar, "Electronically Controllable Explicit Current Output Sinusoidal Oscillator Employing Single VDTA," *ISRN Electronics*, vol. 2012, Article ID 382560, 5 pages, 2012.

[13] R. Sotner, J. Jerabek, N. Herencsar, J. Petrzela, K. Vrba, and Z. Kincl, "Linearly tunable quadrature oscillator derived from LC Colpitts structure using voltage differencing transconductance amplifier and adjustable current amplifier," *Analog Integrated Circuits and Signal Processing*, vol. 81, no. 1, pp. 121–136, 2014.

[14] I. Haritantis, A. Constantinides, and T. Deliyannis, "Wave active filter," *Proceedings of the Institution of Electrical Engineers*, vol. 123, no. 7, pp. 676–682, 1976.

[15] K. Georgia and P. Costas, "Modular filter structures using CFOA," *Radio Engineering*, vol. 19, no. 4, pp. 662–666, 2010.

[16] N. Pandey and P. Kumar, "Realization of resistorless wave active filter using differential voltage current controlled conveyor transconductance amplifier," *Radioengineering*, vol. 20, no. 4, pp. 911–916, 2011.

[17] N. Pandey, P. Kumar, and J. Choudhary, "Current controlled differential difference current conveyor transconductance amplifier and its application as wave active filter," *ISRN Electronics*, vol. 2013, Article ID 968749, 11 pages, 2013.

[18] M. Bothra, R. Pandey, N. Pandey, and S. K. Paul, "Operational trans-resistance amplifier based tunable wave active filter," *Radioengineering*, vol. 22, no. 1, pp. 159–166, 2013.

[19] H. Singh, K. Arora, and D. Prasad, "VDTA-based wave active filter," *Circuits and Systems*, vol. 5, no. 5, pp. 124–131, 2014.

[20] N. Pandey, P. Kumar, and S. K. Paul, "Voltage differencing transconductance amplifier based resistorless and electronically tunable wave active filter," *Analog Integrated Circuits and Signal Processing*, vol. 84, no. 1, pp. 107–117, 2015.

[21] H. Wupper and K. Meerkotter, "New active filter synthesis based on scattering parameters," *IEEE Transaction on Circuit and System*, vol. 22, no. 7, pp. 594–602, 1975.

[22] A. A. M. Shkir, "10kHz, lpw power, 8th order eliptic band—pass filter employing CMOS VDTA," *International Journal of Enhanced Research in Science Technology & Engineering*, vol. 4, no. 1, pp. 162–168, 2015.

[23] M. E. Van Valkenburg and R. Shaumann, *Design of Analog Filters*, Oxford University Press, Oxford, UK, 2001.

[24] Y. Xi and H. Peng, "Realization of lowpass and bandpass leapfrog filters using OAs and CCCIIs," in *Proceedings of the International Conference on Management and Service Science (MASS '09)*, Wuhan, China, September 2009.

[25] M. V. Katageri, M. M. Mutsaddi, and R. S. Mathad, "Comparative study of LC ladder active filter using OTA and current conveyor," *International Journal of Advanced Computer and Mathematical Sciences*, vol. 3, no. 3, pp. 321–325, 2012.

[26] R. Schaumann, "Simulating lossless ladders with transconductance-C circuits," *IEEE Transactions on Circuits and Systems II: Analog and Digital Signal Processing*, vol. 45, no. 3, pp. 407–410, 1998.

[27] V. Novotny and K. Vrba, "LC ladder filter emulation by structures with current conveyor," in *Proceedings of the 4th WSEAS International Conference on Signal Processing, Computational Geometry & Artificial Vision (ISCGAV '04)*, Tenerife, Spain, December 2004.

[28] A. Câmpeanu and J. Gal, "LC-ladder filters emulated by circuits with current controlled conveyors and grounded capacitors," in *Proceedings of the International Symposium On Signals, Circuits and Systems (ISSCS '07)*, vol. 2, Iaşi, Romania, July 2007.

[29] T. S. Rathore and U. P. Khot, "CFA-based grounded-capacitor operational simulation of ladder filters," *International Journal of Circuit Theory and Applications*, vol. 36, no. 5-6, pp. 697–716, 2008.

[30] P. K. Sinha, A. Saini, P. Kumar, and S. Mishra, "CFOA based low pass and high pass ladder filter—a new configuration," *Circuits and Systems*, vol. 5, no. 12, pp. 293–300, 2014.

17

Impact of Band Nonparabolicity on Threshold Voltage of Nanoscale SOI MOSFET

Yasuhisa Omura

ORDIST, Grad. School of Sci. & Eng., Kansai University, 3-3-35 Yamate-cho, Suita, Osaka 564-8680, Japan

Correspondence should be addressed to Yasuhisa Omura; omuray@kansai-u.ac.jp

Academic Editor: Mingxiang Wang

This paper reconsiders the mathematical formulation of the conventional nonparabolic band model and proposes a model of the effective mass of conduction band electrons including the nonparabolicity of the conduction band. It is demonstrated that this model produces realistic results for a sub-10-nm-thick Si layer surrounded by an SiO_2 layer. The major part of the discussion is focused on the low-dimensional electron system confined with insulator barriers. To examine the feasibility of our consideration, the model is applied to the threshold voltage of nanoscale SOI FinFETs and compared to prior experimental results. This paper also addresses a model of the effective mass of valence band holes assuming the nonparabolic condition.

1. Introduction

In the last 3 decades, silicon-on-insulator (SOI) MOSFETs have been attracting attention because of their high short-channel effect immunity [1] and excellent potential with regard to future nanoscale devices [2]. Therefore, many studies on quantum confinement effects have been performed. The author predicted that the threshold voltage (V_{th}) of ultrathin-body (UTB) SOI MOSFET rises as SOI body thickness (t_S) is reduced due to the quantum-mechanical confinement effect [3]. Several studies have demonstrated experiments [4] that, they believe, demonstrate the effect of quantum confinement. However, the author demonstrated that the apparent rise of V_{th} of UTB SOI MOSFETs is due to more than the quantum effect; it includes the semiclassical effect [5]. Other experiments strongly suggested that a simple parabolic band model is not appropriate in analyzing V_{th} of nanoscale SOI MOSFETs [6, 7]; in these experiments, the threshold voltage was lower than the simulation values that assume the parabolic conduction band (X bands for Si) for $t_S < 5\,nm$. The possibility of determining the limits of the conventional parabolic band approximation is a motivation of this study. Recently, a couple of research articles

demonstrated the importance of band nonparabolicity in the analysis of the transport characteristics of nanoscale materials [8, 9], where the first principle calculation and the tight-binding method are used to compute the electronic states. In practical situations, however, demand an analytical closed form for the quantized energy levels and an effective mass tensor is requested for electronic device designs, since they greatly reduce the time costs.

This paper reconsiders the mathematical formulation of the conventional nonparabolic band model. This paper examines whether some perturbations can be added to the conventional model for convenience. In the following discussion, this paper focuses on a low-dimensional electron system confined by insulator barriers. We discuss the impact of the nonparabolic conduction band in Si on the effective mass and propose an analytical expression for the effective mass of electrons including the conduction band nonparabolicity. The model is applied to the threshold voltage of nanoscale SOI FinFETs, and its validity is examined. By examining the mathematical basis for the effective mass of electrons that have conduction band nonparabolicity, this paper also illuminates a model for the effective mass of holes having valence band nonparabolicity.

2. Modeling Nonparabolic Band Structure

When an isotropic band structure is assumed, it is conventionally known that its form can be expressed generally as [10]

$$E_{n\text{-par}}\left(1 + \alpha E_{n\text{-par}}\right) = E_{\text{par}} = \frac{\hbar^2 k^2}{2m^*}, \tag{1}$$

where m^* is the effective mass of electrons in the isotropic band, α is the nonparabolic band factor, $E_{n\text{-par}}$ is the energy of electrons in the nonparabolic band, and E_{par} is the electron energy expression in the form of the parabolic band scheme. Here we reconsider how the realistic effective mass for device analysis should be estimated; the above simplified nonparabolic band model is assumed for the conduction band electrons of a three-dimensional quasi-free electron system.

When an external field effect is taken account of, we must examine whether the following formulation is theoretically valid or not:

$$E_{n\text{-par}}\left(1 + \alpha E_{n\text{-par}}\right) = E_{\text{par}} + \beta E_{\text{ext}}, \tag{2}$$

where E_{ext} is the perturbation energy corresponding to the 1st-order perturbation generated by the external electric field and β is the perturbation factor. When (2) is valid, (2) can be rewritten as

$$E_{n\text{-par}} = \frac{\sqrt{1 + 4\alpha\left(E_{\text{par}} + \beta E_{\text{ext}}\right)} - 1}{2\alpha}. \tag{3}$$

In order to examine the availability of (3), the right-hand side of (3) is changed into Tayler's power series.

$$E_{n\text{-par}} = \frac{1}{2\alpha}\left\{\left(1 + \sum_{n=1}^{\infty} c_n\left(E_{\text{par}} + \beta E_{\text{ext}}\right)^n\right) - 1\right\}$$

$$= \frac{1}{2\alpha}\left\{\left(1 + 2\alpha\left(E_{\text{par}} + \beta E_{\text{ext}}\right)\right.\right. \tag{4}$$

$$\left.\left. + \sum_{n=2}^{\infty} c_n\left(E_{\text{par}} + \beta E_{\text{ext}}\right)^n\right) - 1\right\},$$

where c_n is the expansion coefficient and $c_1 = 2\alpha$. If (4) is meaningful, $E_{n\text{-par}}$ should be the eigenvalue of the corresponding Schrödinger's equation. That is, (4) should be equivalent to the following operator representation.

$$\widehat{H}_{n\text{-par}} = \frac{1}{2\alpha}\left\{\left(1 + 2\alpha\left(\widehat{H}_{\text{par}} + \beta\widehat{H}_{\text{ext}}\right)\right.\right.$$

$$\left.\left. + \sum_{n=2}^{\infty} c_n\left(\widehat{H}_{\text{par}} + \beta\widehat{H}_{\text{ext}}\right)^n\right) - 1\right\} = \widehat{H}_{\text{par}} + \widehat{H}', \tag{5}$$

where all suffice and correspond to those in (4) and $\widehat{H}'$ is the perturbation term that is expressed as

$$\widehat{H}' = \beta\widehat{H}_{\text{ext}} + \frac{1}{2\alpha}\sum_{n=2}^{\infty} c_n\left(\widehat{H}_{\text{par}} + \beta\widehat{H}_{\text{ext}}\right)^n. \tag{6}$$

When we can assume that the amplitude of $\widehat{H}'$ is smaller than that of $\widehat{H}_{\text{par}}$ [11], we approximately have

$$E_{n\text{-par}}(k) \approx E_{\text{par}}(k) - E_C + \beta E_{\text{ext}} + \frac{1}{2\alpha}\sum_{n=2}^{\infty} c_n\left(E_{\text{par}}(k)\right.$$

$$\left. - E_C + \beta E_{\text{ext}}\right)^n = \frac{1}{2\alpha}\left\{\left(1\right.\right.$$

$$+ 2\alpha\left(\left(E_{\text{par}}(k) - E_C + \beta E_{\text{ext}}\right)\right) \tag{7}$$

$$\left.\left. + \sum_{n=2}^{\infty} c_n\left(E_{\text{par}}(k) - E_C + \beta E_{\text{ext}}\right)^n\right) - 1\right\}$$

$$= \frac{\sqrt{1 + 4\alpha\left(\left(E_{\text{par}}(k) - E_C + \beta E_{\text{ext}}\right)\right)} - 1}{2\alpha}.$$

Equation (7) has the same form as (3). However, as the above formulation indicates, the meaning of (7) differs from that of (3) because (7) has been examined with the Hamiltonian operator representation in spite of the approximations used. Provided that prominent nonlinear effects are not significant, it can be concluded that (7) holds important physical meaning in terms of evaluating nonparabolic band effects on transport characteristics.

For the two-dimensional electron system, (7) can be rewritten as [11]

$$E_{n\text{-par},n}(k)$$

$$= \frac{\sqrt{1 + 4\alpha\left(\left(E_n(k_n) - E_C + \beta E_{\text{ext}}\right) + E_{\text{free}}(k_\parallel)\right)} - 1}{2\alpha}, \tag{8}$$

for the nth subband attributed to the specific conduction band, where E_n is the subband energy level of electrons, k_n is the discrete wavenumber of the subband labeled "n" in the quantum well, and $E_{\text{free}}(k_\parallel)$ is the energy component of the transport direction. Equation (8) is valid only for $E_1 - E_C \gg E_{\text{ext}}$ based on the same assumption used for (7). Actually, it is easily found that E_{ext} is much smaller than $E_1 - E_C$ for electric fields less than 10^7 V/cm and $t_S < 10$ nm [10]; therefore, (8) is applicable to the analysis of nanoscale MOSFET characteristics.

The result described above yields an expression for the effective mass tensor ($m^*_{n\text{-par},ij}$) for two-dimensional electron systems around the subband bottom as in [14]. Before using (8) for direct calculations, we review the conventional idea of the effective mass. We must recall the fact that electronic states with finite dimensions are inherently discrete; the difference from a very small size material is just the magnitude of Δk_i between adjacent states. Naive calculation discards the term of the group velocity, but this is erroneous in terms of physics. The correct calculation result is given as

$$
\frac{1}{m_{n\text{-par},ij}} = \frac{1}{\hbar^2} \frac{\partial^2 E_{n\text{-par},n}(k)}{\partial k_j \partial k_i} = \Big(1
$$
$$
+ 4\alpha \left(\left(E_j(k_j) - E_C + \beta E_{ext}\right) + E_{free}(k_\parallel)\right)\Big)^{-1/2}
$$
$$
\cdot \frac{1}{m_{par,ij}} - 2\alpha \Big(1 \tag{9}
$$
$$
+ 4\alpha \left(\left(E_j(k_j) - E_C + \beta E_{ext}\right) + E_{free}(k_\parallel)\right)\Big)^{-3/2}
$$
$$
\cdot v^c_{par,j} v^c_{par,i},
$$

where "j" means confinement direction and $v^c_{par,j}$ is the "*effective group velocity*" along the direction labeled by "j." Given that label "i" indicates the transport direction, we have

$$
v^c_{par,i} = \frac{1}{\hbar} \frac{\partial E_{free}(k_\parallel)}{\partial k_i}. \tag{10}
$$

Given that label "j" represents the confinement direction, we have

$$
v^c_{par,j} = \frac{1}{\hbar} \frac{\partial \left(E_j(k_j) + \beta E_{ext}\right)}{\partial k_j} = \frac{\hbar k_j}{m^*_{jj}} + \frac{1}{\hbar} \frac{\partial \beta E_{ext}}{\partial k_j}, \tag{11}
$$

where k_j is the discrete wavenumber and m^*_{jj} is the effective mass along the confinement direction. The 2nd term of the right-hand side is the perturbation term raised by the external field. When the semiconductor layer thickness is of the order of nanometers, the contribution of the 2nd term of (11) is quite small. Thus, the effective mass tensor ($m^*_{n\text{-par},ij}$) value of low-dimensional electron systems can be calculated around the subband bottom [15]; the important point is the fact that the effective mass of electrons is larger by the factor of $(1 + 4\alpha(E_j(0) - E_C + \beta E_{ext}))^{1/2}$ than that estimated assuming a parabolic band.

3. Calculation Results of Effective Mass of Low-Dimensionality Electrons

Assuming a thin Si layer, we calculated effective mass values (m_l/m_0 and m_t/m_0) for the (001) surface, where it is assumed that 2-fold and 4-fold X-band electrons are confined along the $\langle 001 \rangle$ axis. Physical parameters assumed in calculations are summarized in Table 1. Figure 1 shows the effective mass of electrons occupying the ground state as a function of Si layer thickness (t_S), where the nonparabolicity factor α is assumed to be $0.5\,\text{eV}^{-1}$ and, for comparison, 1st principle calculation results [8, 9] are also shown; the impact of band nonparabolicity on the effective mass of conduction band electrons appears when $t_S < 5$ nm [8, 9]. The present model successfully reproduces the 1st principle calculation results. Figure 2(a) shows the effective mass of electrons occupying the 1st excited state as a function of Si layer thickness. For comparison,

TABLE 1: Physical parameters assumed in calculations [12, 13].

Notations	Values	Comments
m_l^*/m_0	0.92	Longitudinal mass of electrons
m_t^*/m_0	0.19	Transverse mass of electrons
m_{hh}^*/m_0	0.49	Heavy holes
m_{lh}^*/m_0	0.16	Light holes
m_0	9.1×10^{-31} [kg]	Free electrons
n_i (@300 K)	9.7×10^9 [cm^{-3}]	Bulk Si
E_g (@300 K)	1.1 [eV]	
Permittivity of Si (ε_S)	$12\varepsilon_0$ [F/cm]	
Permittivity of SiO$_2$ (ε_{OX})	$3.8\varepsilon_0$ [F/cm]	
Permittivity of vacuum (ε_0)	8.9×10^{-14} [F/cm]	

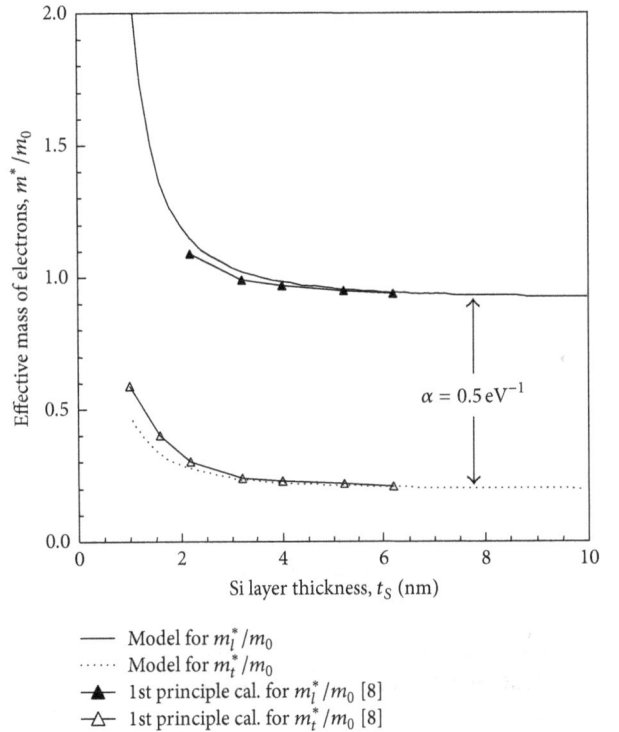

FIGURE 1: Calculated effective mass values of electrons occupying the ground state as a function of semiconductor layer thickness [11]. It is assumed that $\alpha = 0.5\,\text{eV}^{-1}$ in order to calculate the nonparabolic band effect.

1st principle calculation results [8, 9] are also shown in Figure 2(a). The present model does not reproduce the 1st principle calculation results, so the conventional value of α is not appropriate. We, therefore, varied the value of α until the calculated curves fitted the 1st principle calculation results. Figure 2(b) shows the calculation results of the effective mass value of conduction band electrons occupying the 1st excited state, including the nonparabolicity effect, where it is assumed that $\alpha = 0.1\,\text{eV}^{-1}$ for m_l/m_0 and $\alpha = 0.05\,\text{eV}^{-1}$ for m_t/m_0. It

(a)   (b)

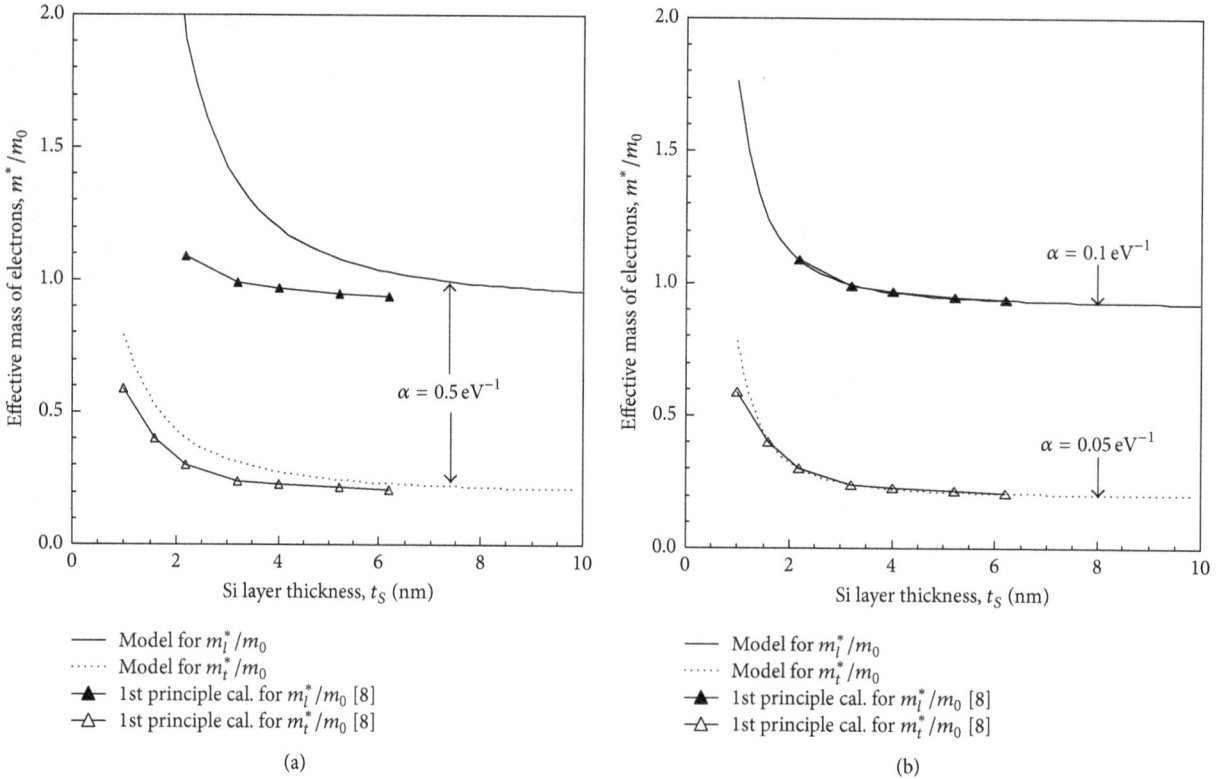

FIGURE 2: Calculated effective mass values of electrons occupying the 1st excited state as a function of semiconductor layer thickness [11]. (a) Effective mass dependence on Si layer thickness. It is assumed that $\alpha = 0.5\,\text{eV}^{-1}$ in order to calculate the nonparabolic band effect. (b) Effective mass dependence on Si layer thickness. It is assumed that $\alpha = 0.1\,\text{eV}^{-1}$ for m_l/m_0 and $\alpha = 0.05\,\text{eV}^{-1}$ for m_t/m_0 in calculating the nonparabolic band effect.

is seen that the present model for effective mass successfully reproduces the theoretical simulation results. It is strongly suggested that the reason why the value of α for the 1st excited state is smaller than expected stems from the approximation that stripped higher-order expansion terms from (7).

4. Applying Effective Mass Model to Express Threshold Voltage of Nanoscale MOSFETs

Since the threshold voltage (V_{th}) of ultrathin-body SOI MOSFETs is effectively ruled by the lowest energy level [3], it is easily anticipated that the impact of band nonparabolicity on V_{th} is very significant. The author already clarified that the rise in threshold voltage stems from not only the quantum-mechanical mechanism [3] but also the semiclassical mechanism [5]. When it is assumed that the threshold voltage rising stems from the quantum confinement [3, 6, 15], V_{th} for ultrathin-body n-channel SOI MOSFET can be expressed as [3]

$$V_{\text{th}} = V_{\text{FB}} + \phi_{\text{sth}} + \frac{Q_B}{C_{\text{ox}}}, \tag{12a}$$

$$\phi_{\text{sth}} = \phi_{F,QM} + \left(\frac{E_G}{2q}\right) + \frac{E_{n1} - E_C}{q}, \tag{12b}$$

where V_{FB} is the flat-band voltage, C_{ox} is the gate oxide capacitance per unit area, Q_B is the depletion charge density, $\phi_{F,QM}$ is the "effective Fermi potential" including quantum effects (defined later), E_G is the bandgap energy, and E_{n1} is the ground-state level energy of confined electrons in the conduction band. Fundamentally, the above expression for the threshold voltage of ultrathin-body SOI MOSFETs should yield the threshold voltage at low temperatures because the Fermi-Dirac function has a step-like function at low temperatures. At room temperature, the Fermi-Dirac function is thermally deformed, and the Fermi level does not simply define the threshold voltage; finite inversion layer charges are needed at the onset of the threshold [5]. From Boltzmann's approximation, at room temperature we have [16]

$$\begin{aligned}\phi_{F,QM} &= \frac{E_i - E_F}{q} \\ &= \frac{1}{q}\left\{E_i - E_{p1} + k_B T \ln\left(\frac{N_A t_S}{q D_{\text{osp}} k_B T}\right)\right\},\end{aligned} \tag{13}$$

where D_{osp} is the density of states of two-dimensional valence-band holes, E_{p1} is the ground-state level energy of confined holes in the valence band, and N_A is the doping density; this expression is available for $E_a - E_{p1} \gg k_B T$, where E_a is the acceptor level energy. This expression is approximately derived from the following expression for intrinsic carrier density (n_i^*), which should be called "*effective intrinsic carrier density*", in the quantum well defined at room temperature [16].

$$n_i^* (T) = \left\{ \frac{\left(D_{osn} D_{osp} \right)^{1/2} k_B T}{t_S} \right\} \exp \left[-\frac{E_G^*}{2k_B T} \right], \quad (14)$$

where D_{osn} is the density of states of two-dimensional conduction band electrons; E_G^* is the "*effective bandgap energy*" of the quantum well and can be expressed as

$$E_G^* = E_{n1} - E_{p1}. \quad (15)$$

In calculating V_{th} of a p-channel SOI MOSFET, we can also apply the present nonparabolic band model to holes (see the Appendix section).

On the other hand, V_{th} for a quantum-wire SOI fin field-effect transistor (FinFET) is expressed as [15, 17]

$$\phi_{sth} = \phi_{F,QM} + \left(\frac{E_G}{2q} \right) + \frac{E_{n,ij} - E_C}{q}, \quad (16a)$$

$$E_{n,ij} - E_C = \frac{h^2}{8} \left(\frac{i^2}{m^*_{nz,i} t_S^2} + \frac{j^2}{m^*_{nx,j} w_S^2} \right), \quad (16b)$$

where h is Planck's constant, t_S is fin height, w_S is fin width, $E_{n,ij}$ is the quantum level energy, and $m^*_{nx,j}$ (or $m^*_{nz,i}$) is the effective mass of electrons occupying the corresponding subband. In (16b), it is assumed that the Si wire is confined along x-axis and z-axis.

Calculated V_{th} dependence on fin width of SOI FinFET two-dimensionally confined with (001) and (011) surfaces [6] and on SOI layer thickness of thin SOI MOSFET [7] is shown in Figure 3, where the top surface of the Si fin body is (001) and the side surface is (011). Physical parameters assumed in the calculations are summarized in Table 1. In this calculation of threshold voltage of FinFET, it is assumed that electrons of the Si body can be represented as quasi-two-dimensional system because the value of t_S is larger than the minimal value of w_S. The conventional parabolic band model overestimates the quantum confinement effect in the range defined by sub-5-nm t_S. This suggests that the nonparabolicity should be taken into account when estimating the V_{th} values of nanoscale SOI MOSFETs. On the other hand, experimental data of FinFET for $t_S > 5$ nm take values higher than two curves. It is anticipated, as one possibility, that fabrication-induced local variation of SOI layer thickness influences the threshold voltage of devices [6, 7, 18].

Ge-on-insulator (GOI) devices are now attracting attention from the viewpoints of high-speed device applications. However, it is known that Ge demonstrates stronger conduction band nonparabolicity than Si. Therefore, the discussion given here is critical when considering nanoscale GOI device characteristics [19].

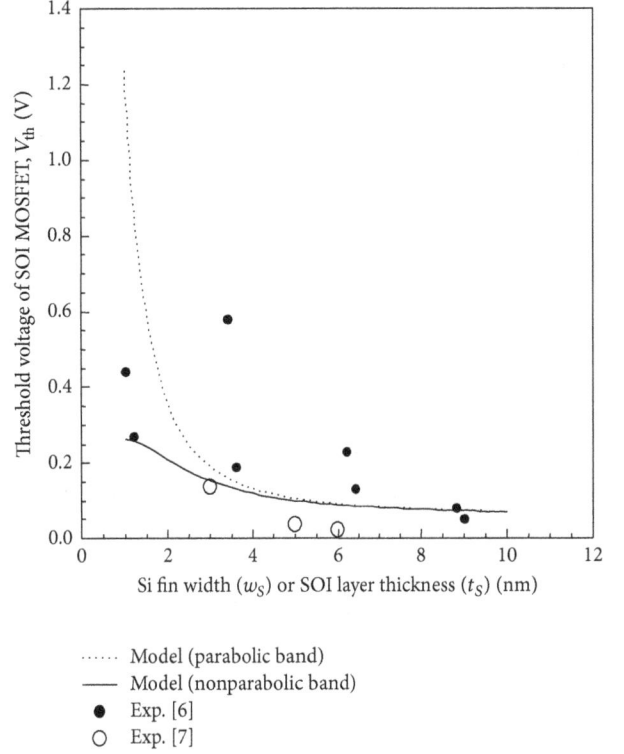

FIGURE 3: V_{th} dependence on fin width or SOI layer thickness of Si SOI MOSFET device. The devices shown in [6] are two-dimensionally confined with (001) and (011) surfaces. It is assumed that the gate oxide layer is 3 nm thick, the body silicon layer is 7 nm thick for [6], and $\alpha = 0.5\,\text{eV}^{-1}$. The devices shown in [7] are two-dimensionally confined with (001) surface. It is assumed that the gate oxide layer is 50 nm thick, the body silicon layer is ranging from 3 nm to 6 nm thick for [7], and $\alpha = 0.5\,\text{eV}^{-1}$. It is also assumed that $m^*/m_0 = 0.916$ for (001) Si surface and $m^*/m_0 = 0.314$ for (011) Si surface.

5. Conclusion

This paper reconsidered the mathematical formulation of the conventional nonparabolic band model. Since the conventional simplified model for band nonparabolicity does not include the perturbation created by the external potential effect, we examined whether such perturbations could be added to the conventional model for convenience. When the perturbation energy is smaller than the unperturbed energy, the insertion of a perturbation term into the conventional expression for the nonparabolic band model is valid; it was confirmed that this approximation is acceptable given a sub-10-nm-thick Si layer surrounded by an SiO_2 layer. The major discussion focused on the low-dimensional electron system confined by an insulator barrier. For the purpose of verifying this consideration, we addressed the influence of band nonparabolicity on the threshold voltage of Si-based nanoscale SOI FinFETs; calculation results yielded by the proposed model were compared to experimental results and the validity of the model proposed here was confirmed.

Appendix

Modeling Nonparabolicity of Valence Band Holes

In the valence band, it is well known that the nonparabolic band energy dispersion relation of holes can be approximately expressed as [10]

$$E_{\text{hole,nonpara}}(k) = \frac{\hbar^2}{2m_0}\left(Ak^2\right.$$
$$\left.\pm\left\{B^2k^4 + C^2\left(k_x^2k_y^2 + k_y^2k_z^2 + k_z^2k_x^2\right)\right\}^{1/2}\right), \tag{A.1}$$

where A, B, and C are constants with values of 4.22, 0.78, and 4.80, respectively [10]. When the system is confined along

z-axis, k_z is discrete and its notation should be replaced with $k_{z,n}$ ($n = 1, 2, 3\ldots$). Equation (A.1) must be replaced with the following:

$$E_{\text{2D-hole,nonpara}} = \frac{\hbar^2}{2m_0}\left(A\left(k_x^2 + k_y^2 + k_{z,n}^2\right)\right.$$
$$\pm\left\{B^2\left(k_x^2 + k_y^2 + k_{z,n}^2\right)^2\right.$$
$$\left.\left.+ C^2\left(k_x^2k_y^2 + k_y^2k_{z,n}^2 + k_{z,n}^2k_x^2\right)\right\}^{1/2}\right). \tag{A.2}$$

Therefore, the effective mass of holes ($m_{h,zz}^*$) in the valence band can be expressed as

$$\frac{1}{m_{h,zz}^*\left(k_x, k_y, k_{z,n}\right)} = \frac{1}{\hbar^2}\frac{\partial^2 E_{\text{2D-hole,nonpara}}(k)}{\partial k_{z,n}^2} = \frac{1}{m_0}\left(A \mp \frac{\left\{2B^2\left(k_x^2 + k_y^2 + k_{z,n}^2\right)k_{z,n} + C^2\left(k_y^2 + k_x^2\right)k_{z,n}\right\}^2}{2\left\{B^2\left(k_x^2 + k_y^2 + k_{z,n}^2\right)^2 + C^2\left(k_x^2k_y^2 + k_y^2k_{z,n}^2 + k_{z,n}^2k_x^2\right)\right\}^{3/2}}\right.$$
$$\left.\pm \frac{2B^2\left(k_x^2 + k_y^2 + 3k_{z,n}^2\right) + C^2\left(k_y^2 + k_x^2\right)}{2\left\{B^2\left(k_x^2 + k_y^2 + k_{z,n}^2\right)^2 + C^2\left(k_x^2k_y^2 + k_y^2k_{z,n}^2 + k_{z,n}^2k_x^2\right)\right\}^{1/2}}\right). \tag{A.3}$$

At the bottom of valence band, $k_x = k_y = 0$, and we have

$$\frac{1}{m_{h,zz}^*\left(0, 0, k_{z,n}\right)} = \frac{1}{m_{h,zz}^*} = \frac{1}{m_0/(A \pm B)}. \tag{A.4}$$

From (A.4), we have $m_{h,zz}^* = 0.20m_0$ for the light hole and $m_{h,zz}^* = 0.29m_0$ for the heavy hole. Thus, when we use those effective mass values, the ground-state level energy $E_{p,1}$ is given by

$$E_V - E_{p,1} = \frac{\hbar^2\pi^2}{2m_{h,zz}^*t_S^2}. \tag{A.5}$$

Equation (A.4) is an approximate expression for the effective mass of holes without any t_S dependence. However, this is not an appropriate result because $m^*_{h,zz}$ values are not accurate for both light holes and heavy holes as shown above. Accordingly, we have to derive a new t_S-dependent expression for the effective mass of holes.

Our solution is to rewrite (A.1) as a function of parabolic band energy $E_{\text{holes,para}(k)}$:

$$E_{\text{hole,nonpara}}(k) = \frac{m_{h,\text{para}}^* E_{\text{hole,para}}(k)}{m_0}\left(A \pm \left\{B^2\right.\right.$$
$$+ \frac{C^2}{2}\left(1\right.$$
$$\left.\left.\left.- \frac{E_{x,\text{hole,para}}^2 + E_{y,\text{hole,para}}^2 + E_{z,\text{hole,para}}^2}{E_{\text{hole,para}}^2(k)}\right)\right\}^{1/2}\right), \tag{A.6}$$

where we assume that

$$E_{\text{hole,para}}(k) = \frac{\hbar^2k^2}{2m_{h,\text{para}}^*} \tag{A.7}$$
$$= E_{x,\text{hole,para}} + E_{y,\text{hole,para}} + E_{z,\text{hole,para}}.$$

Since we have $(E_{x,\text{hole,para}}^2 + E_{y,\text{hole,para}}^2 + E_{z,\text{hole,para}}^2)/E_{\text{hole,para}}^2(k) \approx 1/3$ in (6), (6) can be approximately rewritten as

$$E_{\text{hole,nonpara}}(k) \cong \frac{m_{h,\text{para}}^* E_{\text{hole,para}}(k)}{m_0}\left(A\right.$$
$$\pm\sqrt{B^2 + \frac{C^2}{2}} \mp \frac{1}{\sqrt{B^2 + C^2/2}}$$
$$\left.\cdot\frac{E_{x,\text{hole,para}}^2 + E_{y,\text{hole,para}}^2 + E_{z,\text{hole,para}}^2}{2E_{\text{hole,para}}^2(k)}\right). \tag{A.8}$$

Here, we consider the case of a two-dimensional hole system in which the confinement is along z-axis; from (A.7), we have

$$E_{z,\text{2D-hole,para}}\left(k_{z,n}\right) = \frac{\hbar^2k_{z,n}^2}{2m_{h,\text{para}}^*} \equiv \frac{\hbar^2n^2\pi^2}{2m_{h,\text{para}}^*t_S^2}, \tag{A.9}$$
$$\left(n = 1, 2, 3\ldots\right),$$

where a parabolic band hole system is assumed for simplicity; it is assumed that $m_{h,\text{para}}^*$ is the parabolic band mass of holes;

that is, $m_{h,\text{para}}^{*} = \left(m_{lh}^{*\,3/2} + m_{hh}^{*\,3/2}\right)^{2/3}$. Combining (A.8) with (A.9) yields the following expression for the two-dimensional hole system.

$$E_{2D\text{-hole,nonpara}}(k) \cong \frac{m_{h,\text{para}}^{*} E_{2D\text{-hole,para}}(k)}{m_0} \left(A \right.$$

$$\pm \sqrt{B^2 + \frac{C^2}{2}} \mp \frac{1}{\sqrt{B^2 + C^2/2}} \qquad \text{(A.10)}$$

$$\left. \cdot \frac{E_{x,\text{hole,para}}^2 + E_{y,\text{hole,para}}^2 + E_{z,2D\text{-hole,para}}^2}{2E_{2D\text{-hole,para}}^2(k)} \right).$$

In a practical two-dimensional system, carriers are usually confined by barriers and the external field (F_{ext}). Therefore, we have to introduce a perturbation (E_{ext}) to represent the effect of the external field into (A.10). Accordingly, we have

$$E_{2D\text{-hole,nonpara}}(k) \cong \frac{m_{h,\text{para}}^{*} E_{2D\text{-hole,para}}(k)}{m_0} \left(K_1 \mp K_2 \right.$$

$$\left. \cdot \frac{E_{x,\text{hole,para}}^2 + E_{y,\text{hole,para}}^2 + \left(E_{z,2D\text{-hole,para}} + E_{\text{ext}}\right)^2}{2E_{2D\text{-hole,para}}^2(k)} \right),$$

$$E_{2D\text{-hole,para}}(k) = E_{x,\text{hole,para}} + E_{y,\text{hole,para}} + E_{z,2D\text{-hole,para}} \quad \text{(A.11)}$$

$$+ E_{\text{ext}},$$

$$K_1 = A \pm \sqrt{B^2 + \frac{C^2}{2}},$$

$$K_2 = \frac{1}{\sqrt{B^2 + C^2/2}},$$

where E_{ext} stands for the external energy stemming from the external field (F_{ext}). This expression is physically reasonable when $E_{\text{ext}} \ll E_{z,2D\text{-hole,para}}$ [20].

Competing Interests

The author declares that there are no competing interests.

Acknowledgments

A part of this study was partially conducted by MEXT-Supported Program for the Strategic Research Foundation at Private Universities, "creation of 3d nano-micro structures and its applications to biomimetics and medicine," 2015–2019.

References

[1] J. P. Colinge, *Silicon-on-Insulator Technology: Materials to VLSI*, Springer, New York, NY, USA, 3rd edition, 2004.

[2] J.-P. Colinge, Ed., *FinFETs and Other Multi-Gate Transistors*, Springer, New York, NY, USA, 2008.

[3] Y. Omura, S. Horiguchi, M. Tabe, and K. Kishi, "Quantum-mechanical effects on the threshold voltage of ultrathin-SOI nMOSFET's," *IEEE Electron Device Letters*, vol. 14, no. 12, pp. 569–571, 1993.

[4] Y. Ishikawa, T. Ishihara, T. Tsuchiya, and M. Tabe, "XPS and I-V studies on quantum mechanical effects in ultrathin si layer of SOI structure," in *Proceedings of the Abstract Silicon Nanoelectronics Workshop*, pp. 14–15, Kyoto, Japan, June 2001.

[5] Y. Tamara and Y. Omura, "Empirical quantitative modeling of threshold voltage of sub-50-nm double-gate silicon-on-insulator metal-oxide-semiconductor field-effect transistor," *Japanese Journal of Applied Physics, Part 1: Regular Papers and Short Notes and Review Papers*, vol. 45, no. 4, pp. 3074–3078, 2006.

[6] H. Majima, H. Ishikuro, and T. Hiramoto, "Experimental evidence for quantum mechanical narrow channel effect in ultra-narrow MOSFET's," *IEEE Electron Device Letters*, vol. 21, no. 8, pp. 396–398, 2000.

[7] T. Ernst, S. Cristoloveanu, G. Ghibaudo et al., "Ultimately thin double-gate SOI MOSFETs," *IEEE Transactions on Electron Devices*, vol. 50, no. 3, pp. 830–838, 2003.

[8] K. Nehari, N. Cavassilas, J. L. Autran, M. Bescond, D. Munteanu, and M. Lannoo, "Influence of band structure on electron ballistic transport in silicon nanowire MOSFET's: an atomistic study," *Solid-State Electronics*, vol. 50, no. 4, pp. 716–721, 2006.

[9] P. V. Sushko and A. L. Shluger, "Electronic structure of insulator-confined ultra-thin Si channels," *Microelectronic Engineering*, vol. 84, no. 9-10, pp. 2043–2046, 2007.

[10] B. K. Ridley, *Quantum Processes in Semiconductors*, Clarendon, Oxford, UK, 2nd edition, 1988.

[11] Y. Omura, "Extension of analytical model for conduction band non-parabolicity to transport analysis of nano-scale metal-oxide-semiconductor field-effect transistor," *Journal of Applied Physics*, vol. 105, no. 1, p. 014310, 2009.

[12] S. M. Sze and K. K. Ng, *Physics of Semiconductor Devices*, Wiley InterScience, 3rd edition, 2007.

[13] C. Jacoboni and L. Reggiani, "The Monte Carlo method for the solution of charge transport in semiconductors with applications to covalent materials," *Reviews of Modern Physics*, vol. 55, no. 3, pp. 645–705, 1983.

[14] S. Jin, M. V. Fischetti, and T.-W. Tang, "Modeling of electron mobility in gated silicon nanowires at room temperature: surface roughness scattering, dielectric screening, and band nonparabolicity," *Journal of Applied Physics*, vol. 102, no. 8, Article ID 083715, 2007.

[15] Y. Omura, T. Ishiyama, M. Shoji, and K. Izumi, "Quantum Mechanical Transport Characteristics in Ultimately Miniaturized MOSFETs/SIMOX," in *Proceedings of the 10th International Symposium on SOI Technology and Development*, vol. PV96-3, pp. 199–205, Electrochemical Society, March 1996.

[16] Y. Omura, *Soi Lubistors*, chapter 10, IEEE/Wiley, 1st edition, 2013.

[17] R. Granzner, F. Schwierz, and V. M. Polyakov, "An analytical model for the threshold voltage shift caused by two-dimensional quantum confinement in undoped multiple-gate MOSFETs," *IEEE Transactions on Electron Devices*, vol. 54, no. 9, pp. 2562–2565, 2007.

[18] Y. Omura and M. Nagase, "Low-temperature drain current characteristics in sub-10-nm-thick SOI nMOSFET's on SIMOX (separation by IMplanted OXygen) substrates," *Japanese Journal of Applied Physics, Part 1: Regular Papers & Short Notes & Review Papers*, vol. 34, no. 2B, pp. 812–816, 1995.

[19] L. Pantisano, L. Trojman, J. Mitard et al., "Fundamentals and extraction of velocity saturation in Sub-100 nm (110)-Si and (100)-G," in *Proceedings of the Abstract International Symposium on VLSI Technology, Systems, and Applications*, pp. 52–53, 2008.

[20] Y. Omura, "Modeling hole effective mass of Si modulated by external field," in *Proceedings of the 15th Silicon Nanoelectronics Workshop (SNW '10)*, pp. 65–66, Honolulu, Hawaii, USA, June 2010.

18

Abnormal Capacitance Increasing at Elevated Temperature in Tantalum Capacitors with PEDOT:PSS Electrodes

Qifeng Pan ⓘ,[1,2] Qiao Liu ⓘ,[1] Yuanjiang Yang,[2] and Dongbin Tian[2]

[1]*College of Big Data and Information Engineering, Guizhou University, P.O. Box 550025, Guiyang, Guizhou, China*
[2]*Xinyun Electronics Components Corporation, P.O. Box 550018, Guiyang, Guizhou, China*

Correspondence should be addressed to Qiao Liu; liuqiao1955@163.com

Academic Editor: Gerard Ghibaudo

Due to the importance of capacitance temperature stability in precise analog circuit applications, capacitance instability at elevated temperature of 125°C was investigated in tantalum capacitors with PEDOT:PSS counter electrodes. Capacitance-voltage measurement supposed that residual ions in the PEDOT:PSS dispersion caused an accumulation of charges at the dielectric-cathode interface which contributed to an increase in the dielectric constant and resulted in the capacitance increasing at high temperature. Based on the hypothesis, water wash process was applied and capacitance dropped significantly at high temperature. This study shows that an additional water wash process is necessary to improve the capacitance temperature stability after each dispersion dip step.

1. Introduction

Due to much lower equivalent series resistance (ESR) and more benign failure mode than the incumbent MnO2-cathode technology, conductive polymer tantalum capacitors are expanding their market share in both the commercial world and military world. The first tantalum and aluminum polymer capacitors were introduced into the markets in early 1990s [1, 2]. At that time, tantalum capacitor with conductive polypyrroles cathode outperformed polyanlines and polythiophenes [3]. Because the processing of 3,4-ethylenedioxythiophene (EDOT) is much simpler than other conducting polymers and it is not classified as toxic chemical like pyrrole, Poly(3,4-ethylenedioxythiophene) (PEDOT) becomes the material of first choice for solid electrolytic capacitors.

Tantalum polymer capacitors are not without faults. It was found that the dominant failure mechanism after high temperature life testing may be capacitance loss and ESR increasing rather than dielectric breakdown [4]. At elevated temperature, the thermal stability of PEDOT is a major argument to substitute the much more stable MnO2 in tantalum electrolytic capacitors. The first generation of conductive polymer tantalum capacitors was limited to using

temperatures of 105°C because it was found that some percentage of the devices would suffer steadily increasing ESR at higher temperature [5].

During the last decades, researchers from tantalum capacitor manufacturers had been dedicated to improve the thermal stability of tantalum polymer capacitors. It was learned that principle mechanism of ESR increasing after high temperature lifetime operation is oxidation of the conductive polymer [5]. As the deterioration mechanism of ESR shift was well understood, new 125°C and 85°C/85%RH capable of tantalum capacitors were introduced by leading tantalum capacitor manufacturers [5–7].

Besides ESR increasing and capacitance loss after long-term operating at high temperature, tantalum polymer capacitors also suffer electrical parameter deviation at elevated temperature, such as capacitance instability and DC leakage current increasing. The capacitance of tantalum polymer capacitors can shift dramatically with the change of temperature and dc or ac bias. For example, the capacitance shift of polymer technology at 125°C is up to the range of +30% to +50%, while the predecessor MnO2 remains within +15%. For electrical parameter instability of tantalum polymer capacitors does not cause catastrophic failure in field applications, it draws less attention from researchers.

However, applications such as VCOs, PLLS, RF PAs, and low-level analog signal chains are very sensitive to noise partially contributed by capacitance shift of decoupling capacitors on the power supply rail [8]. Thus, it is important to investigate the phenomenon of capacitance shift with the change of temperature, especially at elevated temperature.

2. PEDOT:PSS Dispersion

Deposition methods for PEDOT cathode used in electrolytic tantalum capacitors include chemical oxidative polymerization or in situ polymerization, electrochemical oxidative polymerization, and conducting polymer dispersions.

In the in situ polymerization, a monomeric precursor of the conducting polymer is polymerized by an oxidizer. Iron salt like Fe(III) toluenesulfonate is commonly used as oxidizer in in situ polymerization. The monomer and oxidizer can be brought either sequentially or as premixed reactive solution. There are several major disadvantages in in situ polymerization process. First, the processing time is long because for every cycle two more dips are necessary. Moreover, residuals of the oxidizer and monomer involved in the polymerization reaction can cause surface charge at the Ta2O5/PEDOT interface, affecting the barrier and resulting in high dc leakage and low breakdown voltage [9].

In electrochemical polymerization process, a monomeric precursor of the conductive polymer is polymerized at an electrode. During the polymerization, ionic dopants from the electrolyte are incorporated into the polymer. For capacitor application, first, an auxiliary electrode layer has to be deposited on the surface of the insulating dielectric. The auxiliary electrode is contacted with an external electrode. Because of the large inner surface and the high aspect ratio of the small pores, polymer built up in the porous anode body by electrochemical polymerization is quite difficult. Another disadvantage of the electrochemical process is the more sophisticated technical setup.

In order to overcome disadvantages of electrochemical or in situ polymerization and to further simplify the manufacturing process, a nanoscale conducting polymer PEDOT:PSS dispersions for the formation of the cathode layer within the porous structure of electrolytic capacitors was developed. The new technology allows for the direct deposition of the cathode layer by simple coating steps without any polymerization. Since no chemical polymerization takes place during the deposition of the conductive polymer, there are no side products like iron salts, which have to be washed out or could deteriorate the performance in the finished product [5]. The leakage current and break-down voltage of the dielectric is not deteriorated, thus leading to the introduction of a new line of high-voltage tantalum polymer capacitors with long-term reliability [10–12].

3. Experimental

3.1. Fabrication of Tantalum Capacitors. Tantalum powder with specific charge per volume from $15,000\,\mu$ Q/g to $23,000\,\mu$ Q/g was pressed with tantalum wires into rectangular pellets, with a 5.8g/cm^3 green density. The pellets were then sintered in vacuum around $1450°C$ for 30 min. The tantalum anodes were anodized in a dilute aqueous solution of 0.01 mol % phosphoric at $65°C$ and then were annealed at $400°C$ for 30 min. The formation voltage was 124 volts for $50V47\,\mu$ F and 153 volts for $63V33\,\mu$ F, respectively. After annealing, a reformation process was performed at the formation voltage for 60 min. In this study, Clevios P with PEDOT:PSS ratio (w/w) of 1:2.5 was applied by dipping the Ta/Ta_2O_5 pellets into a waterborne dispersion and subsequent drying in air and at room temperature and then at $130°C$. The particle size d_{50} of PEDOT particles in the dispersion is less than 30 nm. Before dipping into the dispersion, porous pellets were first vacuum-pumped. The purpose of pumping process is to evacuate air that filled in small pores in porous tantalum anodes and enable the dispersion to penetrate into the pellets more easily. In the prepolymerized process, 4–6 cycles were applied to provide a maximum coverage of the tantalum pentoxide dielectric with the PEDOT particles inside and outside the porous tantalum pellets. After the coating process of PEDOT:PSS, a graphite layer was coated on the pellets and then pasted with a thin silver layer and assembly to lead frame. The capacitors were finally encapsulated with epoxy before measurement.

3.2. Measurement. Prior to starting to test, capacitors were dried at $+125°C\pm5°C$ for 30 min and then restored to room temperature. Capacitance of 13 samples was measured at $25°C\pm3°C$ and $125°C\pm5°C$, respectively. Capacitors were brought to thermal stability at each temperature. Capacitance measurements were performed with Agilent E4980A Precision LCR Meter. Test frequency was 120 Hz ±5 Hz. The magnitude of the ac voltage was 1.0 volt root mean square (rms) and the dc bias voltage was 2.2 volts. Capacitance-voltage measurement technique was used to further investigate the abnormal increasing in capacitance. Capacitance of the samples was measured at 2.2V, 10V, 15V, 20V, 25V, and 30V bias conditions, respectively.

4. Results and Discussion

In this study, abnormal capacitance increasing at elevated temperature of $125°C$ was observed for polymer tantalum capacitors without water wash process as shown in Figure 1. Compared with the initial value at $25°C$, capacitance change at $125°C$ is in the range of 71% to 77% for $63V33\,\mu$ F with $15,000\,\mu$ C/g anodes 153V formation and even as high as over 102% to 122% for $50V47\,\mu$ F with $23,000\,\mu$ C/g anodes 124V formation. The capacitance change is reversible upon the temperature range from $25°C$ to $125°C$ and there is no hysteresis observed.

In Figure 2, capacitance-voltage measurement shows that the average capacitance change of all samples for $50V47\,\mu$ F decreases dramatically with the bias voltage increasing. At low bias condition (2.2V), the capacitance change at high temperature was about 110%. Interestingly, as the bias voltage increased to 30V, capacitance shift rolled off to the level of 20%. The effects of bias voltage on capacitance were also found to be similar in $63V33\,\mu$ F and other type capacitors with PEDOT:PSS electrodes.

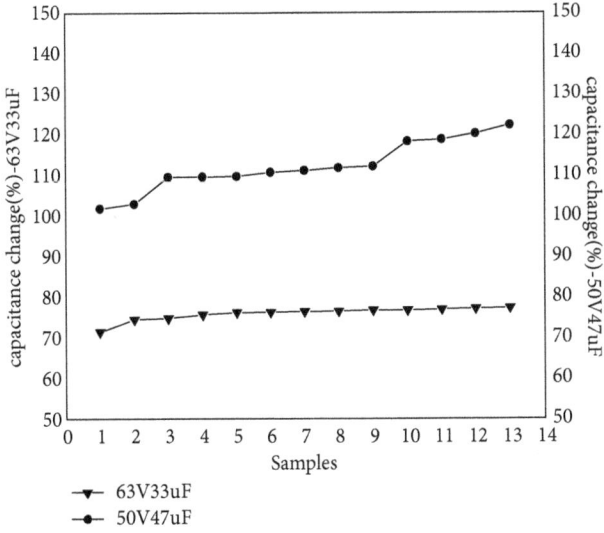

FIGURE 1: Capacitance change in percentage at 125°C without water wash process for polymer tantalum capacitors with 15,000 μ C/g anodes 153V formation for 63V33 μ F and 23,000 μ C/g anodes 124V formation for 50V47 μ F.

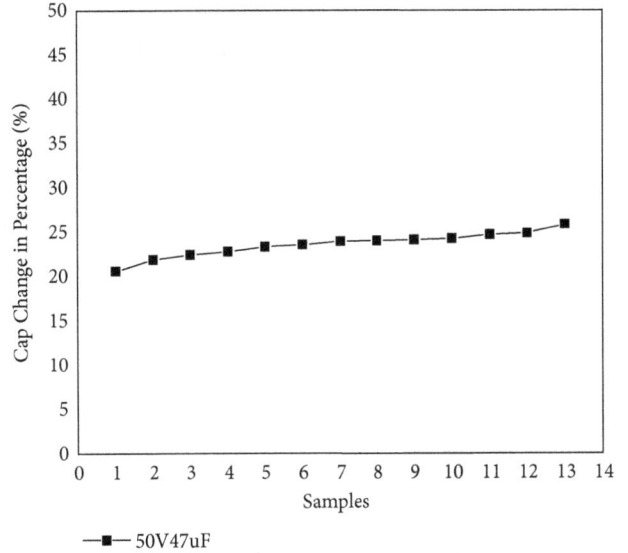

FIGURE 2: Capacitance-voltage characteristic of 50V47 μ F tantalum capacitors at 125°C.

In Figure 3, the capacitance change of all samples with water wash process for 50V47 μ F decreases compared to samples without water wash steps measured at 2.2V of standard bias condition. The effects of water wash process on capacitance stability were also found to be similar in 63V33 μ F and other type capacitors with PEDOT:PSS counter electrode. In this study, the introduction of an additional water wash process after each dispersion dip step is to wash out residual ions with the PEDOT:PSS film.

According to capacitor theory, capacitance is determined by the following formula:

$$C = \frac{\varepsilon_{\circ}\, \varepsilon \cdot A}{d} \qquad (1)$$

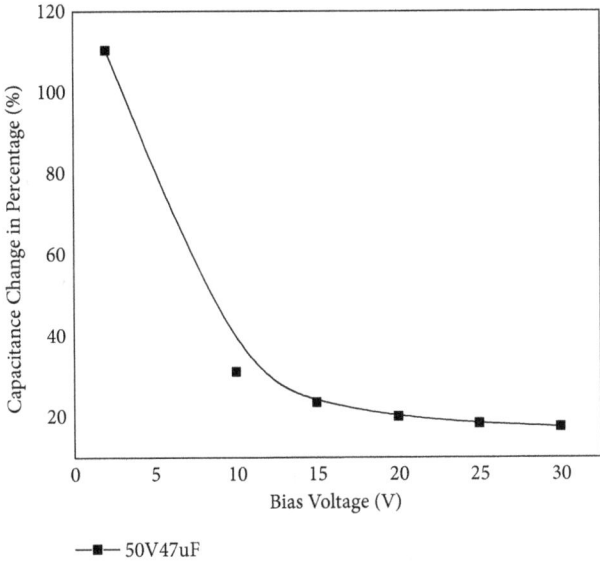

FIGURE 3: Capacitance change of 50V47 μ F at 125°C with water wash steps.

where ε_0 is the dielectric constant for free space (8.855 x 10^{-12} Farads / m), ε is the dielectric constant for tantalum pentoxide (about 27), A is the surface area in m^2, and d is the dielectric thickness in m. In these parameters, ε_0 is a constant and does not vary with temperature and the surface area A is predetermined by the tantalum powder used and the anode dimension which depends less on temperature. Furthermore, the thickness d of the anodic oxide tantalum film is directly proportional to the formation voltage with the coefficient in the range of 1.6 nm/V to 2.0 nm/V [13]. For the coefficient of temperature expansion (CTE) of tantalum pentoxide is in the order of 10^{-5}/K, the thickness of the dielectric film will only change in the same order [14]. Thus, thickness of tantalum pentoxide is not the main cause for the increase in capacitance. Furthermore, the inherent variation of the dielectric constant for tantalum pentoxide with temperature is considerably stable within 10% at 150°C and would not cause abnormal capacitance increase [15]. These conclusions are verified by tantalum capacitors with manganese oxide electrode whose capacitance shift at 125°C is generally less than 15%.

Based on experiment, Freeman, et al. introduced a model that explained the mechanism of capacitance dependency with temperature [15]. According to the model, the PEDOT layer can shrink upon cooling and expand upon heating and resulted in the changes of the effective surface area which contribute to the overall device capacitance. However, the abnormal capacitance increase observed in this study cannot be explained by the surface area model. First, the capacitance measured at 125°C was as high as 1.7 to 2 times compared to the value at room temperature and exceeded the wet capacitance (1.1 times the capacitance of the finished device at room temperature) measured after the dielectric formation. Practically, surface area change would not cause such a large relative change in capacitance. More importantly, strong capacitance dependency on voltage was observed at elevated

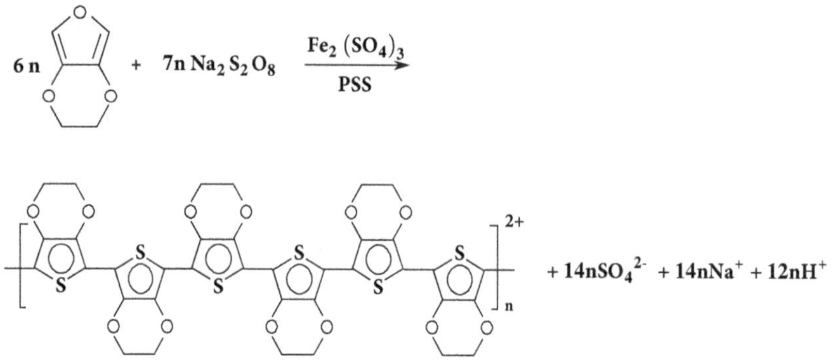

FIGURE 4: Reaction scheme for the PEDOT synthesis using Sodium peroxodisulfate as an oxidant.

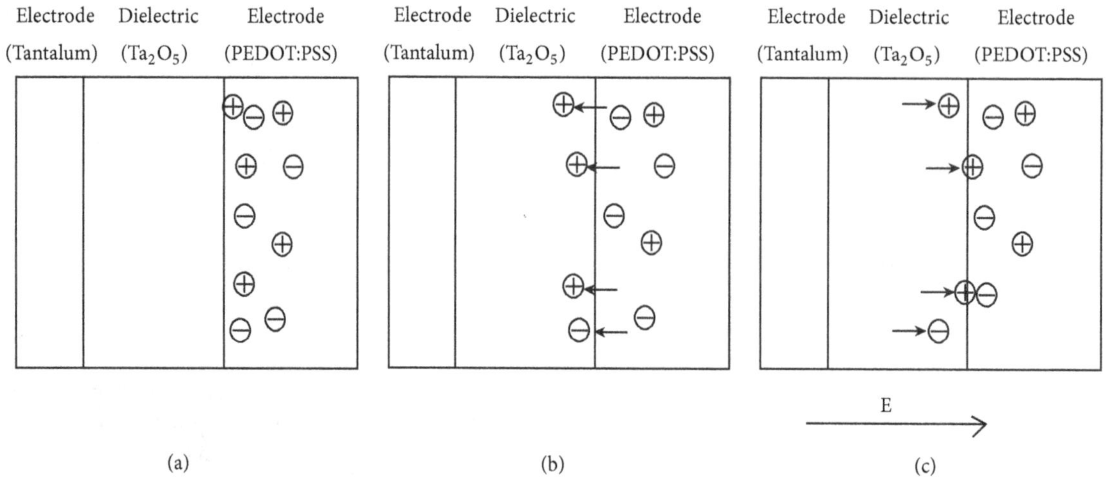

FIGURE 5: Interfacial polarization of tantalum capacitors with PEDOT:PSS electrode.

temperature. As seen in Figure 2, the capacitance decreased dramatically as the biased voltage increased. Obviously, the effective surface area theory cannot explain the capacitance-voltage relationship. It is believed that there are other mechanisms that may contribute to the abnormal capacitance changes.

Fundamentally, dielectric constant is a measure of dielectric polarization under an electric field. There are electronic, ionic, orientational and interfacial polarizations [16]. For tantalum pentoxide, ionic polarization is the main polarization mechanism [16]. Based on dielectric theory, both electronic and ionic polarizations are less dependent on temperature, and orientational polarization is inversely proportional to the temperature. Thus, it is supposed that the abnormal increase in capacitance may be caused by interfacial polarization due to the introduction of conductive polymer PEDOT:PSS.

The PEDOT:PSS complex was synthesized by mixing the polystyrene sulfonic acid with PEDOT. For Clevios P dispersion, the molar ratio of thiophene groups to sulfonic acid groups is 1:1.19, which corresponds to a weight ratio of 1:2.5. During the oxidative polymerization of EDT monomer in the presence of PSS, sodium peroxodisulfate is used in combination with an Fe(III) salt as a catalyst. As shown in Figure 4, the reaction mixture turns more acidic as the reaction progresses, since each mol of EDOT releases

two moles of protons [3]. Theoretically, there are residuals such as protons, sodium, sulfate, Fe(III), and other species in the PEDOT:PSS complex after the completion of the synthesis.

At elevated temperature, impurities such as H+, Na+, and Fe3+ ions diffuse toward the interface between PEDOT:PSS cathode and tantalum pentoxide dielectrics to form an accumulation of charges; thus interfacial polarization occurs as depicted in Figure 5. Dielectric materials, although perfect, contain crystal defects, holes, and surface imperfections. As depicted in Figure 5(a), residuals in PEDOT:PSS have an equal number of positive ions and negative ions, but the positive ions are assumed to be far more mobile because they are relatively small. In the absence of electric field or at lower bias condition, positive ions may diffuse into the dielectrics of tantalum pentoxide under high temperature as shown in Figure 5(b). These positive charges accumulate at the interface and attract more electrons to the negative electrode. These additional charges on the negative electrode, of course, appear as an increase in the dielectric constant. Figure 5(c) shows that, at higher bias condition, the diffused positive ions on the dielectric side may drift back to the cathode or combination with electrons and the effect of interfacial polarization is reduced; thus, the capacitance dropped under higher bias condition as shown in Figure 2.

TABLE 1: Characteristics of Clevios P dispersion.

Characteristics	min.	max.	Measured value	Unit
Solid Content	1.2	1.4	1.32	%
pH	1.5	2.5	1.9	
Sodium		500	233	ppm
Sulfate		80	ND	ppm
Iron	/	/	0.64	ppm

In order to verify the main sources that contributed to the capacitance variation, further analysis was done. For Clevios P dispersion used in this study, the main physical and chemical characteristics are shown in Table 1. Although Fe(III) salt was used as catalyst, it can be easily washed out and the residual of Fe ions is almost negligible and not specified in the specification. The main residual species given are sodium and sulfate, and the maximum contamination levels are 500 ppm and 80 ppm, respectively. In order to further analyze residual species, contents of sodium and iron were measured through leaching analysis. The contents of sodium and iron were 233 ppm and 0.64 ppm, respectively, as shown in Table 1. As seen from the characteristics table, the specified value of pH for the dispersion is in the range of 1.5 to 2.5, and the measured value is 1.9, which indicated that the hydrogen ion concentration is high.

From the analysis and measurement, residuals in the PEDOT:PSS complex contributing to the abnormal capacitance increase at high temperature are mainly H^+ and Na^+ ions in tantalum capacitors with PEDOT:PSS electrodes.

Based on the above hypothesis, it can be assumed that an additional water wash process can reduce the content of impurity ions which resulted in the improvement of capacitance temperature stability. As shown in Figure 3, the capacitance change at high temperature decreased dramatically compared to samples without washing steps during PEDOT:PSS dispersion dipping process.

However, though an additional water wash process can improve the capacitance stability, capacitance changes at high temperature in conductive tantalum capacitors with PEDOT:PSS electrode are still higher than the conventional ones with manganese dioxide electrodes. The reasons for this larger capacitance change have already been well explained by the surface area model developed by Freeman et al. [15]. For mechanisms of capacitance dependency with temperature in PEDOT cathode system are complicated, the pore structure of the anode, preparation methods of the dielectrics of Ta2O5, the material of PEDOT used, deposition methods of PEDOT, and so forth can contribute to the capacitance variation. For the new generation tantalum capacitor technology, more efforts may be needed to further improve the capacitance-temperature stability.

5. Conclusions

The abnormal capacitance increase in tantalum capacitors with PEDOT:PSS electrodes at elevated temperature of 125°C

was investigated by capacitance-voltage measurement. It was supposed that residuals such as hydrogen and sodium ions in the PEDOT:PSS dispersion caused an accumulation of charges at the dielectric-cathode interface, which appeared as an increase in the dielectric constant and resulted in the capacitance increase at high temperature. This study shows that an additional water wash process is necessary to improve the capacitance temperature stability after each dispersion dip step.

Conflicts of Interest

The authors declare that there are no conflicts of interest regarding the publication of this paper.

Acknowledgments

The authors would like to acknowledge the support of Xinyun Electronics Components Corporation for providing capacitor samples, testing facilities, and financial funding.

References

[1] Y. Kudoh, S. Tsuchiya, T. Kojima, M. Fukuyama, and S. Yoshimura, "An aluminum solid electrolytic capacitor with an electroconducting-polymer electrolyte," *Synthetic Metals*, vol. 41, no. 3, pp. 1133–1136, 1991.

[2] M. Fukuyama, Y. Kudoh, N. Nanai, and S. Yoshimura, "Materials Science of Conducting Polymers: An Approach to Solid Electrolytic Capacitors with a Highly-Stable Polypyrrole Thin Film," *Molecular Crystals and Liquid Crystals Science and Technology. Section A. Molecular Crystals and Liquid Crystals*, vol. 224, no. 1, pp. 61–67, 1993.

[3] A. Elshchner, S. Kirchmeyer, W. Lovenich, U. Merker, and K. Reuter, *PEDOT: Principles and Applications of an Intrinsically Conductive Polymer. An introduction to theory and applications of quantum mechanics*, Wiley, 1982.

[4] Reed, Characterization of tantalum polymer capacitors, NASA Electronic parts and packaging program, NEPP Tast 1.21.5,Phase 2,2006.

[5] E. Reed, J. Chen, J. Marshall, J. Paulsen, and R. Weisenborn, "New 125∘C capable tantalum polymer capacitors," in *Proceedings of the 17th Passive Components Symposium CARTS Europe*, pp. 24–31, 2003.

[6] J. Young, "Polymer Tantalum Capacitors for Automotive Applications," in *Proceedings of the CARTS International*, pp. 297–311.

[7] J. Petrzilek, M. Biler, and T. Zednicek, "Hermetically sealed conductive polymer tantalum capacitors," in *Proceedings of the CARTS International 2014*, USA, April 2014.

[8] G. Morita, "Capacitor Selection Guidelines for Analog Devices, Inc., LDOs," Tech. Rep., http://www.analog.com/media/en/technical-documentation/application-notes/AN-1099.pdf.

[9] Y. Freeman, W. R. Harrell, I. Luzinov, B. Holman, and P. Lessner, "Electrical characterization of tantalum capacitors with poly(3,4-ethylenedioxythiophene) counter electrodes," *Journal of The Electrochemical Society*, vol. 156, no. 6, pp. G65–G70, 2009.

[10] U. Merker, W. Lovenich, and K. Wussow, "Conducting polymer dispersions for high-capacitance tantalum capacitors," in *Proceedings of the 20th Passive Components Symposium CARTS Europe*, pp. 21–26, Bad Homburg, Electronic Components, Assemblies & Materials Association, 2006.

[11] U. Merker, W. Lovenich, and K. Wussow, "Tuning conducting polymer dispersions for high-CV tantalum capacitors," in *Proceedings of the in. Proceedings of the 21st Passive Components Symposium CARTS Europe*, Barcelona, 2007.

[12] J. Young, J. Qiu, and R. Hahn, "High voltage tantalum polymer capacitors," in *Proceedings of the 28th Symposium for passive electronics, CARTS-USA 2008*, pp. 185–195, USA, March 2008.

[13] J. P. Manceau, S. Bruyere, S. Jeannot, and A. Sylvestre, "Current instability, permittivity variation with frequency, and their relationship in ta 2 o 5, capacitor," *IEEE Transactions on Device Materials Reliability*, vol. 7, no. 2, pp. 315–323, 2007.

[14] Y. Freeman, P. Lessner, and E. Jones, "Low de-rating reliable and efficient Ta/MnO2 capacitors," in *Proceedings of the CARTS International 2012*, pp. 19–28, USA, March 2012.

[15] Y. Freeman, I. Luzinov, R. Burtovyy et al., "Capacitance stability in polymer tantalum capacitors with PEDOT counter electrodes," *ECS Journal of Solid State Science and Technology*, vol. 6, no. 7, pp. N104–N110, 2017.

[16] M. Henini, *Principles of Electronic Materials and Devices*, S. O Kasap, Ed., McGraw-Hill, 2nd edition, 2002.

19

VHDL-AMS Simulation Framework for Molecular-FET Device-to-Circuit Modeling and Design

Mariagrazia Graziano,[1] **Ali Zahir,**[2] **Malik Ashter Mehdy** ⓘ**,**[1] **and Gianluca Piccinini**[1]

[1]*Electronics and Telecommunication Department of Politecnico di Torino, Torino, Italy*
[2]*Department of Electrical Engineering of the COMSATS Institute of Information Technology, Abbottabad, Pakistan*

Correspondence should be addressed to Malik Ashter Mehdy; malik.mehdy@polito.it

Academic Editor: Gerard Ghibaudo

We concentrate on Molecular-FET as a device and present a new modular framework based on VHDL-AMS. We have implemented different Molecular-FET models within the framework. The framework allows comparison between the models in terms of the capability to calculate accurate *I-V* characteristics. It also provides the option to analyze the impact of Molecular-FET and its implementation in the circuit with the extension of its use in an architecture based on the crossbar configuration. This analysis evidences the effect of choices of technological parameters, the ability of models to capture the impact of physical quantities, and the importance of considering defects at circuit fabrication level. The comparison tackles the computational efforts of different models and techniques and discusses the trade-off between accuracy and performance as a function of the circuit analysis final requirements. We prove this methodology using three different models and test them on a 16-bit tree adder included in Pentium 4 that, to the best of our knowledge, is the biggest circuits based on molecular device ever designed and analyzed.

1. Introduction

Emerging future nanoelectronics is being extensively explored for several applications [1–6]. More specifically molecular devices can play an important role both for sensing and for computational applications with huge advantages in terms of integration capabilities, functional density, and performance. The importance of molecular electronics has led many groups to study the transport properties of conjugated molecular structures [7–9]. Recently, molecular electronics has gained a great interest from both applied electronics and theoretical point of view [1, 2, 10–16]. Here we focus on molecular-FET (MOL-FET); however, our methods can be extended to any other molecular device. Methodologies on molecular device modeling are fragmented, it is extremely difficult to compare models, the scenario offers only a few examples on the application of MOL-FET in circuits and systems for the purpose of computation and sensing, and especially the literature does not offer a thorough analysis of impact of MOL-FET models at circuit level in terms of flexibility and computation requirements.

The novel contributions of the work to advance the state-of-the-art are the following. (1) We present a modular framework (Figures 1(a)–1(c)) based on VHDL-AMS, which is extended version of our previous work [17]. VHDL-AMS is a derivative of VHDL (VHSIC Hardware Description Language) and includes analog and mixed-signal extensions (AMS). Our modular framework enables within the same simulation engine and in the same conditions comparing different MOL-FET models at device level in terms of their capability to generate the *I-V* characteristics comparable to those used as a reference. Here we refer to atomistic simulations in Atomistix ToolKit (ATK) [18, 19] as later explained (ATK Figure 4), but in general any reference data, for example, experimental values, could be included. (2) In this case, three models, developed by the authors and partially new, are being compared and discussed in the following sections: (a) a new semiempirical look-up table model (SE-LUT, Figure 5); (b) a new transmission spectrum based model (TSB Figure 7) fully described in [20] and evolved with respect to [21]; (c) FET-based equivalent model (FBE Figure 6), partially based on [22] and partially new.

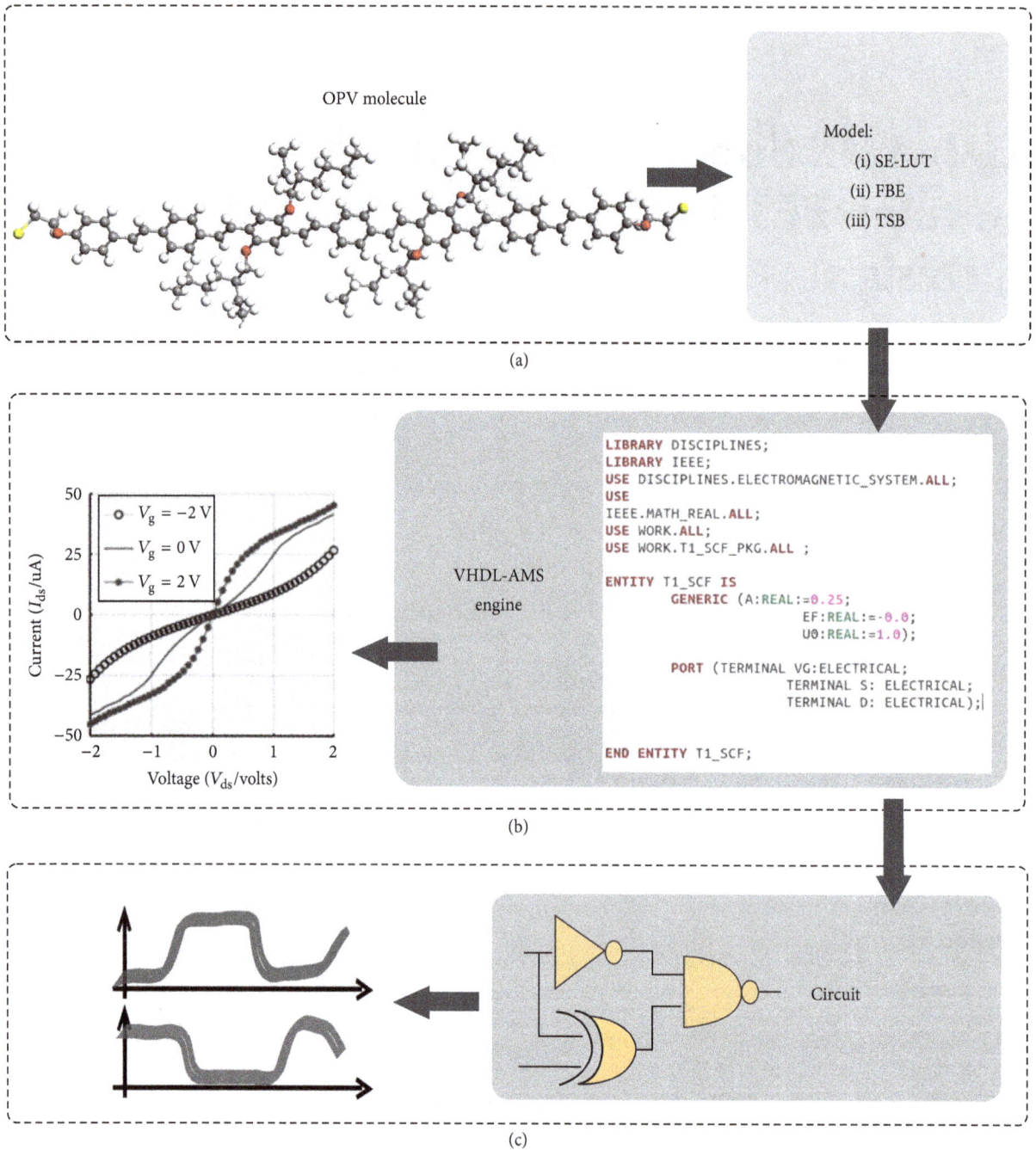

(a)

(b)

(c)

FIGURE 1: Flow of methodology. (a) The OPV molecule coupled to the electrodes through oxygen linkers is simulated in ATK for I-V characteristics. The conduction through the molecule is calculated at different bias conditions by the three models: semiempirical look-up table SE-LUT, transmission spectrum based (TSB) model, and FET-based equivalent model FBE. (b) VHDL-AMS based engine elaborates the results obtained by the three models, generates the corresponding I-V characteristics, and provides the description of MOL-FET in VHDL-AMS. (c) Logic circuits based on MOL-FET can be hierarchically implemented. Correct circuit behavior and performance can be evaluated by the simulation of the circuits.

The I-V characteristics from three models are compared to the reference ATK data (Figure 4) and show an excellent agreement. (3) Another fundamental novelty of this work consists in the possibility of verifying the correct MOL-FET behavior when used in a logic circuit. Several are the possible topologies; in this case we limit to a complementary P-N-MOLFET crossbar that we use to design elementary logic cells. (4) The framework allows analyzing the proper circuit behavior and verifying to what extent it is possible to use each elementary logic cell in a complex architecture organized as

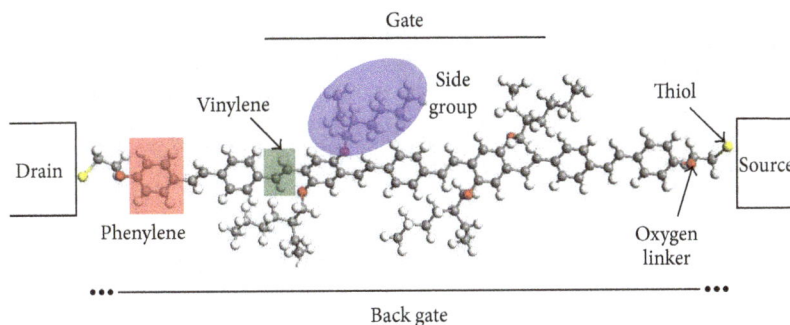

FIGURE 2: The structure of the device with different parts of molecule mentioned.

a cascade of basic crossbars. Here we implement as a benchmark circuit a complex up-to-date 16-bit tree adder (sparse tree used in Pentium 4 family). Steps 1 and 2 have already been presented in our previous work [17] while steps 3 and 4 are the novel contributions of this work. We also demonstrate here the capability of our methods to capture the impact of the possible variations during fabrication. Steps 3 and 4 evidence (i) whether the molecule used has the correct properties for being used in realistic circuits; (ii) the impact of the different technological choices (i.e., molecules, gate coupling) on the circuit behavior; (iii) the capabilities at circuit level offered by the different modeling features of the involved models, that is, the sensitivity to parameters and the capability to capture physical quantities, among which are the technological variations due to the immaturity and the extreme variability of molecular electronics; (iii) the models computational requirements both in terms of simulation time and in terms of setting-up of all the data and parameters necessary when a new molecule or condition is taken into consideration.

To the best of the authors' knowledge *this is the first contribution allowing simulating complex circuits based on molecular devices and on a fan of possible related models whose characteristics and impact can be studied at circuit level.* The rest of the paper is organized as follows: in Section 3 we present the VHDL-AMS framework while Section 4 describes the three models used. In Section 5 we use an example logic gate (EXOR) to discuss the circuit organization and its behavior and to describe the models capabilities. Section 6 finally presents the architecture of the adder and details the simulations results and discusses the performance.

2. Device Structure and Behavior

The molecule under study Oligo Phenylene Vinylene (OPV) is composed of repeating unit phenylene ($-C_6H_4-$, the benzene ring with two hydrogen atoms removed) linked by vinylene ($-C_2H_2-$, ethene with two hydrogen atoms removed) (Figure 2). The electrons are delocalized over the length of molecule because of its conjugated structure so these molecules are efficient in charge transport. The conduction depends on the length of the molecule and anchoring group between the molecule and electrodes. The side groups do not have noticeable impact on the transport properties of this device [23]. We have used thiol as anchoring group to connect with Au electrodes, which is common

in molecular devices. The oxygen in alkyl-oxygen linker group (which links the molecule with electrode through thiol) decouples the molecule from the electrodes which results in the reduction of conductance [24]. Due to its high electronegativity, oxygen attracts the shared electrons in the polar covalent bond with its neighboring carbon atoms. So the negative charge accumulated around the oxygen leads to the decoupling between the molecule and the electrodes. We can see in Figure 4, with no gate voltage applied, a large value of drain to source bias is needed for drain current to flow (high drain to source threshold voltage). Without the linker group the conjugated molecule has good transport properties so if it was directly connected to the electrodes threshold voltage would be very low and it would behave like a molecular wire. Further information about the device chemistry can be found in [23, 24].

3. Simulation Framework

VHDL-AMS [25] is a superset of VHDL and supports the use of digital constructs together with electrical quantities, algebraic constraints, and differential equations. It allows not only the simulation of mixed-signal electronic components, but also mixed-technology systems, ranging from optics to mechanics and chemistry to thermodynamics. It is also suitable for the modeling of molecular systems.

An important property of the simulator supporting VHDL-AMS, in our case ADMS [26], is its capability of using different levels of abstraction for the description of different blocks in the system, based on the focus needed for the different blocks. This makes it possible to design using top-down methodology in which a functionality test is performed by the behavioral description of components and then, by refining the circuit, performance can be improved. ADMS [26] enhances this feature specifically as it allows the simulation of high level VHDL-AMS architectures together with the spice level netlists in the same simulation environment.

The original VHDL language allows discrete system modeling but is not capable of the description of continuous characteristics which is necessary for today's designs. The need arises in case of digital circuits when submicron effects play a role in the performance of the circuits or in case of analog and mixed-signal design. This limitation can be overcome by VHDL-AMS which in addition to the discrete features also supports the implementation of continuous

```
...
architecture behavior of inv is
componet MOLFET
Port (Terminal G,D,S:Electrical);
end component;
...
T1:MOLFET port map(G=>in,S=out,D=ref)
T2:MOLFET port map(G=>in,S=vdd,D=out)
end architecture;
```

(a)

```
.subckt MOLFET NVD NVS NVG
...
.MODEL CMOSN NMOS (VTO=0.2...)
.MODEL CMOSN2 NMOS (VTO=0.1...)
*.param Vg {3.0}
...
ERD NRD NVG1 N1 0 {BETA}
MRD N2 NRD N1 0 CMOSN L=1u W=32u
.ends
```

(b)

```
...
use work.TSB_pkg
entity MOLFET is
...
end entity MOLFET
architecture behav of MOLFET is
...
begin
I=TSB(α_g,V_ds,V_g...);
end architecture behave;
--Function TSB defined in TSB-pkg
Function TSB(α_g,V_ds,V_G...)
--Parameters definition
--TS calculation by SCF
--Current from Landauer's formula
Return Ids;
End Function TSB;
```

(c)

FIGURE 3: A short description of VHDL code of the framework, (a) inverter circuit implemented using the MOLFET device, (b) FBE model, and (c) TSB model description. After defining the device in models we can use it as VHDL component in any circuit ranging from inverter to P4 adder.

models based on mathematical equations. In [27–29] complex mixed-signal electronics and telecommunication systems have been designed using VHDL-AMS as an effective simulation language. In [30, 31] VHDL-AMS is suggested to model complex electronic systems with a varying hierarchical depth and interacting inside the automotive environment.

The framework based on VHDL-AMS here proposed is flexible because different MOL-FET models can be included in it as a parallel module (see the following section) and I-V characteristics can be compared. This structure describes a circuit based on crossbar and allows the analysis of output waveforms. Depending on the type of model different features of ADMS are exploited, as the implementation of algebraic and differential equations (TSB model), the use of numerical data approximation (LUT model), or the reference to SPICE level simulation (FBE model). Once the models have been defined as the VHDL-AMS packages we can define the MOL-FET device based on these models and use it as a component in any circuit (Figure 3).

4. MOL-FET Models

4.1. Reference Atomistic Simulations. The MOL-FET models here adopted are referred for comparison to atomistic simulations obtained using Atomistik Toolkit [18]. The OPV molecule analyzed here has been realized and studied in [23] and it was studied in [24] to use it in an application. The molecular-FET consists of dithiolated molecule with oxygen linker on both sides (Figures 1(a) and 4). This linker group decouples the molecule from electrode due to its electronegative nature which leads to narrow sharp peaks of transmission spectrum

(TS). The molecule OPV7 (OPV with seven phenylene rings) is simulated in ATK connecting the two extremes to gold electrodes as in Figure 4 named herein drain (D) and source (S) subjected to different Vds bias voltages. A Gate (Vg) voltage is also applied, the gate being a reference substrate for the molecular system. Figure 4 on the right also shows the current-voltage characteristics of the OPV7 molecule with oxygen linkers at different gate voltages. N-type behavior in conduction can be observed as the positive gate voltage enhances the current [23]. Later the TS related to this set of simulations will also be shown and discussed in comparison with the TS model.

4.2. Semiempirical Look-Up Table (SE-LUT) Model. The method used in this case is based on a set of given I-V characteristics, at different bias conditions (Vg, Vds), obtained from atomistic simulations. The tabulated data represents the current values for different Vds in rows and those for different Vg in columns (see Figure 5). For every gate voltage, the bias voltage is changed by a constant step of 0.04 V. Of course the smaller the step the higher the number of simulations to be obtained but also the bigger the final resolution achieved. The set of current values for different bias points and gate voltages is provided as the input to simulation engine. The simulation engine finds the current value corresponding to the specific Vds and Vg values in the table using a binary search algorithm. For the intermediate values of external voltages which are not present in the table, current is computed using bilinear interpolation. Due to the nature of I-V function for this device, we could prove that I-V characteristics can be well reproduced, using bilinear

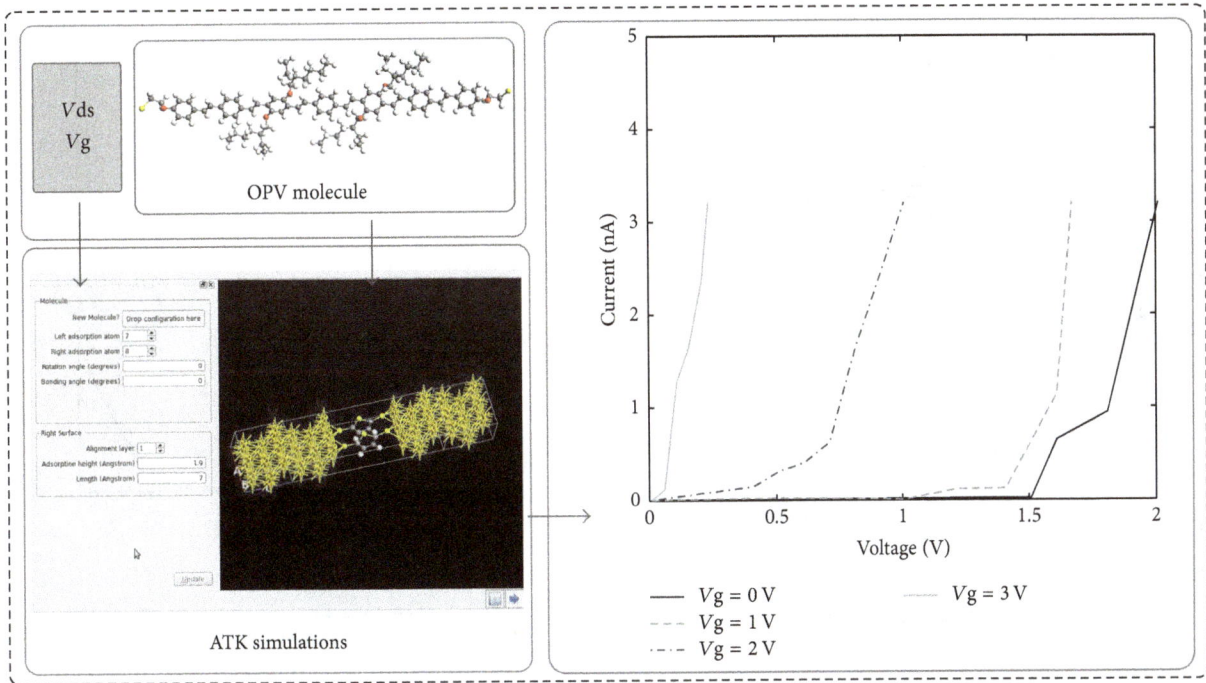

FIGURE 4: The current-voltage characteristics: given ATK simulation obtained at different bias voltages (Vds) and gate voltages (Vg) for OPV7 molecule.

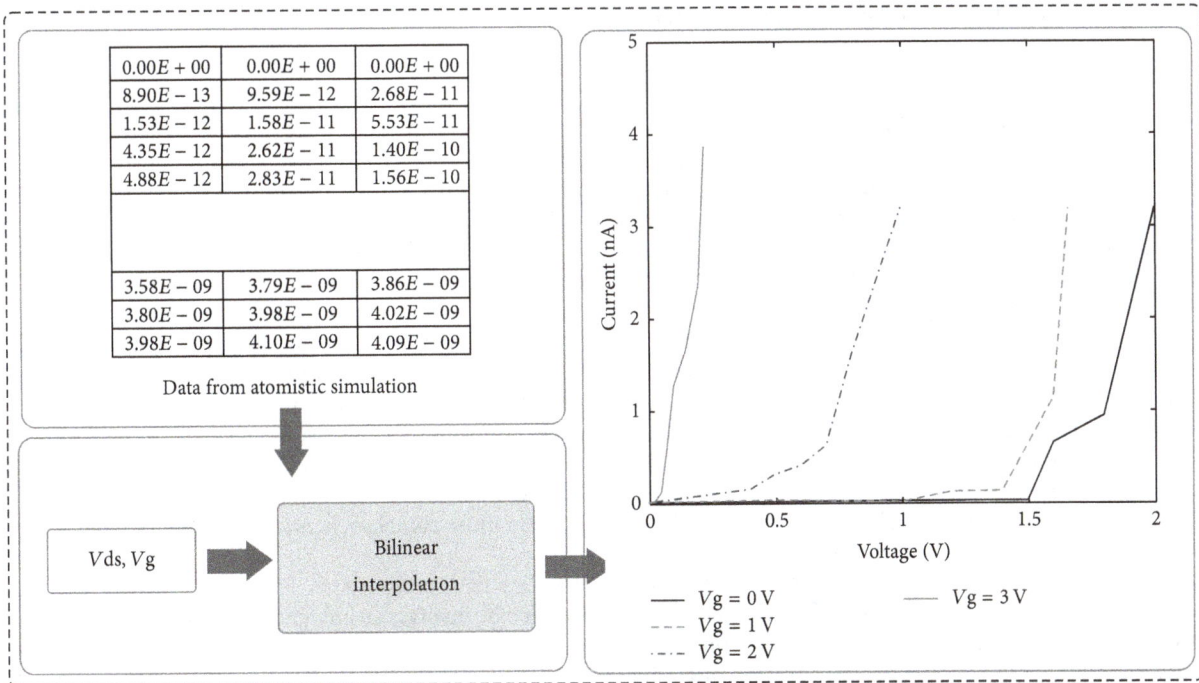

FIGURE 5: Semiempirical look-up table (SE-LUT): results obtained from ATK simulations for specific bias points (Vg, Vds) organized in tabular form; at the intermediate bias points current is obtained by bilinear interpolation.

interpolation. This model has the advantage of simplicity and can be easily implemented but it does not provide any insights on physical behavior of the device. The obtained currents are coherent with the ATK reference simulations, also in cases of Vds where interpolation has to be adopted.

4.3. FET Based Electrical Equivalent (FBE) Model. In this case the model is based on [22] but reckoned in this paper for input data from different source. The method is based on an equivalent circuit model as in Figure 6. In the framework this means that ADMS has a spice-like circuit description as input

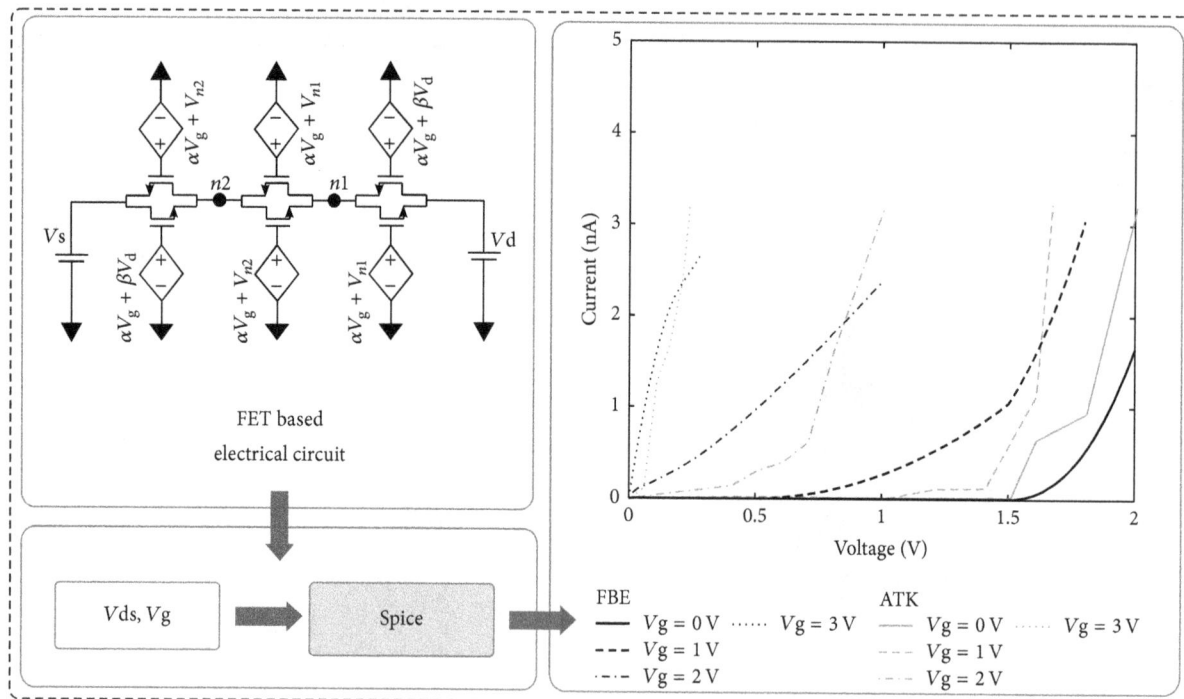

FIGURE 6: Electrical equivalent model (FBE): based on level 1 FET spice model. The phenyl ring with ethylene chain connected to it is modeled with M1 transistor (with V_{th} = 0.1) because of delocalized pi-bonds and M2 transistor (with V_{th} = 0.2) is used to model oxygen with only sigma bonds. Two parallel NFETs are used to model the effect of potential of neighboring atoms. Corresponding IV characteristics for different gate voltages are shown on the right (black lines: FBE; grey lines: ATK).

and calls ELDO, the spice engine included in ADMS tool, for the simulation.

The nature of the bond between atoms and external potential is two parameters which mainly influence the electron transport in the molecules. In our model these effects are represented by an equivalent circuit made of FET transistors and voltage-dependent voltage sources. The phenyl ring connected with the ethylene chain [24] is modeled with M1 transistor having Vth = 0.1 because of the delocalized pi-bonds in Figure 6. And a high electronegative oxygen atom with only sigma bonds is modelled with M2 transistor (Vth = 0.2), values adopted from [22]. In both cases, in order to take into account the effect of potential of neighboring atoms in a chain, two identical parallel NFETs are adopted. A voltage divider models the insulating behavior of alkane chains. The gate to molecule electrostatic coupling is introduced into the model by a coupling coefficient herein referred to as α (see [22] for further details on the different contributions). The I-V curves are shown in Figure 6 on the right, although the shape of the curves are not very similar to ATK results but it gives acceptable results in terms of order of magnitude of the currents. As the simulation engine uses lumped circuit elements to model the current so some differences with respect to ATK can be observed in the I-V characteristics. In the results sections further discussion on the effectiveness of this method will be given.

With respect to the LUT model, this equivalent model provides us with better insights on the physical behavior of the MOL-FET. For example, the gate coupling parameter α

is not present in the SE-LUT model, where, if a different coupling has to be considered, a new set of simulations or of measurements should be obtained and loaded in the system. This partial increment of flexibility is obtained at the price of a more complex description. Moreover, an initial fitting process is required to detect the electrical equivalence. If, then, a different molecule or different linkers or the presence of variations due to technology should be taken into account, then a new fitting procedure should be done. Furthermore, this modeling technique requires using a spice-level engine during the simulation that could increase the total simulation time with respect to the pure LUT method. The model can be improved by including more circuit elements to take into account more complex chemical and physical effects.

4.4. Transmission Spectrum Based (TSB) Model. The model in this case involves an analytical approach; that is, it requires the implementation of a set of functions in the VHDL-AMS framework. The idea is sketched in Figure 7. First of all, the transmission spectrum (TS) at equilibrium is calculated. Semiempirical Extended Huckel Theory (SE-EHT) is used to find the TS and other transport properties. Atomistic simulations (in this case using ATK) at equilibrium are carried out once for all. The TS are numerical inputs for the VHDL-AMS framework. These parameters are then used during the simulation execution to estimate voltage-dependent TS at nonequilibrium conditions. The model functions are called during the simulation and on the basis of the bias conditions will output the expected currents. The model is an extended

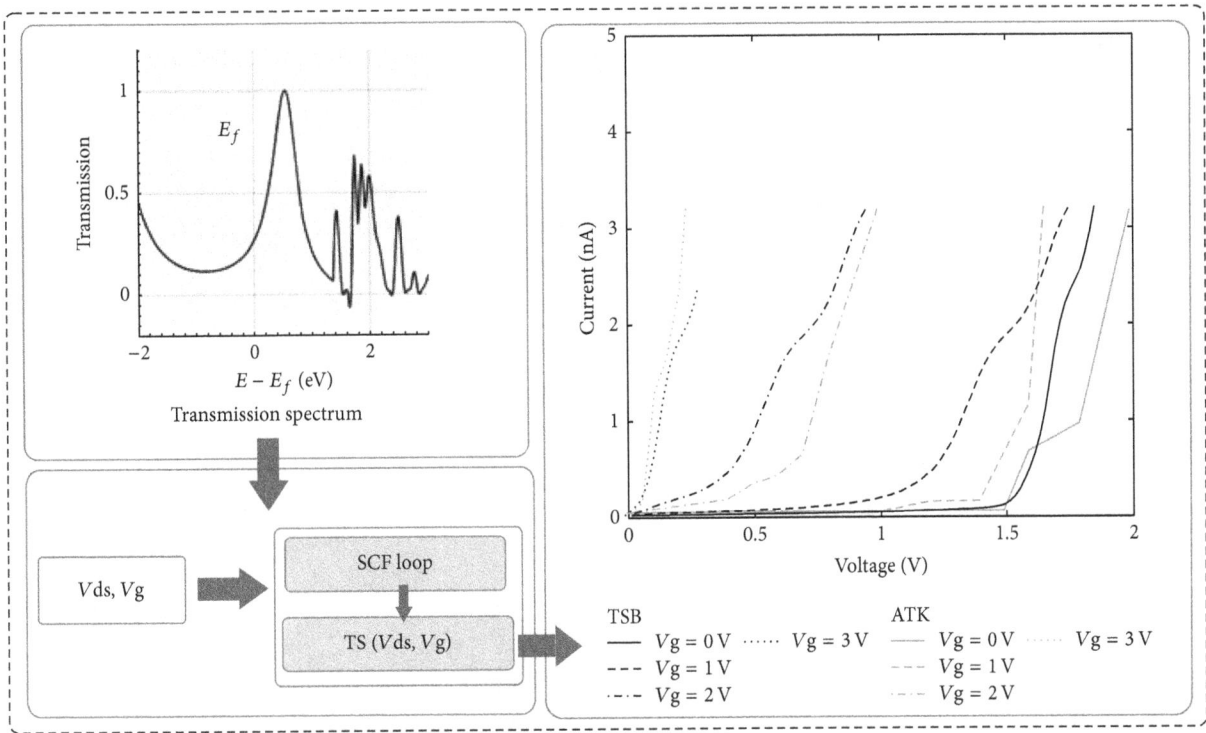

FIGURE 7: Transmission spectrum based (TSB) model: based on a set of equations to compute the quantum transport inside the molecule. The transmission spectrum (TS) at equilibrium is obtained from ATK simulations and given as input to the model. For nonequilibrium conditions, that is, external voltages applied (Vg, Vds), self-consistent field calculations are performed to find the TS and then current is calculated using TS.

version with respect to the one presented in [21]. The charge distribution in the molecule is dependent on the external voltage and changes when voltage is applied. When an external voltage is applied, the charge distribution in the molecular system changes. This causes the modification in terms of shift in the molecular energy levels and thus the TS as it is clear from Figures 8 and 9 where a comparison between this model and atomistic simulations is given in terms of TS. Thus charging effect should be taken into consideration to compute the shift in the TS. In TSB model, self-consistent field method (SCF) is used to estimate the shift in energy levels [1]. After calculating the new TS based on the shifted energy levels, Landauer's formula is used to calculate current from TS. A few assumptions are made: we assume (i) a linear shift in the energy levels at low applied voltages and (ii) that the charging effect modifies all the energy levels uniformly and TS is shifted rigidly on the energy axis. We tested the system in several conditions and verified that these assumptions can efficiently reproduce the transport behavior of the molecular system without losing the accuracy of the system (a detailed discussion on the theory and demonstration of this model is out of the scope of this paper and can be found in our previous paper [20]). The resulting I-V characteristics are shown in Figure 7 on the right. The shape of the IV curves depends strongly on the position of transmission peaks and the area under the peaks. So it is a little different from ATK as the small error in the peaks position and shape can change the magnitude of the current. The shape of the curves from this

model is better than the FBE model and the magnitude of current is also very close to ATK results.

If we compare this model to the previous ones it is easy to understand that this model captures with more details the physical characteristics (see the results section) and provides the option to explore the MOL-FET behavior in different physical conditions. The cost for this model is an initial extraction and analysis of TS and a more complex description, which implies more CPU intensive calculations during circuit simulation. In future the model can be improved by implementing the subtle effects like the evolution in the width of the transmission peaks and the separation between different peaks. This will result in more accurate IV characteristics.

5. From Transistor to Circuit

We designed a small library of logic gates. The circuit topology here chosen is implemented as a crossbar; Figure 10 shows the example of a half adder and a full adder. We used a complementary logic based on both N-type and P-type MTs. Alternatively an N-FET plus resistors could be used but the complementary structure is preferable because of low power consumption. The P-type transistor can be realized using a slightly different procedure. The difference between P-type and N-type is based on the molecular energy levels contributing to the conduction: in the N-type transistor the LUMO levels are involved in conduction, while for the P-type

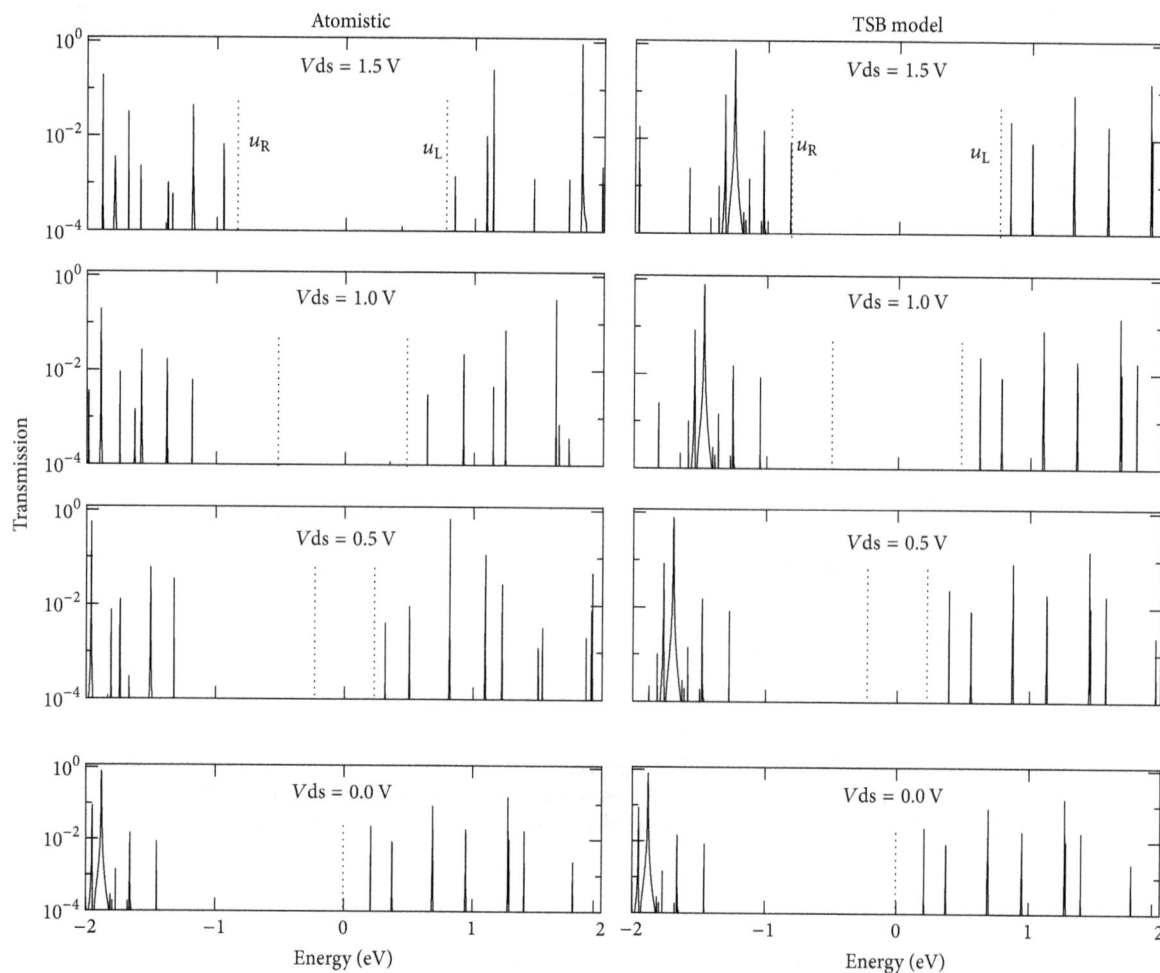

FIGURE 8: Transmission spectrum for different values of Vds (Vg = 0). The dotted lines represent the bias window. Transmission spectrum shifts to new energies when Vds is changed. TSB model shows the same effect using self-consistent field loop.

transistor the HOMO levels contribute to conduction. For the P-type device HOMO level should be near the Fermi level. In this way, when a negative bias is applied on the gate, it shifts the energy levels towards high values and HOMO level contributes to the conduction. This issue can be addressed in two possible ways: selecting the molecule which has HOMO level near Fermi level or using a back-gate to shift the energy levels with respect to the electrodes chemical potential. We used the back gate technique with the same molecule (OPV) as used for the N-type. When we apply an appropriate amount of back gate voltage (V_{BG} = −3.6 V), the energy levels shift with respect to Fermi level. The HOMO level becomes closer to Fermi level and we can use the device as P-type. The back gate is also used for N-type device in order to tune the threshold voltage of the device. In case of N-type device we apply a back gate voltage (V_{BG} = 1 V) to make the LUMO level so close to Fermi level that the conduction becomes possible at low bias conditions (which are stated below).

The different regions of the architectures are highlighted and mentioned in Figure 10(a). Molecular transistors and nanowires are used to construct the P-type and N-type regions. Interconnections enable the communication between the transistors as well as between devices and

inputs/outputs. Back gate electrodes for the N-type and P-type also need interconnection space. A common back gate was used for the whole N-type region and similarly another one for P-type region Figure 10(a). Vertical and horizontal nanowires are connected using resistors at the junctions. A more complex structure like full adder Figure 10(b) is designed and simulated.

Figure 11 shows representative input/output for the input combinations of A, B and Cin. The input signal levels are swept from logic 0 (−0.3 V) to logic 1 (+0.3 V). The value of power supply voltage is 0.3 V and the reference voltage is −0.3 V. The outputs S and C for all three models show that these values have very close high and low values and, moreover, these values are stable after each input changes. The corresponding truth table in Figure 11(b) demonstrates the narrow distribution of the output voltage for both low and high states. This shows the potential to further integrate the device into a larger-scale integrated circuit such as 16-bit full adder in a cascaded configuration (see the next section).

5.1. Impact of Gate Coupling. Transport properties of molecular devices can be significantly effected by the factors like molecular length, gate coupling, and coupling between metal

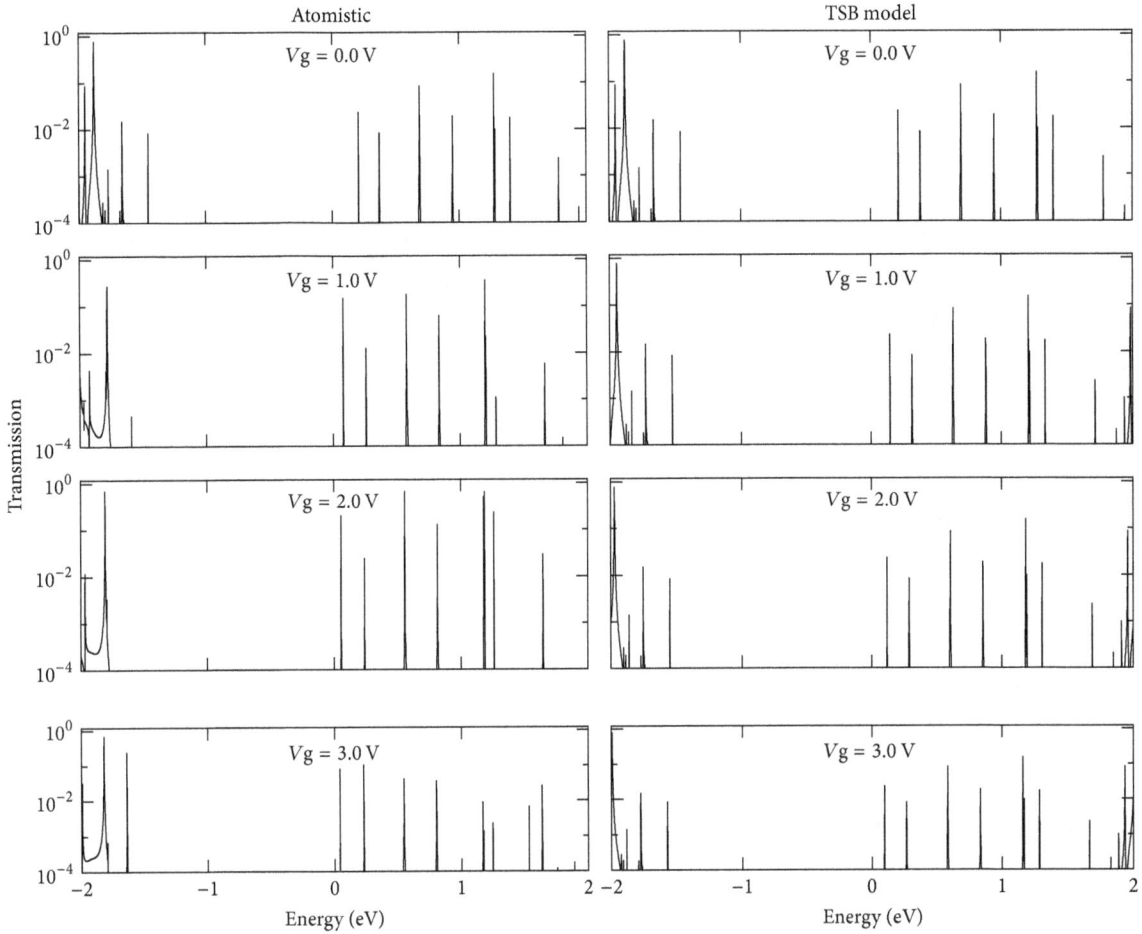

FIGURE 9: Transmission spectrum for different Vg (Vds = 0). The shift in the transmission spectrum can be observed with change in gate voltage.

and molecule. For the case of gate coupling we show in Figure 12 as an example the result for a simple inverter based on a structure not complementary but based on N-FET and resistor as presented in [17] (very similar to the complementary structure, just we do not show the circuit for space reason; please refer to the cited paper for details). We show it in this case because the point is more clear than in the complementary case. The inverter shows a correct logic value at the output evaluation. But the difference can be seen in the form of high voltage level for logic 0 so cascaded gate can not be derived. A different gate coupling can be used to solve this problem. Only FBE and TSB models allow the change in gate coupling. Other physical values like slight changes in energy levels (for example due to change in environment conditions) can only be captured by TSB model by changing the TS according to the change in energy levels. But in case of FBE model the change in energy levels can not be taken into account and it would need new fitting procedure. Our methods can be used to find trade-off between the models according to the simulation requirements and conditions.

5.2. Impact of Fabrication Induced Variations. Increasing length of molecule changes the transmission spectrum (TS) in three ways: (1) the number of transmission peaks in a given energy range is increased; (2) the HOMO-LUMO gap is reduced; (3) width of transmission peaks decreases. The current through device changes as a result of these changes in TS, and thus the circuit behavior also changes. In order to include these variations in the SE-LUT model atomistic simulations for all the values of external voltages are necessary. In case of FBE model initial fitting procedure would be required. The TSB model allows including these variations easily, by analyzing TS and changing metal-molecule coupling. In the following we show some results on the impact of this type of variation.

Figure 13 shows the transmission spectrum for different length of molecule, that is, OPV3 having 3 benzene rings, OPV5 having 5 benzene rings, and OPV7 having 7 benzene rings. As expected, increasing the molecular length changes the TS in two ways. First, it increases the number of transmission peaks. This can be observed if we compare the transmission spectrum of these three molecules. The OPV7 has maximum number of peaks while OPV3 has minimum number of peaks. Second, it reduces the width of the transmission spectrum. It is shown in the encircled area (red) in the figure. The width of the transmission spectrum decreases from OPV3 to OPV7 molecule.

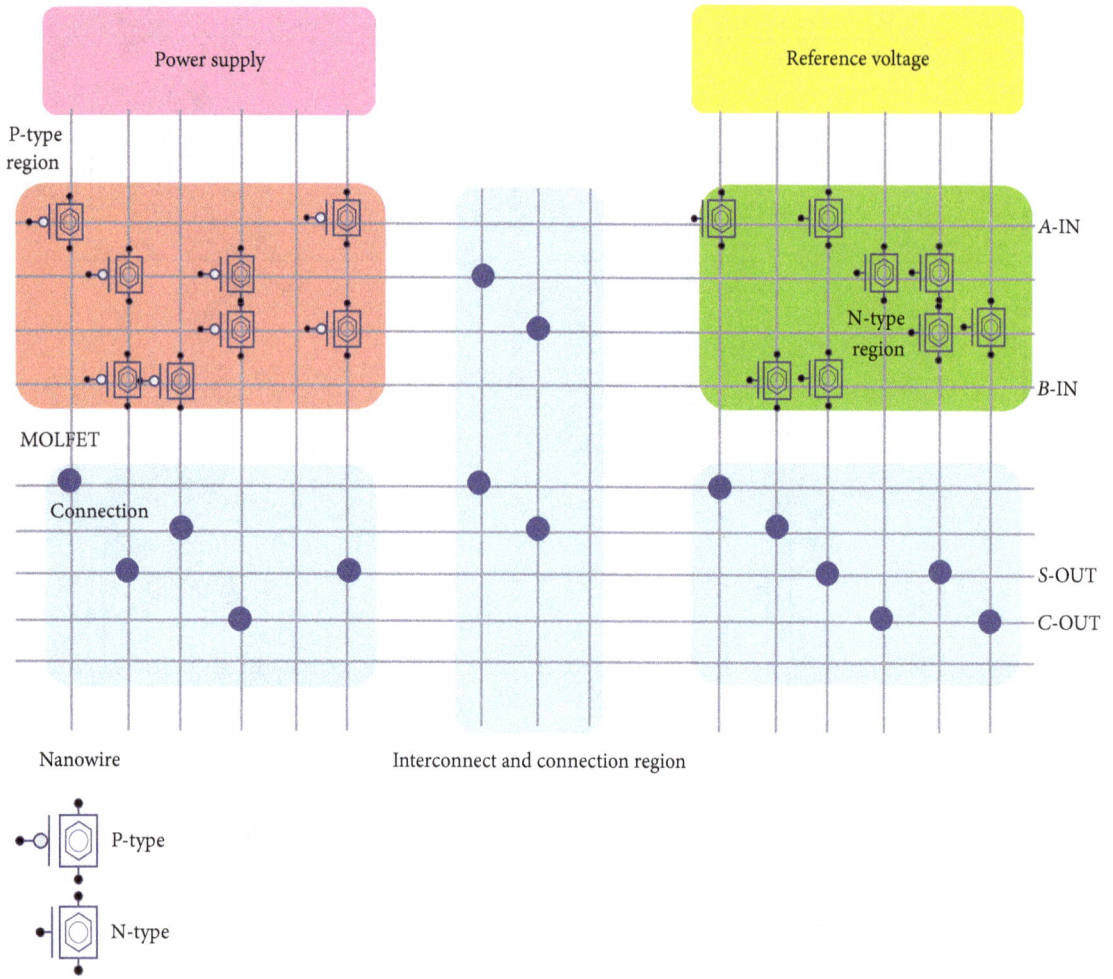

(a)

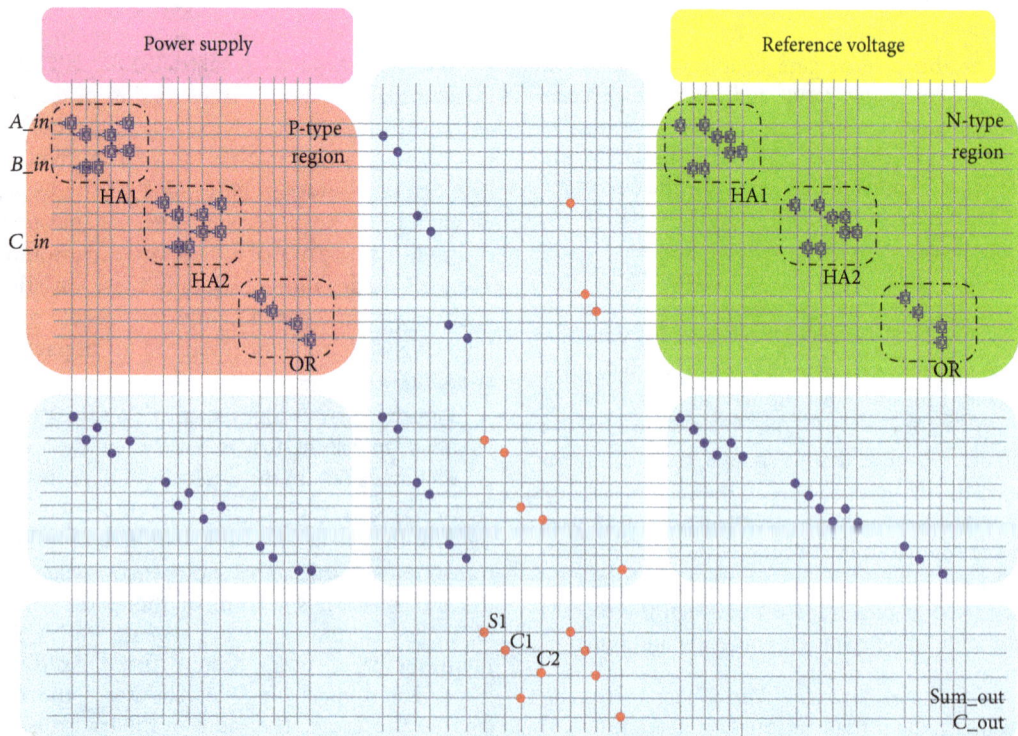

(b)

FIGURE 10: Continued.

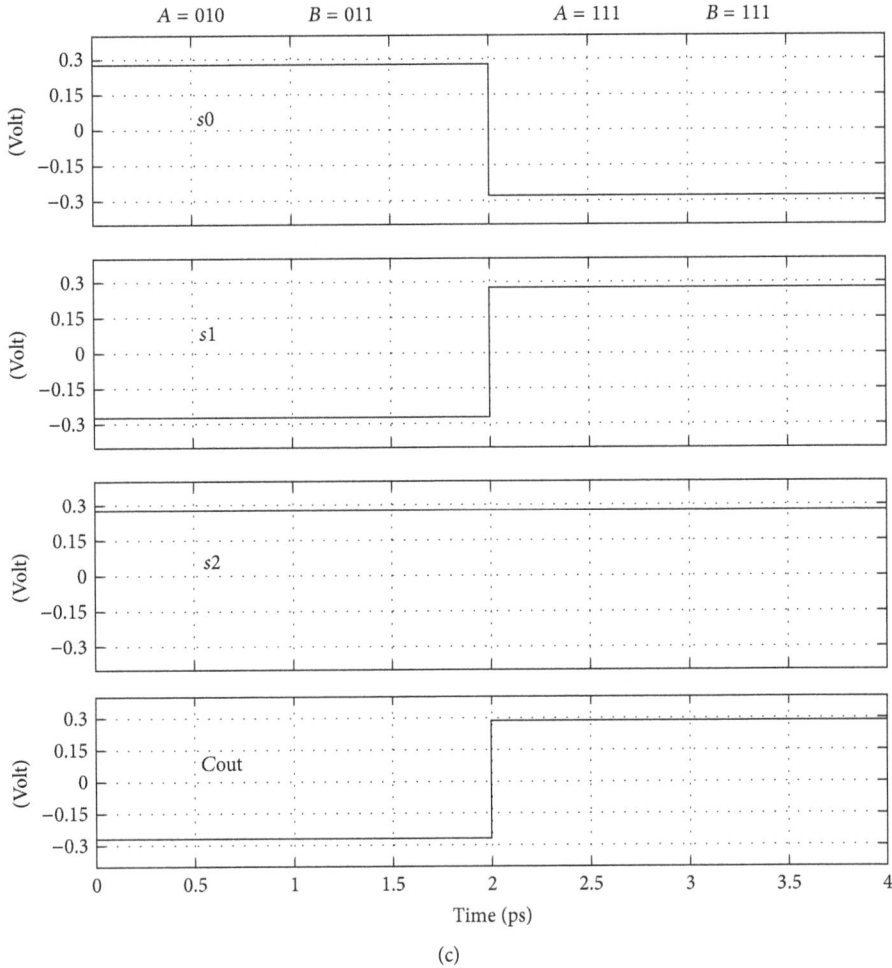

FIGURE 10: Crossbar architecture of a: (a) half adder implemented using complementary logic. The orange color shows a common back gate for P-type region and the green shows that for N-type region. (b) Single-bit full adder, implemented by the connection of two HAs and an OR gate. (c) Output waveforms of a 3-bit ripple carry adder implemented using complementary logic. In this case, the low and high input values of −0.3 V and +0.3 V are used to encode logic "0" and "1," respectively.

Figure 14 shows how the effect of changes in TS due to molecular length impacting the *I-V* characteristics of the molecular system. The highest current is observed in OPV3 molecule and it decreases with increase in length: OPV7 has the smallest current. These simulations were obtained using ATK and the data observed from TS variations were included in the TS model. We were then able to analyze the impact of variations at the circuit level.

Figure 15 shows the implementation of the HA with data found for three different molecules: OPV3, OPV5, and OPV7. Table 1 shows the output voltages for both high and low states. In table input 0 logic represents −0.3 V and input logic 1 represents 0.3 V. As described earlier, the number of transmission peaks in a specific range increases with the increase in molecular length. So when the device is ON there are more numbers of peaks within the bias window and the current increases. When device is OFF there are no peaks within the bias window so increased number of peaks in case of larger molecule does not increase OFF current. On the other hand increasing the molecular length decreases the

TABLE 1: Details on voltage levels for the HA in presence of different molecule length.

Input		Sum			Carry		
B	A	OPV3	OPV5	OPV7	OPV3	OPV5	OPV7
0	0	−169 mV	−183 mV	−295 mV	−293 mV	−295 mV	−299 mV
0	1	179 mV	190 mV	295 mV	−264 mV	−267 mV	−298 mV
1	0	183 mV	193 mV	295 mV	−264 mV	−267 mV	−298 mV
1	1	−195 mV	−206 mV	−195 mV	157 mV	170 mV	295 mV

width of peaks. This leads to decrease in the current of the device with larger molecular length. This phenomenon affects both the ON current and OFF current of the device. Overall effect of increasing the molecular length is to decrease the OFF current and the ON current. But the reduction in OFF current is more than the reduction in ON. So the ratio between ON current and OFF current increases which improves the device performance with increase in molecular length.

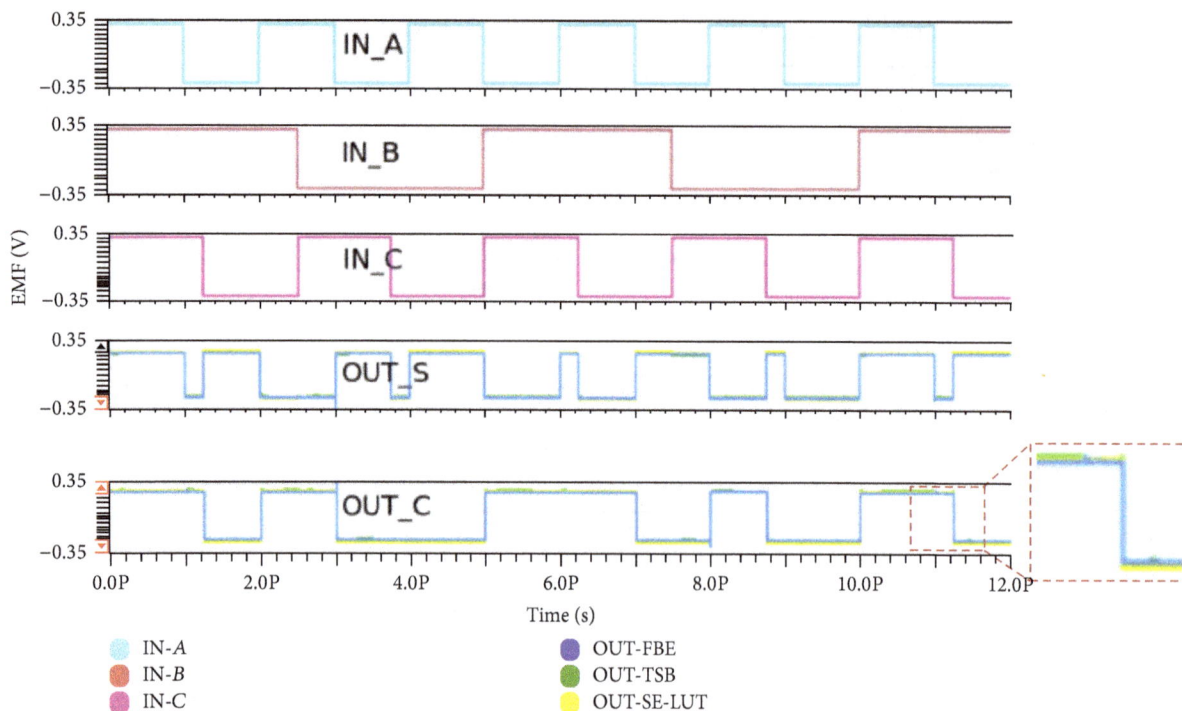

A	B	C_in	Sum TSB_Model	Sum FBE_Model	Sum FBE_Model
0	0	0	0 (V: −0.222)	0 (V: −0.240)	0 (V: −0.235)
0	0	1	1 (V: −0.221)	1 (V: 0.238)	1 (V: 0.222)
0	1	0	1 (V: 0.223)	1 (V: 0.239)	1 (V: 0.236)
0	1	1	0 (V: −0.225)	0 (V: −0.240)	0 (V: −0.236)
1	0	0	1 (V: 0.23)	1 (V: −0.240)	1 (V: −0.224)
1	0	1	0 (V: −0.221)	0 (V: −0.240)	0 (V: −0.236)
1	1	0	0 (V: −0.236)	0 (V: −0.240)	0 (V: −0.234)
1	1	1	1 (V: 0.220)	1 (V: 0.238)	1 (V: 0.220)

FIGURE 11: Logic behavior of the full adder outputs (Sum and Carry) with respect to input *A*, *B* and Carry-in. The output results of all three models are comparable: green line for TSB, blue line for SE-LUT, and yellow line for FBE.

Clearly the molecular length also affects the behavior of the circuit as the voltage levels of the circuits change with molecular length. These kinds of variations are highly important considering the infancy of this technology. Being able to easily include them in the simulation framework is fundamental to execute realistic simulations. These examples were included here just to show the importance of the point and the following section follows the mainstream of demonstrating the framework capabilities in terms of complexity of the description and of the performance evaluation. We

are currently working on the extension of the framework for including process variations in a Monte-Carlo-like fashion exploiting the potentials of the TS model. Results will be presented in a future paper.

6. Architecture

The analysis of a complex architecture based on a cascade of crossbars can evidence (i) the effect of the different technological choices (e.g., gate coupling, molecules) on

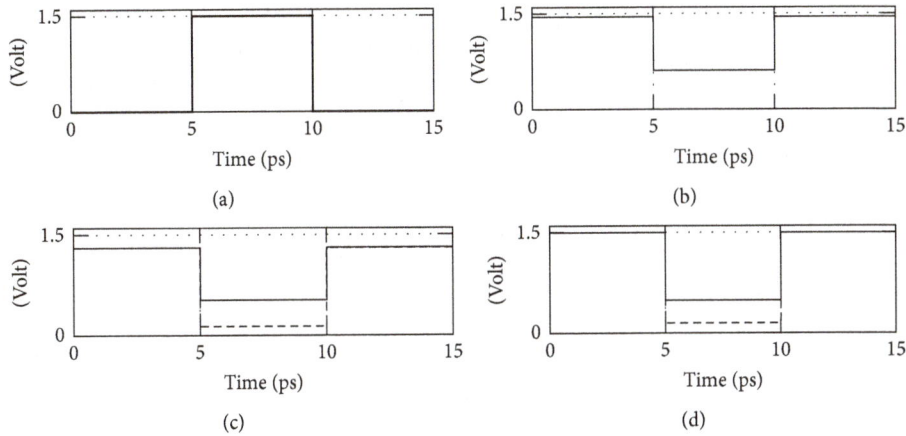

FIGURE 12: Input and output waveforms of inverter: (a) input data. Output data obtained by (b) SE-LUT models, (c) FBE model, and (d) TSB model.

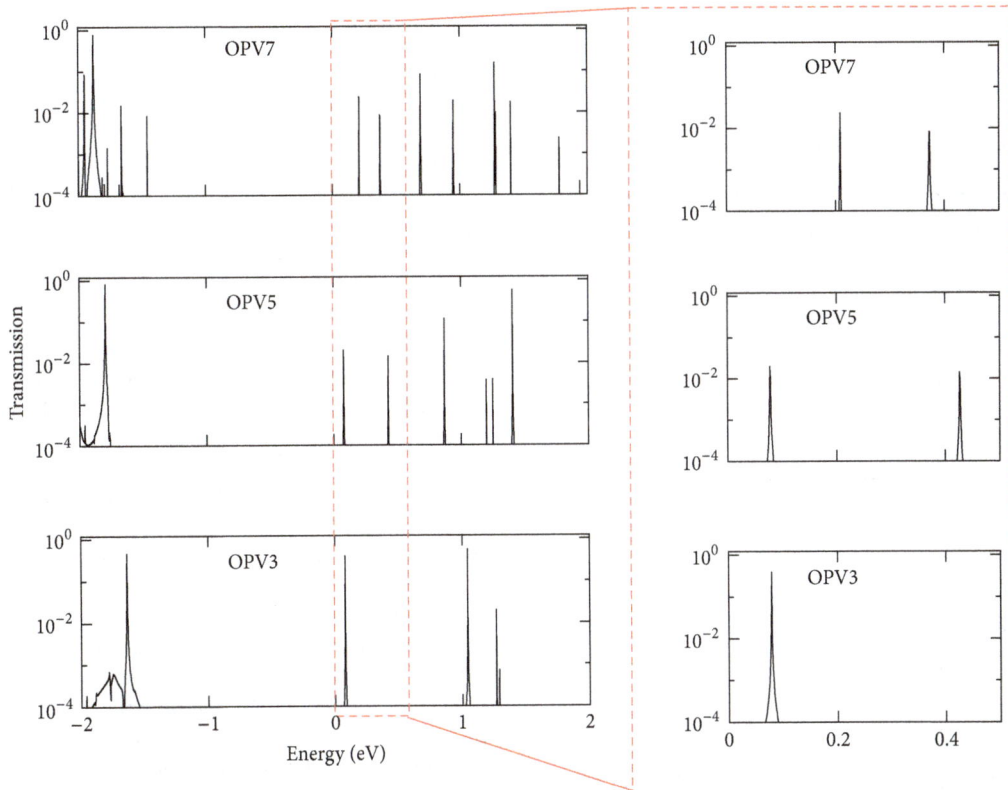

FIGURE 13: TS for molecules of different length OPV3, OP5, and OPV7.

the behavior of circuit, (ii) the capabilities provided by the different models at circuit level, and (iii) the models computational requirements.

The Pentium 4 adder is used here as reference architecture.

6.1. Pentium 4 Adder Implementation. Figure 16 shows the general block diagram of the P4 adder. It is based on two substructures: a carry select adder for sum generator and a sparse tree for carry generator. The Sum generator of 16-bit P4 adder is divided into 4 groups of 4-bit carry select adder. Each

group is organized in two stacked identical 4-bit ripple carry adders (RCA), each generating 4-bit sum and an outgoing carry. One RCA assumes that the incoming carry into the group is 0, while other assumes that it is 1. These two RCAs generate the output in parallel. The final output of the group is selected from the value assigned by sparse tree of the carry generator by means of 4-bit multiplexers.

The carry generator block of P4 adder consists of a sparse tree as shown in Figure 16. This architecture computes every fourth carry in the adder and provides the carry-in signal for individual carry select adder block, effectively

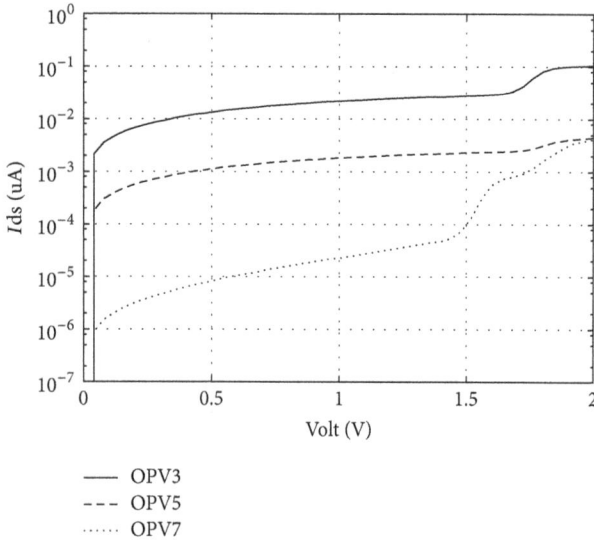

FIGURE 14: Impact of molecule length on *I-V* curves.

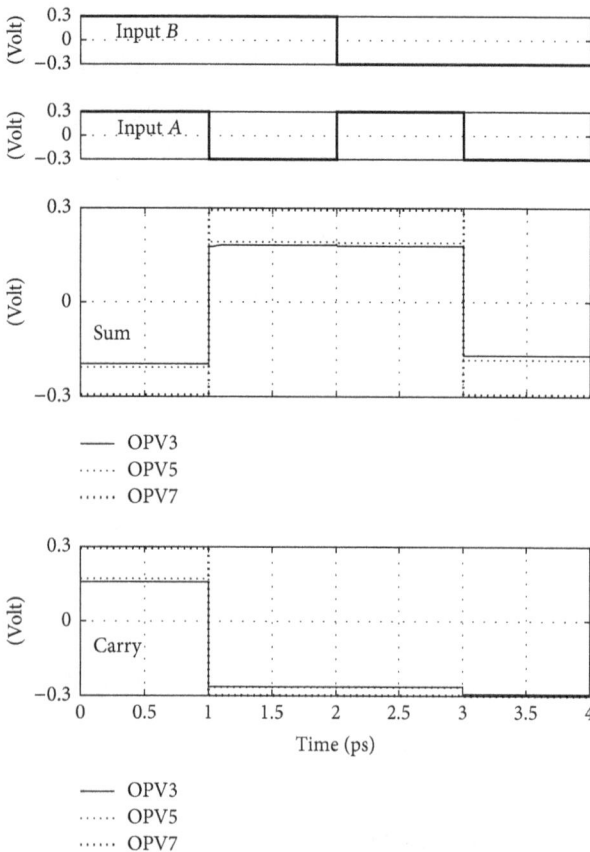

FIGURE 15: HA waveforms in presence of different molecule length.

reducing the critical path of the carry signal. In this way it is possible to work on all the carry generator groups of 4-bits without the need to wait for the carry resulting from previous groups. The tree is made up of a pseudo-regular structure of building blocks. The blocks used to implement the tree are as follows: (1) PG network: it computes propagation (Pin) and

generation (Gin) signals using input signals; (2) PG block: it computes propagation (Pi) and generation (Gi) signals using input signals; (3) G block: it computes only generation (Gi) signals.

Figure 17 shows the output of the 4-bit carry select adder. The P4 adder is based on two substructures, carry generator and sum generator. Sum generator of 16-bit adder is divided into 4 groups of 4-bit carry select adder CSA. Figure 17 is used to show the proper functionality of the CSA block. The logic level of Sum [2] and sum [3] (3rd and 4th bit of sum) of three models are also shown in the figure. Good voltage levels are maintained which shows the potential to integrate the device into large scale.

For performance analysis, the result of 16-bit sparse tree and the full adder is shown in Figure 18. As an example, input values and their corresponding sum outputs are shown by red lines. Simulation results show the correct operation of the adder. The output voltage levels are also maintained. For a comparison, the output waveforms obtained by all three models are placed together which shows very small difference with respect to each other.

To analyze the voltage levels at different nodes, we select nodes N1 to N9 in the longest chain of the sparse tree as shown in Figure 19(a). Voltage level of each selected node (N1 to N9) obtained by three models is compared: TSB, SE-LUT, and FBE (Figure 19(b)). The input values are set such that on each node we get logic "0" output (−300 mV). For FBE model voltage varies from −245 mV to −265 mV, for TBS the variation is between −225 mV and −300 mV, and for SE-LUT it is between −215 mV and −295 mV. Initially, the swing of voltage level increases from moving to higher nodes. However, after node 5 the swing of voltage level becomes stable. This trend can be observed by all three models.

6.2. Timing Comparison. We here want to give a very simple identification of the difference in timing performance for the three models. Timing does not involve only the final circuit simulation time, but also the effort required to setup the model data in order for them to be included in the simulation framework. This analysis gives an idea of which model could be useful depending on the type of data available and the type of analyses requested. We compare the SE-LUT and the TSB models as they are highly different in terms of characteristics on timing and on flexibility in the final results achievable.

Time required to implement P4 adder is evaluated as a function of the number of bias points (NVds) and number of gate voltages (NVg) points of the MOL-TRAN characteristics. For the sake of simplicity hereinafter we consider NVg = NVds = N. The timing analysis of both systems is estimated as follows:

$$T_{\text{SE-LUT}} = T_{\text{Atm}} * N^2 + T_{\text{Arch}} + T_{\text{Algo}},$$
$$T_{\text{TSB}} = (2 * N * T_{\text{Atm}}) + (10 * T_{\text{Arch}}) + T_{\text{Algo}}. \tag{1}$$

T_{Atm} is the time required by atomistic simulations. The time required to compute current at given Vds and Vg depends on the convergence time of the self-consistence field loop. If the convergence occurs early (e.g., within 10 iterations), then the computation time would be less. And if

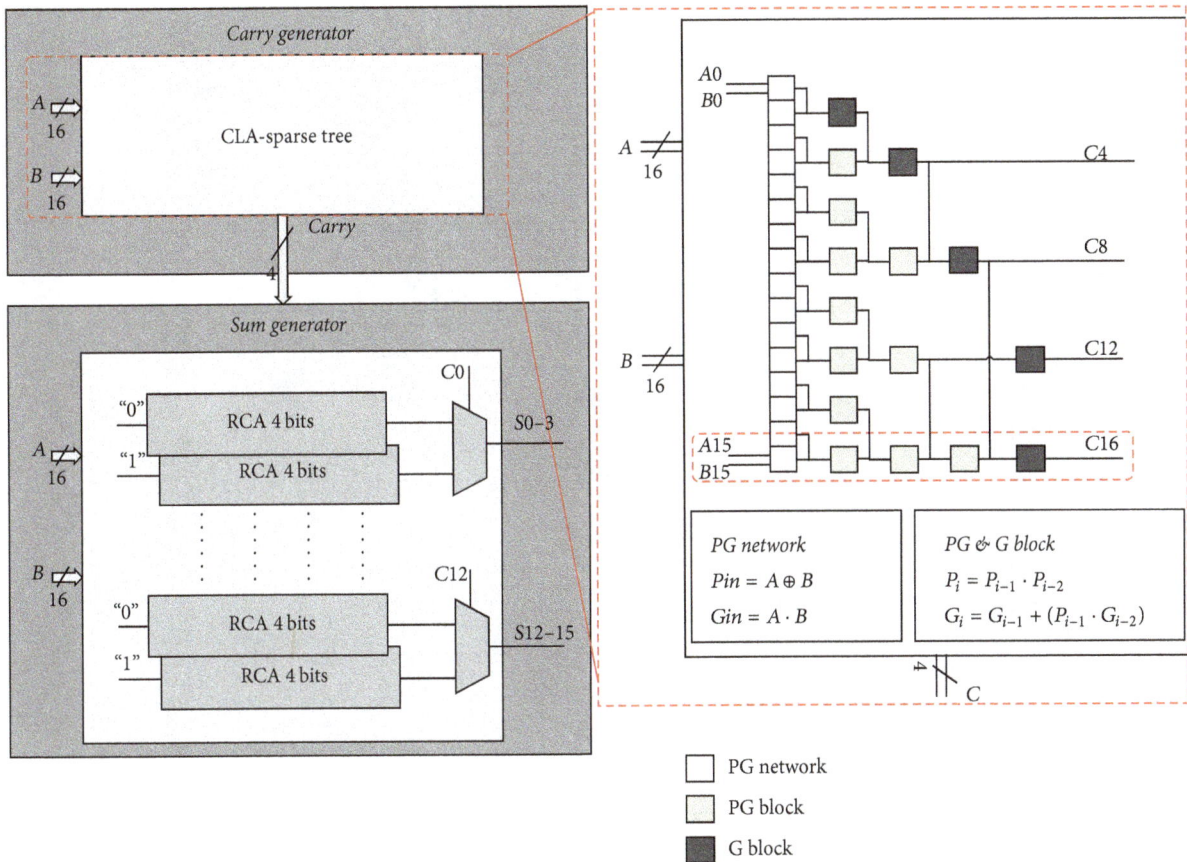

FIGURE 16: The general block diagram of the P4 adder and the carry generator block of P4 adder consists of a sparse tree.

PG network

$Pin = A \oplus B$

$Gin = A \cdot B$

PG & G block

$P_i = P_{i-1} \cdot P_{i-2}$

$G_i = G_{i-1} + (P_{i-1} \cdot G_{i-2})$

☐ PG network

☐ PG block

■ G block

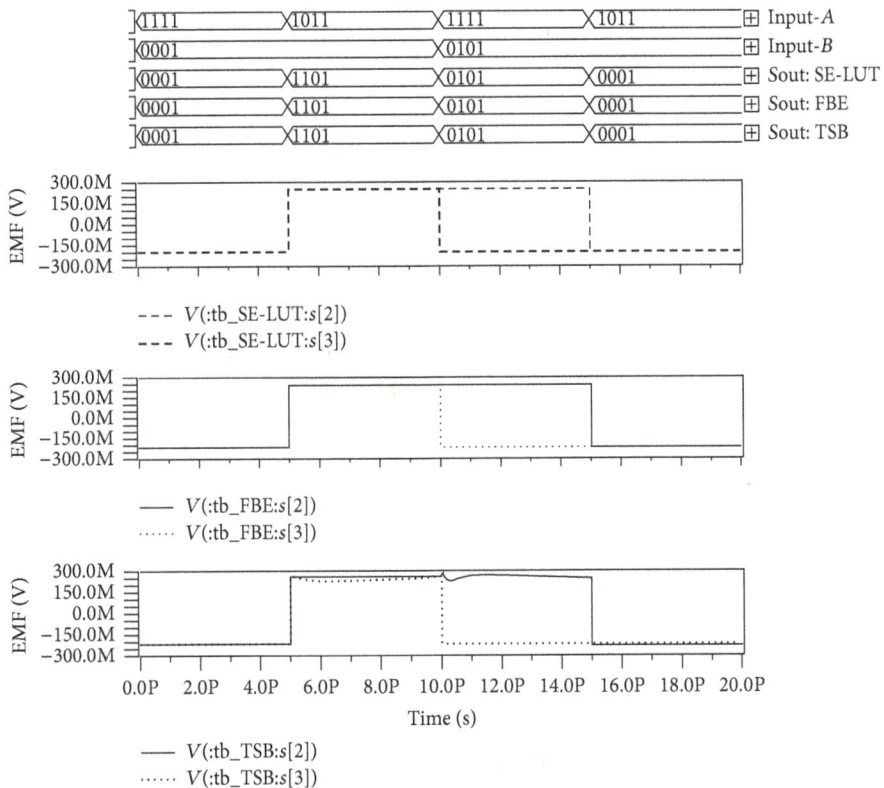

$--- V(:tb\_SE\text{-}LUT:s[2])$
$--- V(:tb\_SE\text{-}LUT:s[3])$

$--- V(:tb\_FBE:s[2])$
$\cdots\cdots V(:tb\_FBE:s[3])$

$--- V(:tb\_TSB:s[2])$
$\cdots\cdots V(:tb\_TSB:s[3])$

FIGURE 17: Carry select adder output waveforms.

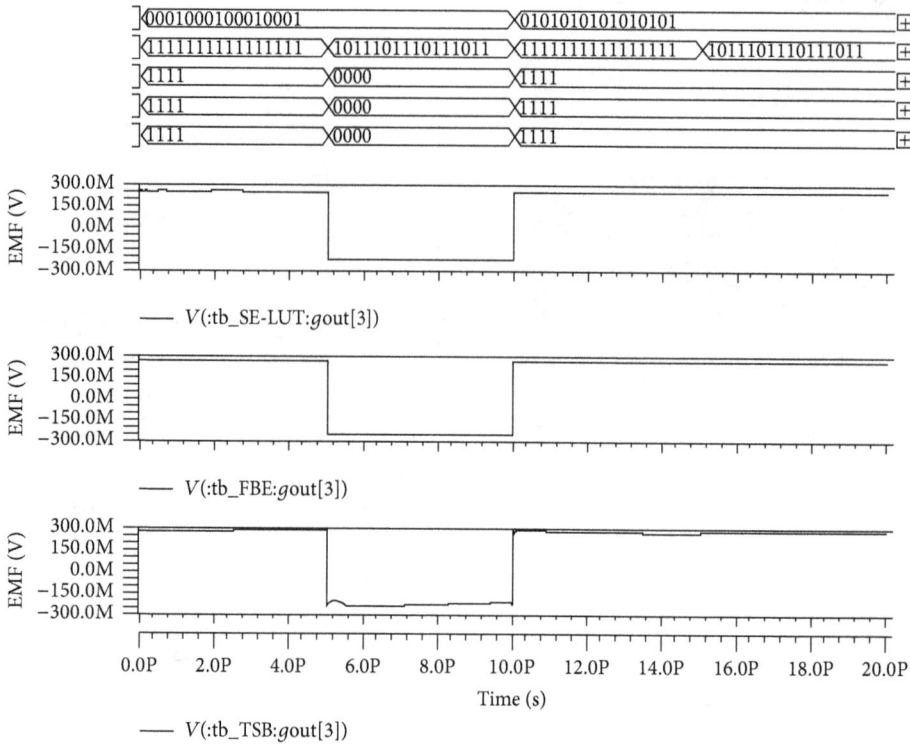

FIGURE 18: Output waveforms for the sparse tree adder.

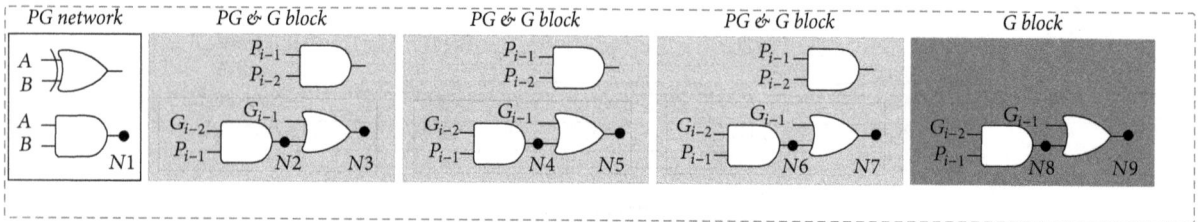

(a)

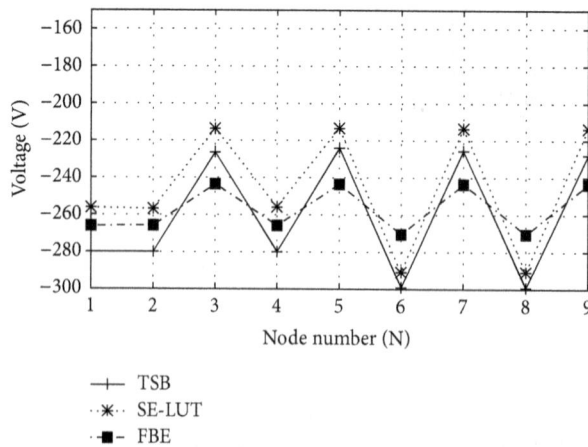

(b)

FIGURE 19: (a) A schematic of sparse tree area highlighted in Figure 16. Voltage level at each selected node (N1 to N9) of the sparse tree.

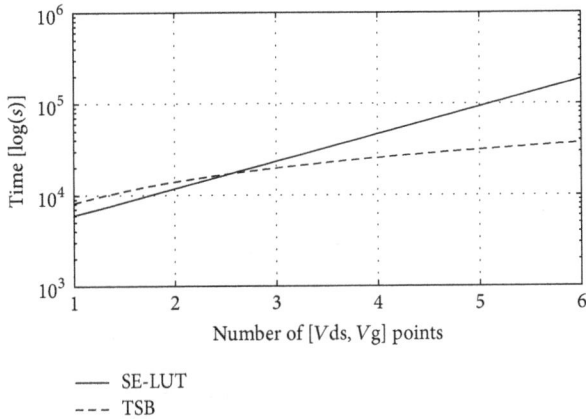

FIGURE 20: Timing comparison of SE-LUT and TSB model.

the number of iterations is large then the time would be large. In our case, we take the average time to calculate current (i.e., total time/total number of Vds and Vg points). In this case, $T_{Atm} = 2891$ s.

T_{Arch} is the time required to implement the architecture in VHDL-ams. In all our simulations, implementation of P4 architecture by semiempirical lookup table requires less time as compared to other models. For transmission spectrum based model the time is about ten times the SE-LUT model, and T_{Arch} (for SE – LUT) = 240 s.

T_{Intr} is the time required for interpolation of SE-LUT model: this time is estimated as about 1 millisecond. T_{Algo} is the time required for implementation of algorithm by TSB model in our case $T_{Algo} > 1.35e - 3$ s.

Using these values in the equations we obtain the trend in Figure 20. The time required to implement the whole set of data and simulation for SE-LUT model increases exponentially, while the time of the TSB increases linearly. Thus, when NVds < 2 SE-LUT requires a smaller amount of total time. However, after NVds > 2 the total time of SE-LUT increases as compared to TBS. Thus in more realistic set of values, the computational demand of SE-LUT model will be much higher as compared to TSB model.

7. Remarks and Conclusions

In this work, we presented a modular framework based on VHDL-AMS for the description of circuits based on molecular-FET. The I-V characteristics obtained by the three models proposed were compared to atomistic simulations. The three models used in the framework are transmission spectrum based model (TSB), FET based equivalent model (FBE), and semiempirical look-up table based model (SE-LUT). Though I-V characteristics showed good results, the three models differ for flexibility and effort required to generate the necessary data. The framework allows the description of complex molecular circuits that have never been previously designed and tested in terms of functionality. In this case we adopted a crossbar based circuit and implemented a 16-bit tree adder reproducing the architecture of the Pentium 4 adder. This is not at all possible using atomistic level simulations. We demonstrated the correct functionality and

estimated the timing requirements to achieve the results if we consider to have a new molecule or new conditions on a previously available molecule. Results show that the SE-LUT model is straightforward for the setup and simple for the simulation but might require very long time in case a realistic set of data is necessary. The other models are more complex to be described and require some initial effort but are more flexible. In particular, TSB model easily allows including parameters variation derived by technological or environmental conditions for the molecule.

We have simulated, for the first time, the molecular transistor on a circuit level using the OPV molecule. Our approach allows fast assessment of the performance overcoming the need of complex quantum simulations of the system with many molecular devices currently not possible. Our methods allow seeing the impact of physics and chemistry of molecular devices on the applications and giving feedback to the technologists. In the future we aim to include the effect of process variations into the models. Some examples of the process variations include different binding strength between anchoring group and electrodes, stacking of molecules, and variation in molecular structure during synthesis.

Conflicts of Interest

The authors declare no conflicts of interest.

Acknowledgments

The authors wish to thank professor Paolo Lugli for his support and for allowing them to use the FBE model.

References

[1] S. Datta, *Quantum Transport: Atom to Transistor*, Cambridge University Press, 2005.

[2] G. Csaba and P. Lugli, "Read-out design rules for molecular crossbar architectures," *IEEE Transactions on Nanotechnology*, vol. 8, no. 3, pp. 369–374, 2009.

[3] A. Cao, E. J. R. Sudhölter, and L. C. P. M. de Smet, "Silicon nanowire-based devices for gas-phase sensing," *Sensors*, vol. 14, no. 1, pp. 245–271, 2013.

[4] A. Chiolerio, P. Allia, and M. Graziano, "Magnetic dipolar coupling and collective effects for binary information codification in cost-effective logic devices," *Journal of Magnetism and Magnetic Materials*, vol. 324, no. 19, pp. 3006–3012, 2012.

[5] M. T. Björk, H. Schmid, J. Knoch, H. Riel, and W. Riess, "Donor deactivation in silicon nanostructures," *Nature Nanotechnology*, vol. 4, no. 2, pp. 103–107, 2009.

[6] M. Vacca, J. Wang, M. Graziano, M. R. Roch, and M. Zamboni, "Feedbacks in QCA: A Quantitative Approach," *IEEE Transactions on Very Large Scale Integration (VLSI) Systems*, vol. 23, no. 10, pp. 2233–2243, 2015.

[7] S. Hong, R. Reifenberger, W. Tian, S. Datta, J. I. Henderson, and C. P. Kubiak, "Molecular conductance spectroscopy of conjugated, phenyl-based molecules on Au(111): the effect of end groups on molecular conduction," *Superlattices and Microstructures*, vol. 28, no. 4, pp. 289–303, 2000.

[8] T. Ishida, W. Mizutani, N. Choi, U. Akiba, M. Fujihira, and H. Tokumoto, "Structural Effects on Electrical Conduction of Conjugated Molecules Studied by Scanning Tunneling Microscopy,"

The Journal of Physical Chemistry B, vol. 104, no. 49, pp. 11680–11688, 2000.

[9] H. C. Seong, B. Kim, and C. D. Frisbie, "Electrical resistance of long conjugated molecular wires," *Science*, vol. 320, no. 5882, pp. 1482–1486, 2008.

[10] F. Zahid, M. Paulsson, E. Polizzi, A. W. Ghosh, L. Siddiqui, and S. Datta, "A self-consistent transport model for molecular conduction based on extended Hückel theory with full three-dimensional electrostatics," *The Journal of Chemical Physics*, vol. 123, no. 6, Article ID 064707, 2005.

[11] A. Mahmoud and P. Lugli, "Designing the rectification behavior of molecular diodes," *Journal of Applied Physics*, vol. 112, no. 11, Article ID 113720, 2012.

[12] M. Graziano, A. Pulimeno, R. Wang, X. Wei, M. R. Roch, and G. Piccinini, "Process variability and electrostatic analysis of molecular QCA," *ACM Journal on Emerging Technologies in Computing Systems*, vol. 12, no. 2, article no. 18, 2015.

[13] A. Pulimeno, M. Graziano, A. Sanginario, V. Cauda, D. Demarchi, and G. Piccinini, "Bis-ferrocene molecular QCA Wire: Ab initio simulations of fabrication driven fault tolerance," *IEEE Transactions on Nanotechnology*, vol. 12, no. 4, pp. 498–507, 2013.

[14] M. A. Reed, H. Song, and T. Lee, "Molecular Transistors," *Emerging Nanoelectronic Devices*, pp. 194–226, 2015.

[15] W. D. Wheeler and Y. Dahnovsky, "Molecular transistors with perpendicular gate field architecture: A strong gate field effect," *The Journal of Physical Chemistry C*, vol. 113, no. 3, pp. 1088–1092, 2009.

[16] T. M. Perrine, R. G. Smith, C. Marsh, and B. D. Dunietz, "Gating of single molecule transistors: Combining field-effect and chemical control," *The Journal of Chemical Physics*, vol. 128, no. 15, Article ID 154706, 2008.

[17] A. Zahir, A. Pulimeno, D. Demarchi et al., "Modular framework for molecular-FET device-to-circuit modeling," in *Proceedings of the 15th IEEE International Conference on Nanotechnology, IEEE-NANO 2015*, pp. 156–159, Italy, July 2015.

[18] Atomistix ToolKit, "Version 12.8.2," QuantumWise A/S, Copenhagen, Denmark, 2012.

[19] J. M. Soler, E. Artacho, J. D. Gale et al., "The SIESTA method for ab initio order-N materials simulation," *Journal of Physics: Condensed Matter*, vol. 14, no. 11, pp. 2745–2779, 2002.

[20] A. Zahir, A. Pulimeno, D. Demarchi et al., "EE-BESD: molecular FET modeling for efficient and effective nanocomputing design," *Journal of Computational Electronics*, vol. 15, no. 2, pp. 479–491, 2016.

[21] A. Zahir, S. A. A. Zaidi, A. Pulimeno et al., "Molecular transistor circuits: From device model to circuit simulation," in *Proceedings of the 2014 IEEE/ACM International Symposium on Nanoscale Architectures, NANOARCH 2014*, pp. 129–134, fra, July 2014.

[22] A. Mahmoud and P. Lugli, "Toward circuit modeling of molecular devices," *IEEE Transactions on Nanotechnology*, vol. 13, no. 3, pp. 510–516, 2014.

[23] A. Mahmoud and P. Lugli, "Atomistic study on dithiolated oligophenylenevinylene gated device," *Journal of Applied Physics*, vol. 116, no. 20, Article ID 204504, 2014.

[24] M. I. Schukfeh, K. Storm, A. Mahmoud et al., "Conductance enhancement of InAs/InP heterostructure nanowires by surface functionalization with oligo(phenylene vinylene)s," *ACS Nano*, vol. 7, no. 5, pp. 4111–4118, 2013.

[25] VHDL Analog and Mixed-Signal Extensions, IEEE Std 1076.1-1999.

[26] ADVance MS (ADMS), Reference Manual, Mentor Graphics, 2013.

[27] E. Christen and K. Bakalar, "VHDL-AMS - a hardware description language for analog and mixed-signal applications," *IEEE Transactions on Circuits and Systems II: Analog and Digital Signal Processing*, vol. 46, no. 10, pp. 1263–1272, 1999.

[28] P. Loumeau and J. F. Naviner, "VHDL-AMS behavioral modelling and simulation of high-pass delta-sigma modulator," in *Proceedings of the BMAS 2005 - 2005 IEEE International Behavioral Modeling and Simulation Workshop*, pp. 106–111, USA, September 2005.

[29] M. R. Casu, M. Crepaldi, and M. Graziano, "A VHDL-AMS simulation environment for an UWB impulse radio transceiver," *IEEE Transactions on Circuits and Systems I: Regular Papers*, vol. 55, no. 5, pp. 1368–1381, 2008.

[30] F. Pêcheux, C. Lallement, and A. Vachoux, "VHDL-AMS and Verilog-AMS as alternative hardware description languages for efficient modeling of multidiscipline systems," *IEEE Transactions on Computer-Aided Design of Integrated Circuits and Systems*, vol. 24, no. 2, pp. 204–224, 2005.

[31] M. Graziano and M. Ruo Roch, "An automotive CD-player electro-mechanics fault simulation using VHDL-AMS," *Journal of Electronic Testing*, vol. 24, no. 6, pp. 539–553, 2008.

20

Design of a Narrow Bandwidth Bandpass Filter using Compact Spiral Resonator with Chirality

Weiping Li,[1,2] **Zongxi Tang,**[1] **and Xin Cao**[1]

[1]*School of Electronic Engineering, University of Electronic Science and Technology of China, No. 2006 Xiyuan Ave, West Hi-Tech Zone, Chengdu 611731, China*
[2]*School of Information Engineering, East China Jiaotong University, No. 88 Shuanggang Road, Nanchang 330013, China*

Correspondence should be addressed to Weiping Li; lwp8277@126.com

Academic Editor: Gerard Ghibaudo

In this article, a compact narrow-bandpass filter with high selectivity and improved rejection level is presented. For miniaturization, a pair of double negative (DNG) cells consisting of quasi-planar chiral resonators are cascaded and electrically loaded to a microstrip transmission line; short ended stubs are introduced to expand upper rejection band. The structure is analyzed using equivalent circuit models and simulated based on EM simulation software. For validation, the proposed filter is fabricated and measured. The measured results are in good agreement with the simulated ones. By comparing to other filters in the references, it is shown that the proposed filter has the advantage of skirt selectivity and compact size, so it can be integrated more conveniently in modern wireless communication systems and microwave planar circuits.

1. Introduction

Design of very compact microwave devices compatible with printed circuit board and monolithic-microwave integrated-circuit fabrication technologies has gained great interest in the last decades. The split-ring resonators (SRRs) and their counterparts, complementary split-ring resonators (CSRRs), are key aspects that propose new design strategies to miniaturize planar microwave circuit [1]. It is demonstrated that, by combining two metal levels at both sides of a dielectric layer connected by vias with an appropriate topology, it is possible to design new resonators with a higher level of miniaturization [2]. These resonators which often exhibit chirality [3] can be used to synthesize left handed structures with neither severe degradation in the quality factor [2] nor causing electromagnetic compatibility (EMC) problems dealing with SRR and CSRR structures [3, 4].

Narrow-bandpass filters (NBPFs) with sharp selectivity and high rejection are increasingly demanded in modern microwave communications systems. In order to miniaturize such filters for circuit integration, many research works

have been done. In [5], composite right/left-handed coplanar waveguide (CRLH-CPW) resonators are used to create a passband. But in general, the parasitic effects of CPW on the coupling gaps usually cannot be neglected. Therefore, the selectivity would be compromised. In [6], CSRRs have been loaded on the top side of the substrate, but a good deal of insertion loss of the passband is caused by the coupling and fringing capacitance brought by this structure. Also, unwanted spurious responses occur. In [7], multisection stepped-impedance resonators are cascaded to create a relative wide passband. However, the resonant mode cannot be easily controlled and the design process is comparatively complex. In [8], electromagnetic bandgap (EBG) is introduced on the ground plane of the filter. Better out-of-band rejection has been achieved and the design process can be more flexible. But in most cases, the loss of the in-band frequency response is inevitably increased by the rejection effect of EBG. Therefore, some performances must be balanced to make compromises. In [9], SRRs are coupled together to create a single passband. Since the structures of SRRs and CSRRs are quite the same, similar problems have

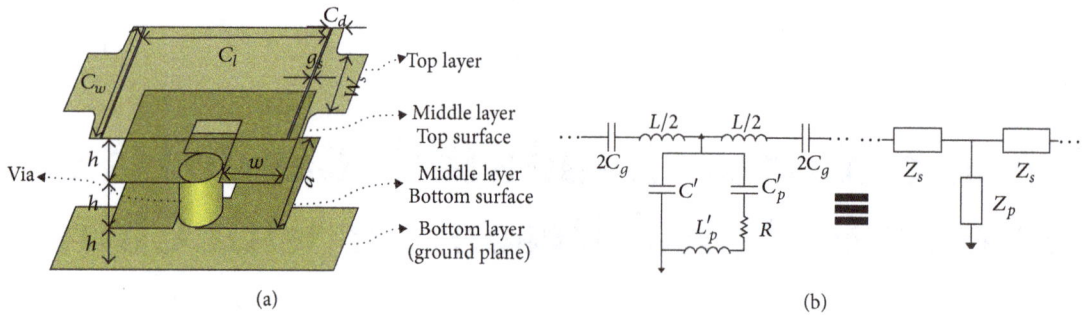

(a)

(b)

FIGURE 1: (a) 3D layout of the DNG cell including quasi-planar chiral resonator (particle) electrically loaded to a microstrip line with series gaps. (b) Lumped element equivalent circuit and equivalent T-model.

been encountered as that with CSRRs. Recently, other types of structures have been proposed to overcome the afore-mentioned disadvantages. In [10–17], composite right/left-handed (CRLH) metamaterial structure is demonstrated. CRLH resonators have positive, negative, and zeroth resonant modes based on the working condition. At zeroth resonant mode, the passband can achieve the highest selectivity with the lowest insertion loss due to the low conductor loss at this special resonant mode. In addition, CRLH resonators are often designed in the form of interdigital or spiral geometric structure, which makes the size more compact. In [18–25], defected ground structure (DGS) has been proposed. DGSs are etched at the bottom plane of the resonators and band-stop frequency responses are achieved. Since DGS is independent of the resonators on the upper layer, the design process can be much more flexible. Out-of-band rejection can be increased by the slow-wave effect of DGS. Moreover, DGS are perfectly compatible with CRLH structures. In order to further miniaturize the size of the microwave circuit, in [26], multilayered quasi-planar structure is proposed. Different electromagnetic structures are packed into different layers of the substrate through weak coupling. Thus, more compact size has been achieved without major influence on the performances of the device. Inspired by these previous literatures, in our work, a novel NBPF with good selectivity and higher level of miniaturization using quasi-planar chiral resonators is proposed. The corresponding equivalent circuit has been analyzed and the results of full-wave simulation and experimental measurement have been presented.

2. Filter Design

Figure 1(a) shows the topology of the double negative (DNG) cell conceptually proposed in [4], consisting of a microstrip line with series gaps and a quasi-planar chiral resonator, electrically loaded to the transmission line. As illustrated in Figure 1(a), this topology can be printed on the surfaces of a three-layer substrate. The circuit model of the cell and its transformed T-model are depicted in Figure 1(b) (the circuit model is clearly described in [4]). In [4] it was demon-strated that this resonator provides negative permittivity in a narrow band after its resonant frequency. On the other hand, series gaps (C_g) between transmission lines expose

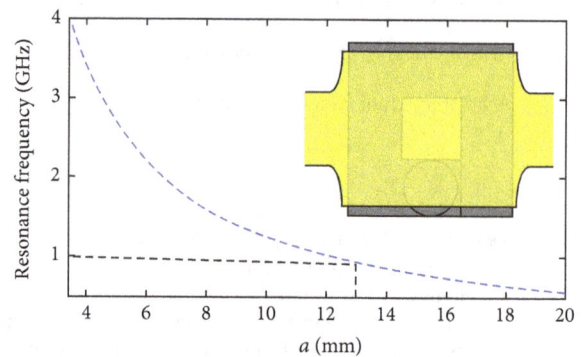

FIGURE 2: The variation of the resonant frequency of the chiral particle with respect to changing "a" (depicted in Figure 1). Other dimension parameters are set to $w \approx a/3$, $W_s = 5.5$ mm, $C_l = 10$ mm, $C_w = 11.4$ mm, and $C_d = 1.4$ mm.

negative permeability before their plasma frequency. Thus, this subwave-length structure with appropriate topology supports backward-wave propagation in a narrow frequency band. Therefore, the main idea behind this work is to use this DNG cell for implementing a compact NBPF.

Because the particle is excited with electrical field perpen-dicular to the substrate, for better coupling, the transmission line above the particle is widened. In addition, this widening prevents gap size from being closer to the limits imposed by the fabrication technology (approximately 0.1 mm). The line is tapered next to the widened section for a better matching.

Particle resonant frequency is determined by its dimen-sions. Among the dimension parameters indicated in Fig-ure 1(a), to the side length of the loop, "a" is the most important one on the resonant frequency. Figure 2 shows the resonant frequency variation of the resonator with respect to changing "a." In the simulation, three layers of Rogers 4003C substrate with relative dielectric constant (ε_r) of 3.55, thickness (h) of 0.813 mm, and loss tangent δ of 0.0027 are used and copper metallization thickness is 35 μm. To realize the passband at 1 GHz, the value of "a" is set to be 12.9 mm.

While setting the gap size (g_s), it can be noticed that larger g_s provides more suitable rejection up to the plasma frequency but results in a higher insertion loss within the

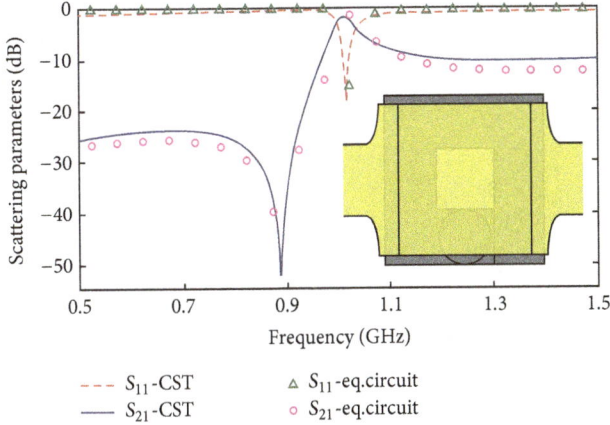

FIGURE 3: Scattering parameters of the DNG cell. The elements of the equivalent circuit are extracted as $C'_p = 4.396$ pF, $L'_p = 7.308$ nH, $R = 0.6\,\Omega$, $C' = 10.24$ pF, $L = 4.182$ nH, and $C_g = 0.8575$ pF.

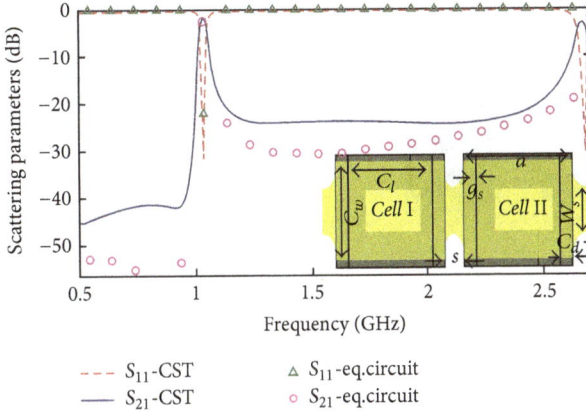

FIGURE 4: Scattering parameters of two cascade DNG cells ($s = 2.1$ mm).

passband. On the other hand, smaller g_s decreases the rejection level within the stopband. In this design g_s is set to be 0.2 mm to lessen the influence caused by fabrication uncertainties.

The results of full-wave electromagnetic and equivalent circuit simulations of the structure shown in Figure 1(a) are depicted in Figure 3, which clearly shows a narrow-passband around 1 GHz (the equivalent circuit parameters of the topology are extracted from equations (14), (15), and (17)-(19) of [4] and given in the caption).

To achieve a deeper rejection and also a sharper pass-band, two DNG cells are cascaded as shown in Figure 4, where cell I and cell II contain a left-handed and a right-handed chiral resonator, respectively. Figure 4 also shows the simulations results. Comparing to the results shown in Figure 3, improvement in filter performance is obvious. Also, the spurious response occurs far beyond twice the resonant frequency of the resonators, and the out-of-band rejection has been improved.

In addition, to expand upper rejection band and improve selectivity, a transmission zero is placed at 2.6 GHz using

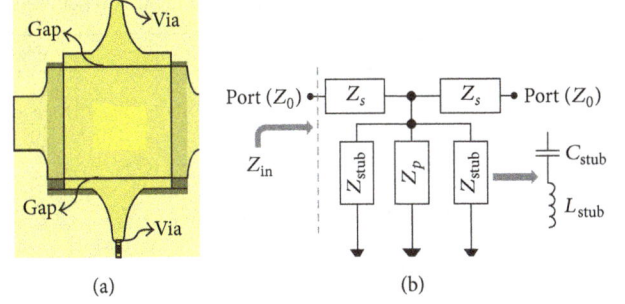

FIGURE 5: A DNG cell with loaded stubs; (a) layout and (b) equivalent circuit.

short ended stubs coupled with gaps to the wide sections of the transmission line as shown in Figure 5(a). In Figure 5(b) the equivalent circuit of a cell loaded with short ended stubs is depicted where the stubs (with coupling gaps) are modeled by LC resonators (Z_{stub}). Appendix describes how to extract the elements of the equivalent circuit, that is, L_{stub} and C_{stub}. These capacitors and inductors can achieve different resonant modes with the change of operating frequency.

3. Experimental Results and Discussions

Three layers of Rogers RO4003C are used to fabricate the proposed filter. The transmission line with gaps and stubs are printed on the top surface of the top layer. Two metal rings are printed on each surface of the middle layer and connected to each other using copper ribbons to form quasi-planar resonators. The bottom layer is ground plane (depicted in Figure 6). Then, the three layers are connected using insulating gel. After gluing, the structure is punctured and the vias of the stubs are soldered with tin. Then, by using the substrate with less thickness, stronger coupling between the upper and lower resonator rings can be obtained with the reducing of resonant frequency, which indicates that further miniaturization could be achieved. Moreover, advanced monolithic fabrication method can mitigate some problems due to the lack of complete alignment among the layers (especially between the first and the second layers) and the substrate loss of the gel.

The results obtained from equivalent circuit simulation, full-wave simulation, and measurements are shown in Figure 7, demonstrating good agreement between them; however, use of copper ribbons instead of vias has effect on the resonant frequency of the resonators which appears by a slight frequency-shift in the measurement results. The small differences between simulated and measured results are mainly due to fabrication inaccuracy. But in general, the measured results are in good agreement with the simulated results.

Finally the proposed filter is compared with the other compact NBPFs reported in the literature in terms of resonator type, center frequency (f_0), fractional bandwidth (FBW), length (λ_g), and insertion loss (IL), where λ_g is the guided wavelength of the electromagnetic waves in the substrate at the resonant frequency. The comparison is shown

TABLE 1: Comparison between this work and other published works.

Ref.	Resonator	f_0 [GHz]	FBW [%]	Length [λ_g]	IL [dB]
[9]	SRR	4.6	1.7	0.53	5.2
[8]	EBG	5	3.7	0.47	2.8
[5]	CRLH-CPW	5	6	0.42	2.5
[6]	CSRR	1	8	0.42	1.5
[7]	Multisection SIR	2.4	10.8	0.24	5.8
This work	Quasi-planar chiral	1	2.2	0.17	5.1

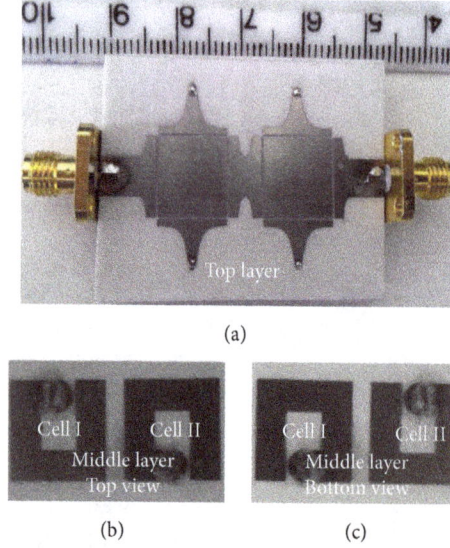

(a)

(b)   (c)

FIGURE 6: The layers of the fabricated filter: (a) top layer, (b) top view of the middle layer, and (c) bottom view of the middle layer (notice that the bottom layer (not shown in this figure) is ground plane).

in Table 1 (which is sorted by filter length in descending order) indicating that the proposed filter with $0.17\lambda_g$ length is the most compact (the operating frequency of the proposed filter is the lowest, but its size is the smallest), yet with FBW = 2.2% the second narrowest. On the other hand, the insertion loss of the filter in this work is still acceptable.

4. Conclusion

In this article, a narrow bandwidth bandpass filter based on the quasi-planar chiral resonators is presented. The proposed filter has the merits of simple design, backward propagation, compact size, and high selectivity. The results of equivalent circuit simulation, full-wave simulation, and measurement are in good agreement. Compared with other similar filters, it could be observed that the proposed filter has the advantages of better size integration and frequency selectivity. The proposed filter is feasible and applicable in modern microwave communication circuits.

Appendix

The input impedance (Z_{in}) indicated in Figure 5(b) at the resonant frequency of the quasi-planar resonator $\omega_{0,p}$ (where

$Z_p = \infty$) can be written as $Z_{in} = [(Z_0 + Z_s)//Z_{stub}/2 + Z_s]$. Hence, Z_{stub} at $\omega_{0,p}$ is derived as

$$Z_{stub}\left(\omega_{0,p}\right)$$
$$= 2\frac{\left[Z_0 + Z_s\left(\omega_{0,p}\right)\right]\left[Z_{in}\left(\omega_{0,p}\right) - Z_s\left(\omega_{0,p}\right)\right]}{Z_0 + 2Z_s\left(\omega_{0,p}\right) - Z_{in}\left(\omega_{0,p}\right)}. \tag{A.1}$$

Z_{stub} is a complex amount whose real part describes losses which is ignored here for simplicity. Considering Figure 5(b), it can be obtained that

$$\mathfrak{I}\left(Z_{stub}\right) = j\omega L_{stub} - \frac{j}{\omega C_{stub}}. \tag{A.2}$$

And the resonant frequency of LC network is

$$\omega_{0,s} = \frac{1}{\sqrt{L_{stub}C_{stub}}}. \tag{A.3}$$

Then, L_{stub} and C_{stub} can be extracted as

$$L_{stub} = \frac{\omega_{0,p}\mathfrak{I}\left(Z_{stub}\left(\omega_{0,p}\right)\right)}{\Delta\omega_0^2}$$

$$C_{stub} = \frac{\Delta\omega_0^2}{\omega_{0,p}\omega_{0,s}^2\mathfrak{I}\left(Z_{stub}\left(\omega_{0,p}\right)\right)}, \tag{A.4}$$

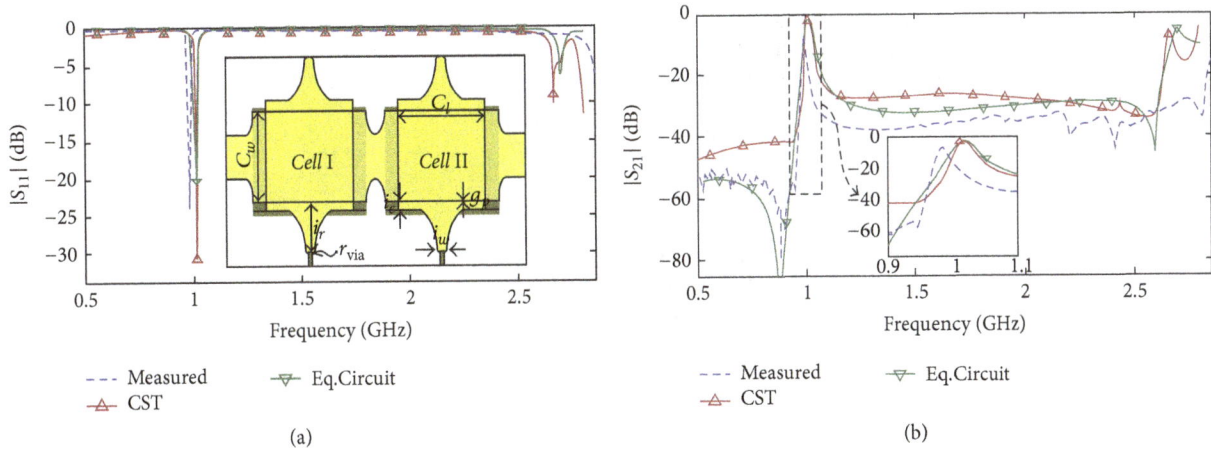

FIGURE 7: Measured, full-wave, and equivalent circuit simulations of (a) $|S_{11}|$ and (b) $|S_{21}|$. Layout parameters are $i_r = 6.5$ mm, $i_c = 1$ mm, $i_w = 1$ mm, $r_{via} = 0.25$ mm, and $g_p = 0.2$ mm. Extracted values of circuit model elements are $L_{stub} = 10.8$ nH and $C_{stub} = 0.35$ pF.

where

$$\Delta\omega_0^2 = \omega_{0,p}^2 - \omega_{0,s}^2. \qquad (A.5)$$

Competing Interests

The authors declare that there is no conflict of interests regarding the publication of this paper.

Acknowledgments

The work is supported by National Natural Science Foundation of China (no. 61563015), Young Foundation of Humanities and Social Sciences of Ministry of Education in China (no. 13YJCZH089), and Young Foundation of Educational Commission of Jiangxi Province of China (no. GJJ14401).

References

[1] R. Markques, F. Martin, and M. Sorolla, *Metamaterials with Negative Parameters: Theory, Design and Microwave Applications*, Wiley Interscience, 2008.

[2] F. Aznar, J. García-García, M. Gil, J. Bonache, and F. Martín, "Strategies for the miniaturization of metamaterial resonators," *Microwave and Optical Technology Letters*, vol. 50, no. 5, pp. 1263–1270, 2008.

[3] R. Marqués, L. Jelinek, and F. Mesa, "Negative refraction from balanced quasi-planar chiral inclusions," *Microwave and Optical Technology Letters*, vol. 49, no. 10, pp. 2606–2609, 2007.

[4] S. M. Hashemi, M. Soleimani, and S. A. Tretyakov, "Compact negative-epsilon stop-band structures based on double-layer chiral inclusions," *IET Microwaves, Antennas and Propagation*, vol. 7, no. 8, pp. 621–629, 2013.

[5] S.-G. Mao, M.-S. Wu, Y.-Z. Chueh, and C. H. Chen, "Modeling of symmetric composite right/left-handed coplanar waveguides with applications to compact bandpass filters," *IEEE Transactions on Microwave Theory and Techniques*, vol. 53, no. 11, pp. 3460–3466, 2005.

[6] J. Bonache, I. Gil, J. García-García, and F. Martín, "Novel microstrip bandpass filters based on complementary split-ring resonators," *IEEE Transactions on Microwave Theory and Techniques*, vol. 54, no. 1, pp. 265–271, 2006.

[7] H. Zhang and K. J. Chen, "Miniaturized coplanar waveguide bandpass filters using multisection stepped-impedance resonators," *IEEE Transactions on Microwave Theory and Techniques*, vol. 54, no. 3, pp. 1090–1095, 2006.

[8] S.-G. Mao and Y.-Z. Chueh, "Coplanar waveguide bandpass filters with compact size and wide spurious-free stopband using electromagnetic bandgap resonators," *IEEE Microwave and Wireless Components Letters*, vol. 17, no. 3, pp. 181–183, 2007.

[9] A. L. Borja, J. Carbonell, V. E. Boria, J. Cascon, and D. Lippens, "A 2% bandwidth C-band filter using cascaded split ring resonators," *IEEE Antennas and Wireless Propagation Letters*, vol. 9, pp. 256–259, 2010.

[10] V. G. Veselago, "The electrodynamics of substances with simultaneously negative values of ε and μ," *Soviet Physics Uspekhi*, vol. 10, no. 4, pp. 509–514, 1968.

[11] M. Studniberg and G. V. Eleftheriades, "A quad-band bandpass filter using negative-refractive-index transmission-line (NRI-TL) metamaterials," in *Proceedings of the IEEE Antennas and Propagation Society International Symposium (AP-S '07)*, pp. 4961–4964, Honolulu, Hawaii, USA, June 2007.

[12] Z.-P. Wang, Z.-Y. Han, and L.-M. Guo, "Miniaturized zeroth-order resonator based on simplified CRLH TL structure," *Microwave and Optical Technology Letters*, vol. 53, no. 4, pp. 848–852, 2011.

[13] X. Q. Lin, Q. Cheng, R. P. Liu, D. Bao, and T. J. Cui, "Compact resonator filters and power dividers designed with simplified meta-structures," *Journal of Electromagnetic Waves and Applications*, vol. 21, no. 12, pp. 1663–1672, 2007.

[14] S. M. Jiang, W. T. Li, X. H. Wang, Q. Y. Song, and X. W. Shi, "A novel method of designing cross-coupled filters through optimization," *Journal of Electromagnetic Waves and Applications*, vol. 23, no. 14-15, pp. 2011–2019, 2009.

[15] N. Dolatsha, M. Shahabadi, and R. Dehbashi, "Via-free CPW-based composite right/left-handed transmission line and a calibration approach to determine its propagation constant," *Journal of Electromagnetic Waves and Applications*, vol. 22, no. 11-12, pp. 1599–1606, 2008.

[16] H.-X. Xu, G.-M. Wang, Q. Peng, and J.-G. Liang, "Novel design of tri-band bandpass filter based on fractal-shaped geometry of a complementary single split ring resonator," *International Journal of Electronics*, vol. 98, no. 5, pp. 647–654, 2011.

[17] J.-Q. Gong, C.-H. Liang, and B. Wu, "Novel dual-band hybrid coupler using improved simplified CRLH transmission line stubs," *Microwave and Optical Technology Letters*, vol. 52, no. 11, pp. 2473–2476, 2010.

[18] Y. Guo and R. Xu, "Ultra-wideband power splitting/combining technique using zero-degree left handed transmission lines," *Journal of Electromagnetic Waves and Applications*, vol. 21, no. 8, pp. 1109–1118, 2007.

[19] J.-Y. Lee, D.-J. Kim, and J.-H. Lee, "High order bandpass filter using the first negative resonant mode of composite right/left-handed transmission line," *Microwave and Optical Technology Letters*, vol. 51, no. 5, pp. 1182–1185, 2009.

[20] C.-H. Tseng and C.-L. Chang, "An image reject mixer with composite right/left-handed quadrature power splitter and if hybrid," *Journal of Electromagnetic Waves and Applications*, vol. 22, no. 11-12, pp. 1557–1564, 2008.

[21] J. He, B.-Z. Wang, and K.-H. Zhang, "Arbitrary dual-band coupler using accurate model of composite right/left handed transmission line," *Journal of Electromagnetic Waves and Applications*, vol. 22, no. 8-9, pp. 1267–1272, 2008.

[22] J.-S. Li and Y.-Y. Zhuang, "Compact microstrip bandpass filter using composite right/left-handed transmission lines," *Microwave and Optical Technology Letters*, vol. 49, no. 8, pp. 1929–1931, 2007.

[23] J. Ju and S. Kahng, "A compact UWB bandpass filter using a center-tapped composite right/left-handed transmission-line zeroth-order resonator," *Microwave and Optical Technology Letters*, vol. 53, no. 9, pp. 1974–1976, 2011.

[24] Q. Zhang and S. N. Khan, "Compact broadside coupled directional coupler based on coplanar CRLH waveguides," *Journal of Electromagnetic Waves and Applications*, vol. 23, no. 2-3, pp. 267–277, 2009.

[25] J. He, B.-Z. Wang, and K.-H. Zhang, "Wideband differential phase shifter using modified composite right/left handed transmission line," *Journal of Electromagnetic Waves and Applications*, vol. 22, no. 10, pp. 1389–1394, 2008.

[26] S.-W. Zhao, Z.-X. Tang, W. Dai, and B. Zhang, "A novel dual-layer high-directivity directional coupler using printed metamaterial MS/NRI coupled-line," *Microwave and Optical Technology Letters*, vol. 52, no. 12, pp. 2706–2708, 2010.

21

Multitransmission Zero Dual-Band Bandpass Filter using Nonresonating Node for 5G Millimetre-Wave Application

Liyun Shi [1,2] **and Jianjun Gao** [1]

[1] School of Information and Science Technology, East China Normal University, 500 South Lianhua Road, Shanghai, China
[2] School of Electronic Information and Electronical Engineering, Shanghai Jiao Tong University, 800 Dongchuan Road, Shanghai, China

Correspondence should be addressed to Liyun Shi; shiliyun525@163.com

Academic Editor: S. M. Rezaul Hasan

A planer millimetre-wave dual-band bandpass filter with multitransmission zeros is proposed for 5G application. This filter includes two dual-mode open-loop resonators. The U-shape nonresonating node is employed to generate an extra coupling path. Finally, a dual-band bandpass filter with five transmission zeros is obtained. The filter is fabricated and measured. Good agreement between simulation and measurement is obtained.

1. Introduction

The fifth-generation mobile communication system (5G) is currently experiencing rapid development. The Third-Generation Partnership Project's (3GPP) Release-16, or "5G phase 2", should be completed in December 2019 [1]. There are three typical usage scenarios for 5G, including Enhanced Mobile Broadband (eMBB), Massive Machine Type Communications (mTC), and Ultrareliable and Low-Latency Communications (uRLLC), which were defined by the International Telecommunication Union-Radio Communication Sector (ITU-R) in 2015. To achieve very high-speed data transmission, the 5G high-frequency systems will have much wider operating bands as well as multiband. As a key passive circuit block in these systems, bandpass filters (BPFs) featuring multiband, high out-of-band power rejection and miniaturization are very attractive [2]. These BPFs efficiently reject out-of-system interferences while preserving processed signals in-band, where a specific narrow frequency band needs to be strongly attenuated. Much effort has been devoted to designing 5G millimetre-wave BPFs with high selectivity, compact size, and low cost, as shown in [2–9]. In [3–5], substrate integrated waveguide (SIW) technology is applied to design millimetre-wave BPFs. SIW technology has the advantage of a high Q-factor and easy integration with other planar circuits. In [6], a magic-T with imbedded

Chebyshev filter response is developed with a multilayer, low-temperature cofired ceramic (LTCC) technology, which is popular due to its low dielectric loss, high material reliability and compatibility, and high level of robustness. However, its application is limited by the thick substrate's large feature size. In recent years, with the rapid scaling down of advanced complementary metal-oxide semiconductor (CMOS) technology, many on-chip millimetre-wave BPFs have been implemented in CMOS technology as shown in [7–9], which could provide good performance but have high cost and difficult fabrication. Compared with the above technologies, printed circuit technology still has advantages for millimetre filter design as the mainstream technology at low frequency, such as its low profile, easy fabrication, and easy integration with other planer circuits.

Bandpass filters with multiband and high selectivity are attractive because they could meet the demand for multiband wireless communication systems. Usually, filtering selectivity could be generally enhanced by producing transmission zeros (TZs) at finite real frequencies [10]. Several methods have been reported to introduce transmission zeros. In [10] an original and simple method of signal-interference source/load coupling is employed to improve filters' selectivity. In [11], special coupling topology is designed to introduce electromagnetic coupling between nonadjacent resonators. An SIW filter with etched capacitive slots at the coupling region is proposed in [12]. These slots could create a negative coupling coefficient but also result in a higher radiation loss.

The nonresonating node concept has recently been applied to synthesize high selectivity single-band planar filters as discussed in [13, 14]. In [14], a stepped-impedance nonresonating node is employed to introduce an extra signal transmission path between the source and the load. The underlying idea is to use these elements to generate alternative interresonator couplings favouring transmission zero creation. In this paper, the realization of high selective dual-band millimetre-wave BPF by means of nonresonating node is approached. Additionally, two dual-mode open-loop resonators are employed in this filter. Dual-mode open-loop resonators are well known for their compact size and have been used for bandpass filter design in [15–17], which only use a single type of resonator and show good performance. Hence, the dual-mode open-loop bandpass filters in millimetre-wave applications are worthy of study.

The aim of this paper is the presentation of a millimetre-wave filter design that achieves a very good trade-off between performance and fabrication cost. The organization of the rest of the manuscript is as follows: Section 2 expounds on the theoretical foundations of dual-band bandpass filters. Nonresonating nodes are analyzed and discussed in Section 3. Finally, the conclusions of this study are described in Section 4.

2. Design of the Dual-Band BPF

Figure 1 shows the detailed configuration of the proposed filter. In fact, this filter can be seen as a constitution of five resonators. Two dual-mode open-loop resonators are denoted by L_1, L_2, L_3, L_4, W_1, and W_3. The nonresonating node is denoted by L_7, L_8, and W_4, which produces extra coupling paths. Compared with traditional feedlines, two 1/4 wavelength resonators are added to the CPW feedlines.

Dual-mode open-loop resonators are well known for their compact size, which is much smaller than the conventional dual-mode loop resonator [15–17]. Dual-mode open-loop filters can excite two nondegenerate modes as well as two controllable finite-frequency transmission zeros, which can be controlled simply by varying the size of perturbation.

Figure 2 shows the simulated resonant frequency response which is plotted for different values of L_2 and W_2. As can be seen from Figure 2(a), when L_2 and W_2 are changed the even mode resonant frequency is effectively shifted, while the odd mode resonant frequency is much less affected. It is also interesting to notice from Figure 2(b) that there are two finite-frequency transmission zeros when the two modes split. One transmission (TZ$_1$) is allocated on the left side of the passband, while the other (TZ$_2$) experiences a significant frequency shifting.

Since the dual-mode resonator is composed of symmetric structures, even-odd mode theory can be adopted to analyze its equivalent circuit structure and its equivalent circuit is given in Figure 3. For even mode excitation, the symmetry plane along the dashed line AA′ is considered as an open end which can be seen from Figure 3(a). The resonator works like

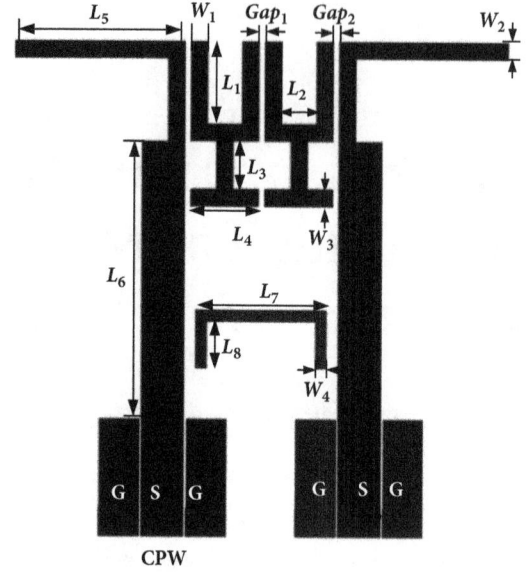

FIGURE 1: Schematic views of the proposed dual-band BPF.

a stepped impendence resonator (SIR) at even mode resonant frequency. The resonant condition can be calculated by (1)-(7):

$$Y_{even} = Y_{up,even} + Y_{low,even} \tag{1}$$

$$Y_{low,even} = \frac{j \tan(\theta_1 f_{e1})}{Z_1} \tag{2}$$

$$Y_{up,even} = \frac{(Q_a - Q_b)}{(P_a - P_b)} \tag{3}$$

where

$$P_a = Z_1 Z_2^2 Z_3 - Z_1 Z_2^3 \tan(\theta_3 f_e) \tan(\theta_4 f_e) \tag{4}$$

$$P_b = -Z_1^2 Z_2 \tan(\theta_2 f_e) \tan(\theta_4 f_e) \\ - Z_3 \tan(\theta_2 f_e) \tan(\theta_3 f_e) \tag{5}$$

$$Q_a = j Z_1 Z_2 \tan(\theta_4 f_e) + j Z_1 Z_3 \tan(\theta_2 f_e) \tag{6}$$

$$Q_b = j Z_2^2 \tan(\theta_3 f_e) \tan(\theta_4 f_e) \tan(\theta_2 f_e) \tag{7}$$

For odd mode excitation, its equivalent circuit is shown in Figure 3(b). The resonator works like a uniform impedance resonator (UIR) at the odd mode resonance frequency. The resonant condition can be described by

$$Y_{odd} = \frac{j(\tan(\theta_1 f_0) \tan(\theta_2 f_0) - 1)}{Z_1} \tag{8}$$

where $\theta_1 = \beta(L_1 + L_3)$, $\theta_2 = \beta(L_2/2)$, $\theta_3 = \beta L_3$, $\theta_4 = \beta L_4$, θ_i, β, Zi are the electrical length, phase constant, and characteristics impedance of resonator, respectively, $f_e = f_{even}/f_r$, $f_o = f_{odd}/f_r$, f_{even}, f_{odd}, f_r present the even mode resonate frequency, odd mode resonate frequency, respectively.

The frequency of two transmission zeros can be obtained when input admittance at even mode or odd mode is zero. It is

(a)

(b)

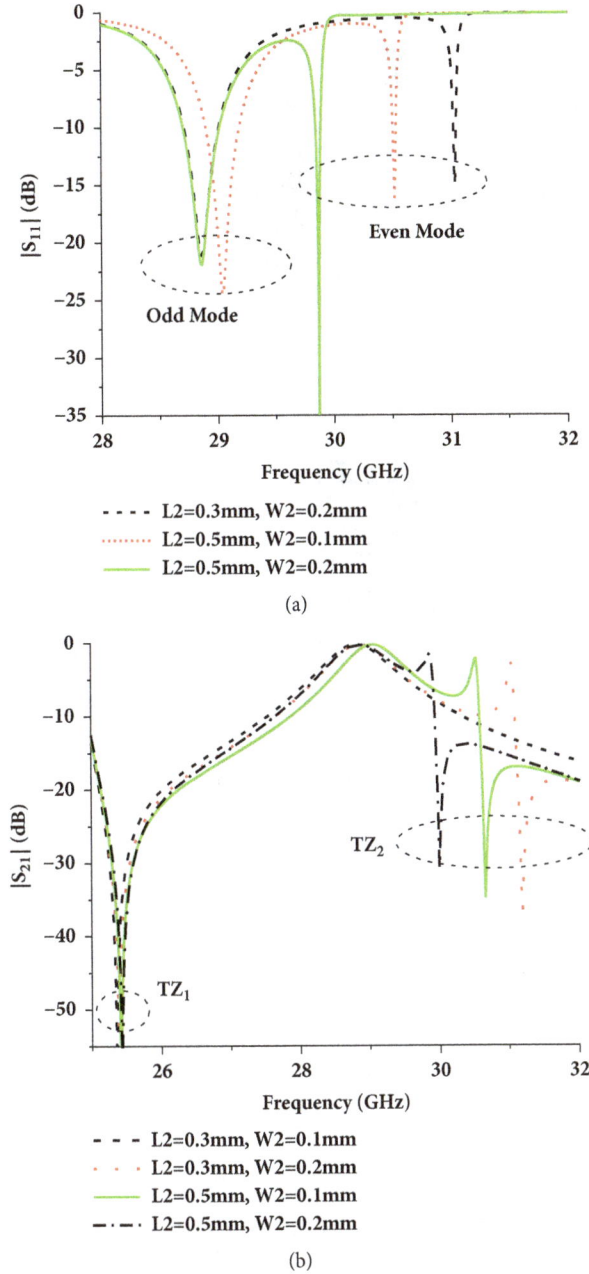

FIGURE 2: Simulated frequency responses of the dual-mode open-loop resonator with different size of L_2 and W_2 (mm): (a) S_{11} responses; (b) S_{21} responses.

noted that both f_{odd} and f_{even} can be estimated approximately by the half-wavelength of this microstrip resonator. They are obtained by (9)-(10):

$$f_{even} = \frac{2c}{(L_1 + L_3 + L_2/2 + L_4/2)\sqrt{\varepsilon_{eff}}} \quad (9)$$

$$f_{odd} = \frac{4c}{(L_1 + L_2/2)\sqrt{\varepsilon_{eff}}} \quad (10)$$

where ε_{eff} is the effective relative permittivity of substrate.

3. Analysis of Nonresonating Node

Nonresonating nodes have been employed to generate transmission zeros because they could provide multiple signal-interaction paths [18, 19]. To explain the nonresonating nodes, an original method based on a low-pass prototype with a nonresonating node is described. Elements of the low-pass prototype nonresonating node are given by the method in [19, 20], to yield a response with prescribed transmissions.

The coupling scheme of a second-order filter with no transmission zero in the cascaded configuration is shown in Figure 4(a). The dark nodes are resonators and the

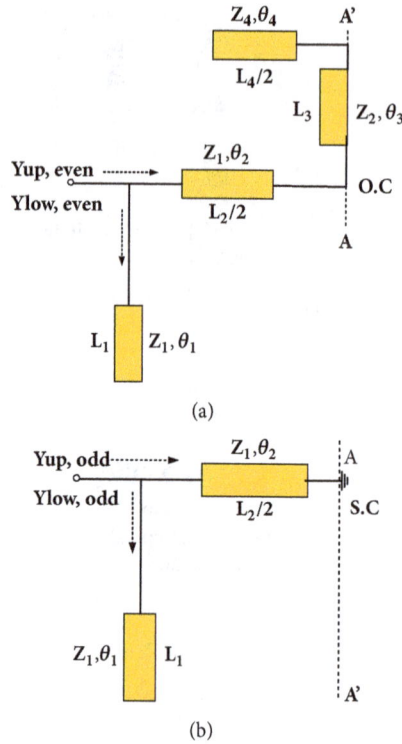

FIGURE 3: Equivalent circuit of different modes: (a) even mode; (b) odd mode.

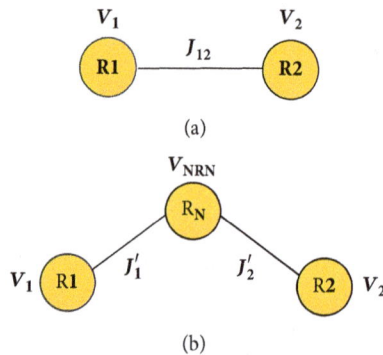

FIGURE 4: (a) First coupling scheme of two-order filter with none transmission zeros; (b) second coupling scheme of two-order filter in-line configuration with nonresonating nodes (patterned circles).

lines connecting them are admittance inverters. A second-order loss-pass prototype which realizes the same response is shown in Figure 4(b). Note the presence of additional nonresonating nodes, shown as the patterned circles in this prototype.

For the two cells in Figure 4 to be equivalent, they must have the same node equations (admittance matrix) or the same normal modes. The node matrix equation of the cell in Figure 4(a) is shown in

$$
\begin{bmatrix} \omega + J_{11} & J_{12} \\ J_{12} & \omega + J_{22} \end{bmatrix} \begin{bmatrix} V_1 \\ V_2 \end{bmatrix} = \begin{bmatrix} e_1 \\ e_2 \end{bmatrix} \tag{11}
$$

where ω is the normalized frequency, VI and V2 are the node voltages, and e1 and e2 are the excitations at the same nodes.

The normalized frequency shifts of the resonators are J_{11} and J_{22}, respectively. On the other hand, the node matrix equation of the cell in Figure 4(b) is given by

$$
\begin{bmatrix} \omega + b'_1 & 0 & J'_1 \\ 0 & \omega + b'_2 & J'_2 \\ J'_1 & J'_2 & J_{NRN} \end{bmatrix} \begin{bmatrix} V_1 \\ V_2 \\ V_{NRN} \end{bmatrix} = \begin{bmatrix} e_1 \\ e_2 \\ 0 \end{bmatrix} \tag{12}
$$

where V_{NRN} is the voltage of the internal nonresonating node which is not externally excited. The normalized frequency shifts of the resonators are b1$'$ and b2$'$, respectively. The susceptance of the NRN is JNRN. If the voltage VNRR is

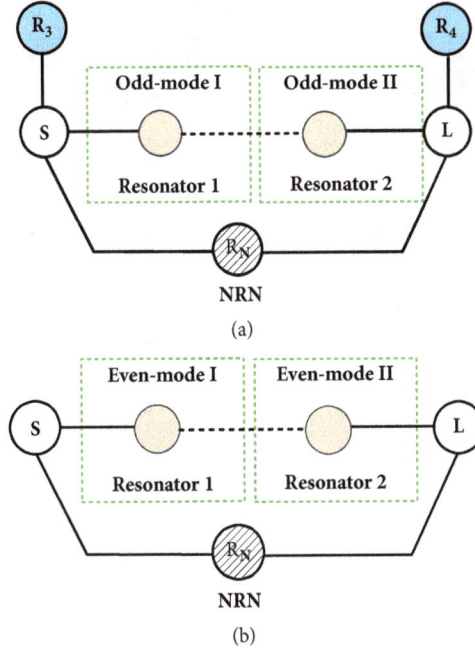

FIGURE 5: Coupling structure of the proposed dual-band bandpass filter. (a) At lower passband. (b) At upper passband.

eliminated from (12), by using the last row, we get the new matrix equation:

$$
\begin{bmatrix}
\omega + b_1' - \dfrac{J_1'^2}{J_N} & -\dfrac{J_1'}{J_N} \\
-\dfrac{J_1' J_2'}{J_N} & \omega + b_2' - \dfrac{J_2'^2}{J_N}
\end{bmatrix}
\begin{bmatrix} V_1 \\ V_2 \end{bmatrix}
=
\begin{bmatrix} e_1 \\ e_2 \end{bmatrix}
\tag{13}
$$

By directly comparing (11) and (13), we get the elements of the coupling matrix of the cascaded cell in Figure 4(a) in terms of the elements of the cell with an internal nonresonating node in Figure 4(b), described by (14)-(16).

$$
J_{11} = b_1' - \frac{J_1'^2}{J_N}
\tag{14}
$$

$$
J_{22} = b_2' - \frac{J_2'^2}{J_N}
\tag{15}
$$

$$
J_{12} = -\frac{J_1'^2}{J_N}
\tag{16}
$$

In our design, one U-shaped nonresonating node is employed to increase source-load coupling paths. The coupling and routing scheme in the inset of Figure 4 can be used to model the configuration of the proposed filter. The coupling structure for the designed dual-band bandpass filters is shown in Figure 5, where resonators 1 and 2 denote two dual-mode open-loop resonators and resonators 3 and 4 denote L-shape feedline resonators. The white circles are source and load; blue circles are resonating nodes; the patterned circle is a nonresonating node; solid lines are main direct couplings; dashed lines are minor cross couplings. Figure 5(a) illustrates

the schemes at a lower resonant frequency and Figure 5(b) is at an upper resonant frequency. Note that Figure 5 gives the main coupling structure; some other very small couplings could also take place in the proposed filter schemes.

Indeed, this nonresonating node is a resonator at certain frequencies in a strict sense; it works as a detuned folded open-ended transmission-line resonator in our design. This particular geometry could help to create a pair of transmissions zeros on both sides of the passband. As seen from Figure 2, the odd mode resonant frequency is located at the lower passband, and the even mode resonant frequency is at the higher passband. Additionally, the two L-shaped feedlines (described by L_5 and W_2) work as two resonators and their resonant frequencies are also at the lower passband. To demonstrate this design concept, a filter without a nonresonating-node is also simulated. Note that the only difference between the two filters is whether there is a nonresonating-node; the rest of the parameters are the same. Performance comparisons are illustrated in Figure 6. It is clear that two extra transmissions zeros are produced with a nonresonating node.

4. Results and Discussion

Based on the above discussion, one dual-band millimetre-wave bandpass filter has been manufactured and measured. The filter is fabricated on substrate Rogers 5880 with permittivity of 2.2, loss tangent of 0.001, and thickness of 0.254 mm, which could suppress the occurrence of high-order modes and the substrate surface-waves. The dimensions of this dual-band filter indicated in Figure 1 are L_1=1.4 mm L_2=0.6 mm, L_3=0.8 mm, L_4=1.2 mm, L_5=3 mm, L_6=6.7 mm, L_8=0.8 mm, W_1 =W_2= W_3 =0.3 mm, and W_4=0.2 mm. The width of this

TABLE 1: Performance comparison between the proposed design and some previously published millimetre-wave filters.

Ref.	Passband (GHz)	Insertion Loss (dB)	Fabrication Technology	Number of TZs	Size (mm²)
[1]	28	0.4	waveguide	2	5,600,000
[2]	28	2.2	SIW	2	93.7s65
[5]	24	4.5	LTCC	1	59.625
[7]	28	2.6	CMOS	2	0.038
This work	28.1/31.1	2.1/2.5	PCB	5	72

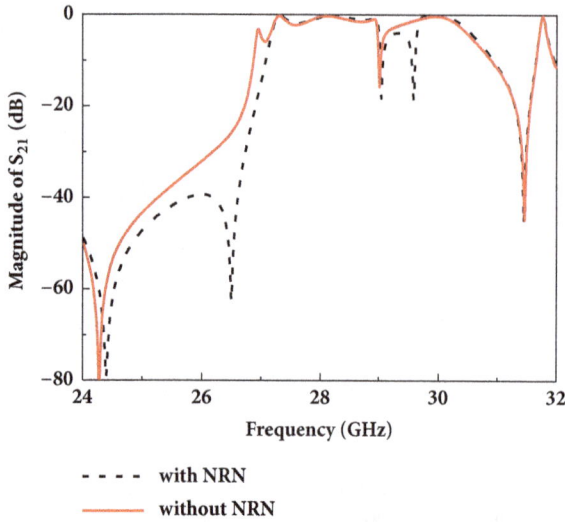

FIGURE 6: Comparisons of simulated S21 of the proposed BPF (solid line: filter without nonresonating node; dash line: filter with measured nonresonating node).

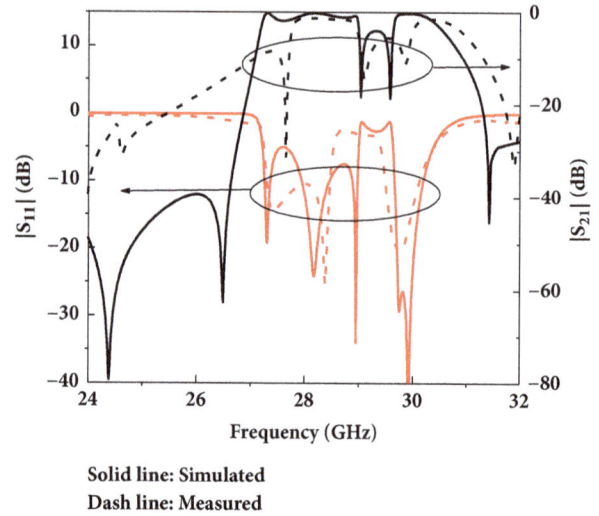

FIGURE 8: Simulated and measured results of the proposed dual-band BPF (solid line: simulated results; dash line: measured results).

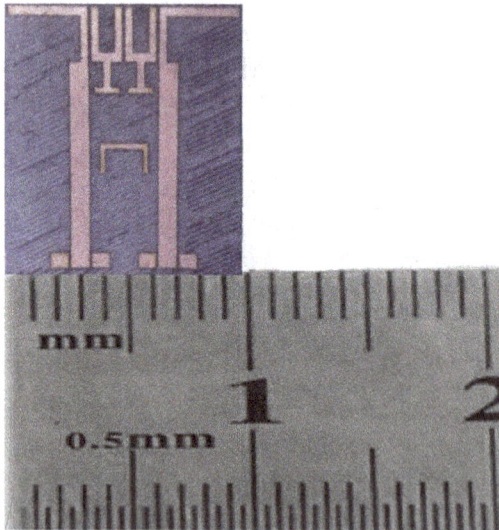

FIGURE 7: Photographs of this fabricated filter.

pair of I/O feedlines is W_1 =1.5 mm. A photograph of the proposed filter is shown in Figure 7. The filter size is 8×9 mm² (1.17λg X 1.04λg, where λg is the waveguide length at 28 GHz).

The measured and simulated results are shown in Figure 8, where the measured result is carried out by a Keysight N5227A network analyzer and Cascade probe station Summit 11K. A good agreement between simulation and measurement results is obtained. Two passbands are centred near 28.1 GHz and 31.1 GHz, with -3 dB fractional bandwidths (FBWs) of 6.4% and 3.8%, respectively. There are five transmission zeros on both sides of the passband, which greatly improve the filter's selectivity. The attenuation slope of the first passband generated at 28.1 GHz is 271 dB/GHz on the left side and 38.25 dB/GHz on the right side; the attenuation slope of each side of the second passband generated is 70.323 dB/GHz and 30.04 dB/GHz, respectively. A little higher insertion loss in the passband and frequency shift can be observed. The deviations between the simulated and measured results are mainly caused by two reasons. One is due to fabrication tolerance; millimetre-wave filters are very sensitive to their fabrication precision. Second, the discontinuity between the RF probe and our filter is not considered in our simulation, which may not be ignored at high frequencies.

To further demonstrate the performance of the proposed dual-band filter, comparisons with previously reported millimetre filters are listed in Table 1. It can be found that the proposed filter has achieved a very good trade-off between selectivity, size, and fabrication cost.

5. Conclusion

A highly selective planer millimetre-wave dual-band BPF is developed by cascading two dual-mode open-loop resonators with a pair of L-shaped feedlines. Two odd modes resonate at the lower passband and two even modes work at the higher passband. The proposed U-shaped nonresonating node is used to generate cross coupling. As a result, five transmission zeros are obtained, which effectively enhances this filter's selectivity. Finally, a millimetre-wave dual-band BPF is fabricated with a measured centre frequency of 28.1 GHz and 31.1 GHz and an FBW of 6.4% and 3.8%, respectively. This design achieves a very good trade-off between performance and fabrication cost.

Conflicts of Interest

The authors declare that there are no conflicts of interest regarding the publication of this paper.

Acknowledgments

This work is partly supported by the National Science and Technology Major Project of the Ministry of Science and Technology of China (under Grant no. 2016ZX03001007) and partly supported by the National Natural Science Foundation of China (under Grant no. 61370008).

References

[1] *Third Generation Partnership Project (3GPP) Release 16*, http://www.3gpp.org/release-16.

[2] J. R. Garai, J. A. Cruz, J. M. Rebollar, and T. Estrada, "In-line pure E-plane waveguide band-stop filter with wide spurious-free response," *IEEE Microwave and Wireless Components Letters*, vol. 21, no. 4, pp. 209–211, 2009.

[3] K. Gong, W. Hong, Y. Zhang, P. Chen, and C. J. You, "Substrate integrated waveguide quasi-elliptic filters with controllable electric and magnetic mixed coupling," *IEEE Transactions on Microwave Theory and Techniques*, vol. 60, no. 10, pp. 3071–3078, 2012.

[4] F. Zhu, W. Hong, J.-X. Chen, and K. Wu, "Cross-coupled substrate integrated waveguide filters with improved stopband performance," *IEEE Microwave and Wireless Components Letters*, vol. 22, no. 12, pp. 633–635, 2012.

[5] S. O. Nassar and P. Meyer, "Pedestal substrate integrated waveguide resonators and filters," *IET Microwaves, Antennas & Propagation*, vol. 11, no. 6, pp. 804–810, 2017.

[6] T.-M. Shen, T.-Y. Huang, C.-F. Chen, and R.-B. Wu, "A laminated waveguide magic-T with bandpass filter response in multilayer LTCC," *IEEE Transactions on Microwave Theory and Techniques*, vol. 59, no. 3, pp. 584–592, 2011.

[7] A.-L. Franc, E. Pistono, D. Gloria, and P. Ferrari, "High-performance shielded coplanar waveguides for the design of CMOS 60-GHz bandpass filters," *IEEE Transactions on Electron Devices*, vol. 59, no. 5, pp. 1219–1226, 2012.

[8] Y. Zhong, Y. Yang, X. Zhu, E. Dutkiewicz, K. M. Shum, and Q. Xue, "An On-Chip Bandpass Filter Using a Broadside-Coupled Meander Line Resonator with a Defected-Ground Structure," *IEEE Electron Device Letters*, vol. 38, no. 5, pp. 626–629, 2017.

[9] S. Chakraborty, Y. Yang, X. Zhu et al., "A broadside-coupled meander-line resonator in 0.13-μm SiGe technology for millimeter-wave application," *IEEE Electron Device Letters*, vol. 37, no. 3, pp. 329–332, 2016.

[10] M. Sanchez-Renedo and R. Gomez-García, "Multi-coupled-resonator dual-band bandpass microstrip filters with non-resonating nodes," in *Proceedings of the 2011 IEEE/MTT-S International Microwave Symposium - MTT 2011*, pp. 1–4, Baltimore, MD, USA, June 2011.

[11] P.-J. Zhang and M.-Q. Li, "Cascaded trisection substrate-integrated waveguide filter with high selectivity," *IEEE Electronics Letters*, vol. 50, no. 23, pp. 1717–1719, 2014.

[12] Z. He, C. J. You, S. Leng, and X. Li, "Compact inline substrate integrated waveguide filter with enhanced selectivity using new non-resonating node," *IEEE Electronics Letters*, vol. 52, no. 21, pp. 1778–1780, 2016.

[13] D. Rebenaque, F. Pereira, J. Gomez Tornero, J. Garcia, and A. Alvarez Melcon, "Two simple implementations of transversal filters with coupling between non-resonant nodes," in *Proceedings of the IEEE MTT-S International Microwave Symposium Digest, 2005.*, pp. 957–960, Long Beach, CA, USA.

[14] S. Cogollos, R. Cameron, R. Mansour, M. Yu, and V. Boria, "Synthesis and Design Procedure for High Performance Waveguide Filters Based on Nonresonating Nodes," in *Proceedings of the 2007 IEEE/MTT-S International Microwave Symposium*, pp. 1297–1300, Honolulu, HI, USA, June 2007.

[15] J.-S. Hong, H. Shaman, and Y.-H. Chun, "Dual-mode microstrip open-loop resonators and filters," *IEEE Transactions on Microwave Theory and Techniques*, vol. 55, no. 8, pp. 1764–1770, 2007.

[16] X. Guan, Y. Yuan, L. Song et al., "A novel triple-mode bandpass filter based on a dual-mode defected ground structure resonator and a microstrip resonator," *International Journal of Antennas and Propagation*, vol. 2013, 2013.

[17] Chia-Mao Chen, Shoou-Jinn Chang, Sung-Mao Wu, Yuan-Tai Hsieh, and Cheng-Fu Yang, "Investigation of Compact Balun-Bandpass Filter Using Folded Open-Loop Ring Resonators and Microstrip Lines," *Mathematical Problems in Engineering*, vol. 2014, Article ID 679538, 6 pages, 2014.

[18] D. Zhang, J. Zhou, and Z. Yu, "Coupling topology of substrate integrated waveguide filter using unequal length slots with non-resonating nodes," *IEEE Electronics Letters*, vol. 53, no. 20, pp. 1368–1370, 2017.

[19] S. Amari and G. Macchiarella, "Synthesis of inline filters with arbitrarily placed attenuation poles by using nonresonating nodes," *IEEE Transactions on Microwave Theory and Techniques*, vol. 53, no. 10, pp. 3075–3081, 2005.

[20] S. Amari, "Direct synthesis of cascaded singlets and triplets by non-resonating node suppression," in *Proceedings of the 2006 IEEE MTT-S International Microwave Symposium Digest*, pp. 123–126, USA, June 2006.

22

Power Device Thermal Fault Tolerant Control of High-Power Three-Level Explosion-Proof Inverter based on Holographic Equivalent Dual-Mode Modulation

Shi-Zhou Xu, Chun-jie Wang, and Yu-feng Peng

College of Electronic and Electrical Engineering, Henan Normal University, Xinxiang, China

Correspondence should be addressed to Shi-Zhou Xu; xushizhousiee@163.com

Academic Editor: Michele Riccio

It is necessary for three-level explosion-proof inverters to have high thermal stability and good output characteristics avoiding problems caused by power devices, such as IGBT, so it becomes a hot and difficult research point using only one control algorithm to guarantee both output characteristics and high thermal stability. Firstly, the simplified SVPWM (Space Vector Pulse Width Modulation) algorithm was illustrated based on the NPC (neutral-point-clamped) three-level inverter, and then the quasi-square wave control was brought in and made into a novel holographic equivalent dual-mode modulation algorithm together with the simplified SVPWM. The holographic equivalent model was established to analyze the relative advantages comparing with the two single algorithms. Finally, the dynamic output and steady power device losses were analyzed, based on which the power loss calculation and system simulation were conducted as well. The experiment proved that the high-power three-level explosion-proof inverter has good output characteristics and thermal stability.

1. Introduction

The high-power three-level inverter has been applied widely in the field of mine hoist, and it shows high performance in application. However, with the diversification and complexity of applications, inverters used in explosion-proof environments need increasingly high requirements of system stability. There are many factors, like internal faults, overload, transients, and so on, that affect the system stability. But for explosion-proof inverters, feedback from the applications of this stage shows that the primary factor affecting the stability of the system is the thermal fault; that is, the power loss of power device is too large, and the cooling system with certain cooling capacity can not meet the cooling needs and results in damage of power devices due to accumulated heat. Fault tolerant control for thermal fault of inverters has received attention paid by scholars at home and abroad, and some progress has been made already.

The power losses of internal power devices during operation are the main heat sources of inverters. So the modeling

analysis of power devices is the major research of thermal analysis, and the topology is shown in Figure 1.

The three-level NPC inverter compared with traditional two-level topology achieves higher transfer density of energy and lower harmonic content of voltage and current. It has been widely used in the field of high-voltage power inverter and now is the main topology of power circuit in the mainstream products from many world-renowned electric manufacturers at this stage, and, therefore, the thermal fault tolerant control based on high-power explosion-proof NPC three-level inverter is the main content of this paper. Several experts and scholars have conducted research in this area with some practical results. Power loss, heat models, and life prediction of power devices were researched in [1] based on three-level NPC inverter, but the thermal models mentioned in the paper are from the datasheets, and the RC equivalent circuit is established accordingly, which causes larger calculation error. The power loss and thermal models were established in [2] based on the platform of ANPC converter and the study of IGBT-Diode encapsulation module, and

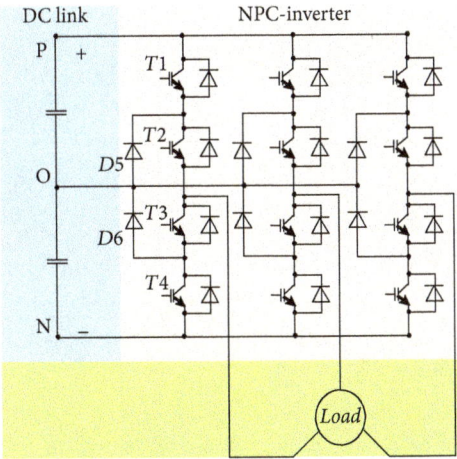

FIGURE 1: Topology of neutral-point-clamped three-level inverter.

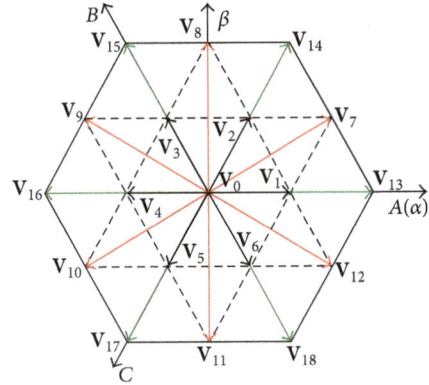

FIGURE 2: Space voltage vector diagram of three- level inverter.

meanwhile, the steady-state thermal models were analyzed under air cooling conditions. In [3], the authors analyzed the thermal stress of IGBT module based on traditional PSCAD/ETMDC internal models and achieved the thermal models in consistence with electrical simulation modes, but the mode established by this method has a large calculation error. As the heat source, power losses of power devices have an effect on the inverter thermal field, which has attracted much attention from experts and scholars. A thermal management method of NPC three-level inverter based on IGCT has been proposed in [4], and failover design is also realized by applying appropriate heat management method of power semiconductor devices considering the possibility of protection strategy failure. With respect to loss distribution and junction temperature of power devices, several sets of MW wind power converter are analyzed under different wind speed scales and parameters [5], and it is claimed that thermal stress of inverter power devices in machine side of DFIG is greater than the grid-side (PMSM) converter's. However, the objects are two-level converters. As for the low voltage ride through (LVRT) operating condition, a novel space vector modulation method based on NPC wind power converter to reallocate the loss during failure is proposed in [6], which reorientates heat load of the power devices and equally splits heat generated during the fault process to reduce the junction temperature of power devices. In [7], the temperature management of H-bridge inverter based on dual-mode control is developed, and the strategy combined quasi-square wave modulation with PWM modulation, with which the power loss can be reduced by 20% to 50%. Thermal analysis has been conducted in various topologies and applications, and temperature management is especially virtual in engineering application. High-power three-level explosion-proof inverters are more sensitive to temperature, when working in terrible underground coal mine, and therefore, an effective control method considering power quality and thermal stability of inverters equally during operation is imperative. Heat-pipe and water-cooling heat dissipation methods for high-power explosion-proof inverters have been used widely at this stage. And fault tolerant capability of the

heatsink is needed to meet the demand of heat tolerance. But it is difficult to increase redundancy of cooling systems due to the limited space under explosion-proof conditions.

Hence, considering modulation strategy, a new modulation algorithm based on holographic equivalent dual mode is proposed in this paper. The calculation can not only adjust dynamic power device losses according to output property, but also achieve optimal thermal management of power devices in different speed zones of hoister. Therefore, heat fault tolerant control for high-power three-level explosion-proof inverter can be realized.

2. Modulation Principle of Holographic Equivalent Dual Mode

Firstly, the principle of SVPWM is introduced simply based on the inverter topology shown in Figure 1. As can be seen in Figure 1 each bridge arm is composed of four anti-parallel diodes S_{xy} ($x = a, b, c; y = 1, 2, 3, 4$) and two clamping diodes D_{xy} ($x = a, b, c; y = 1, 2$), respectively. As clamping diodes, each phase outputs P, O, and N three levels, whose voltages are $+U_{dc}/2$, 0 and $-U_{dc}/2$, respectively. Through combining three-phase output voltages and 27 kinds of space voltage vectors are obtained as shown in Figure 2 [8]. The general equation of space voltage vector is shown as follows:

$$\mathbf{V} = \frac{U_{dc}}{3} \left(S_a + S_b e^{j2\pi/3} + S_c e^{j4\pi/3} \right), \tag{1}$$

where S_a, S_b, and S_c represent the state of phases A, B, and C respectively.

The space voltage vectors depicted in Figure 2 can be divided into four types: zero vector $\mathbf{V}_0$, small vectors $\mathbf{V}_1$–$\mathbf{V}_6$, medium vectors $\mathbf{V}_7$–$\mathbf{V}_{12}$, and long vectors $\mathbf{V}_{13}$–$\mathbf{V}_{18}$. Table 1 lists the space voltage vectors corresponding to the output voltage states.

The diagram of three-level voltage space vectors shown in Figure 2 is resolved to stacked six two-level space vector diagrams, as shown in Figure 3, which are marked by $S = 1 \sim 6$.

Due to the symmetry of the six small hexagons, the algorithm of SVPWM is analyzed based on reference voltage vector through the small hexagon marked with $S = 1$. As

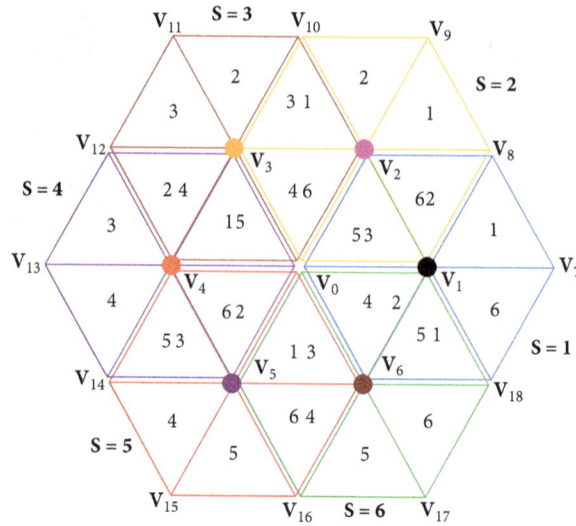

FIGURE 3: Diagram of transforming between three-level space vector to two-level space vector.

TABLE 1: Space voltage sectors and corresponding output voltage states.

Space voltage vectors	Output voltage state
V_0	PPP, OOO, NNN
V_1	POO, ONN
V_2	PPO, OON
V_3	OPO, NON
V_4	OPP, NOO
V_5	OOP, NNO
V_6	POP, ONO
V_7	PON
V_8	OPN
V_9	NPO
V_{10}	NOP
V_{11}	ONP
V_{12}	PNO
V_{13}	PNN
V_{14}	PPN
V_{15}	NPN
V_{16}	NPP
V_{17}	NNP
V_{18}	PNP

shown in Figure 4, the two-level space vector is divided into six triangular sectors, as reference voltage vector V_{ref} located in the first triangular sector, and, according to vectorial resultant rule, the basic vectors V_1, V_7, and V_{13} result in V_{ref}. Vector translation of reference voltages under different S values is shown in Table 2.

Based on the principle of volt-second balance, the relationship can be expressed as follows:

$$\mathbf{V}_{ref}T_s = \mathbf{V}_1 T_1 + \mathbf{V}_{13} T_{13} + \mathbf{V}_7 T_7$$

$$T_s = T_1 + T_{13} + T_7, \tag{2}$$

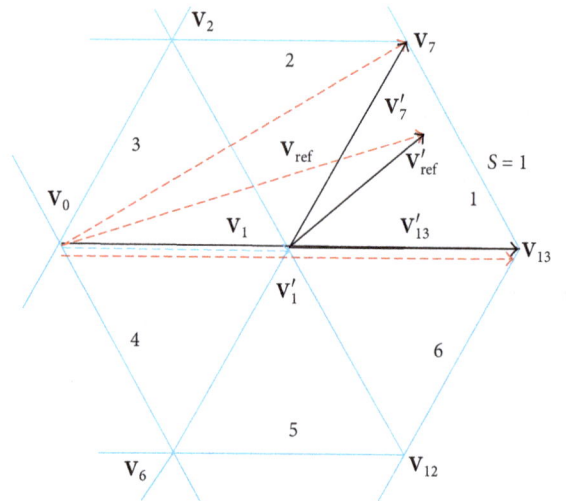

FIGURE 4: Vector translation of reference voltage under $S = 1$.

where $\mathbf{V}_{ref}$ is the reference voltage vector, T_s is the carrier cycle, and T_1, T_{13}, and T_7 are effecting time, respectively.

Introducing translation vector $\mathbf{V}_1$, so

$$(\mathbf{V}_{ref} - \mathbf{V}_1) \cdot T_s = (\mathbf{V}_1 - \mathbf{V}_1) \cdot T_1 + (\mathbf{V}_{13} - \mathbf{V}_1) \cdot T_{13}$$
$$+ (\mathbf{V}_7 - \mathbf{V}_1) \cdot T_7. \tag{3}$$

On the basis of Figure 4, it can be expressed as follows:

$$\mathbf{V}'_{ref} \cdot T_s = \mathbf{V}'_1 \cdot T_1 + \mathbf{V}'_{13} \cdot T_{13} + \mathbf{V}'_7 \cdot T_7, \tag{4}$$

where $\mathbf{V}'_{ref}$, $\mathbf{V}'_1$, $\mathbf{V}'_{13}$, and $\mathbf{V}'_7$ are the shifted voltage space vectors.

As shown in Figure 4, after the shift of voltage, voltage vector $\mathbf{V}'_1$ becomes the origin of the two-level space vector diagram, which is known as zero vector. $\mathbf{V}'_{13}$ and $\mathbf{V}'_7$ are the basic vectors.

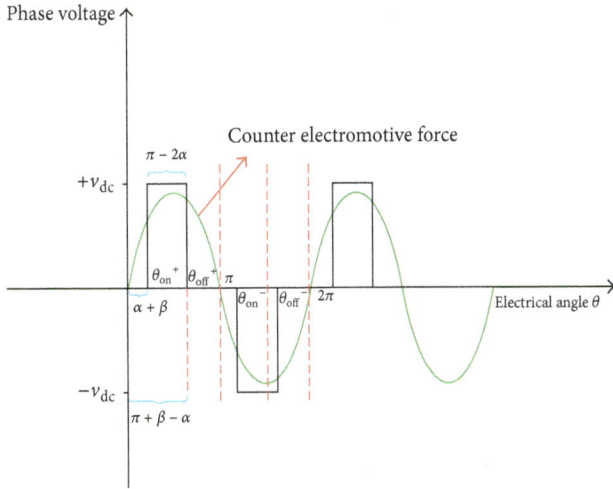

FIGURE 5: Quasi-square-wave voltage and counter electromotive force corresponding to the angle of the basic waveform and turn-off position.

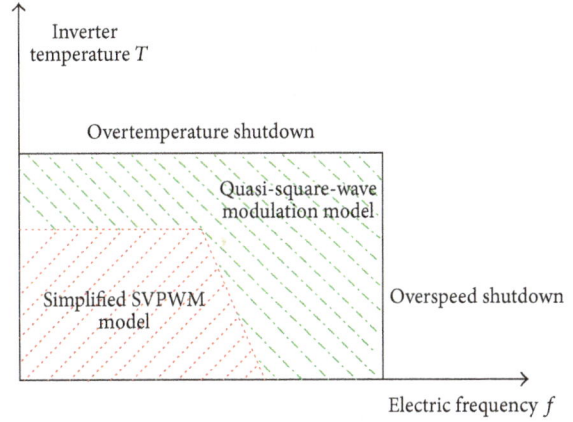

FIGURE 6: Temperature-frequency regional distribution with runtime of dual-mode modulation.

Through the above analysis, it can be found out that the first step is to judge the reference voltage vector, located in two-level space vector label S, according to Figures 2 and 3; then according to Table 2, depending on the value of different S, shift the reference voltage vector correspondingly. Finally, effecting time of basic vector can be calculated and the switching sequence of power devices can be determined as well [9–13].

In Table 2, V_{α_ref}, V_{β_ref} are the reference voltages of space vectors, which are the two basic components of coordinate. V'_{α_ref}, V'_{β_ref} are their corresponding translation voltages.

Fault tolerant control of thermal stability based on holographic equivalent dual mode is a method to switch two holographic equivalent modes to adjust the losses of power devices. The first mode, improved on the basis of traditional PWM modulation, is a simplified SVPWM algorithm based on three-level reference voltage vector transformation. This algorithm can simplify the three-level power device control, reduce the computation time, and improve the quality of the output power quality. But the power loss under this modulation is too large to maintain the thermal stability under continuous working. The second mode is quasi-square wave modulation algorithm, which needs the quasi-square wave voltage in each cycle to switch the power devices on and off twice. As shown in Figure 5, compared with the simplified SVPWM, this algorithm has a much lower power loss of power devices and generates much less heat during operation process. However, the output power quality is much worse and cannot be suitable for some operation conditions asking for high performance control effect. These two kinds of modulation act as the two holographic equivalent modes. In the different periods of hoist, according to the performance requirement of the control motor and different temperature control of power devices, the modulation will be switched between the two modes, which will keep balance between outputting high quality power to ensure the excellent ascension performance of hoist, and reducing power loss of

power devices. On the basis of modulation switching, the thermal stability fault tolerant control during transient failure of cooling system can be realized. The operation time and temperature-frequency regional distribution of dual-mode modulation are shown in Figure 6.

Two kinds of modulation mechanisms of the holographic equivalent modes are as follows:

(1) The simplified SVPWM is based on vector translation of three-level reference voltage: in simplified SVPWM mode, the direct relationship between input voltage command and PWM duty ratio is

$$d(t) = \frac{v(t)}{v_{dc}}. \tag{5}$$

(2) Quasi-square-wave modulation: in quasi-square-wave modulation mode, the relationships between switch-on and -off angles and orthogonal and direct harmonic components are as follows:

$$
\begin{aligned}
v_{n,q} &= \left(\frac{4v_{dc}}{n\pi}\right)\cos(n\alpha)\cos(n\beta), \\
v_{n,d} &= \left(\frac{4v_{dc}}{n\pi}\right)\cos(n\alpha)\sin(n\beta).
\end{aligned}
\tag{6}
$$

Benchmark effective power/torque and reactive power are decided by the value of α and β. Both of them have relationships with reference voltage components as follows:

$$\alpha = \left(\frac{\pi}{4v_{dc}}\right)\cos^{-1}\sqrt{\left(v_{1q}^{2} + v_{1d}^{2}\right)}, \tag{7}$$

$$\beta = \tan^{-1}\left(\frac{v_{1d}}{v_{1q}}\right). \tag{8}$$

3. Simulation and Experiment

The experiment platform is shown in Figure 7, including motor part and inverter part [14–17].

The parameters of winding asynchronous motor and component parameters of inverter main circuit are shown in Tables 3 and 4, respectively.

TABLE 2: Vector translations of reference voltages under different S values.

S	V'_{α_ref}	V'_{β_ref}
1	$V_{\alpha\_ref} - \dfrac{u_{dc}}{3}$	V_{β_ref}
2	$V_{\alpha\_ref} - \dfrac{u_{dc}}{6}$	$V_{\beta\_ref} - \dfrac{u_{dc}}{2\sqrt{3}}$
3	$V_{\alpha\_ref} + \dfrac{u_{dc}}{6}$	$V_{\beta\_ref} - \dfrac{u_{dc}}{2}$
4	$V_{\alpha\_ref} + \dfrac{u_{dc}}{3}$	V_{β_ref}
5	$V_{\alpha\_ref} + \dfrac{u_{dc}}{6}$	$V_{\beta\_ref} + \dfrac{u_{dc}}{2\sqrt{3}}$
6	$V_{\alpha\_ref} - \dfrac{u_{dc}}{6}$	$V_{\beta\_ref} + \dfrac{u_{dc}}{2\sqrt{3}}$

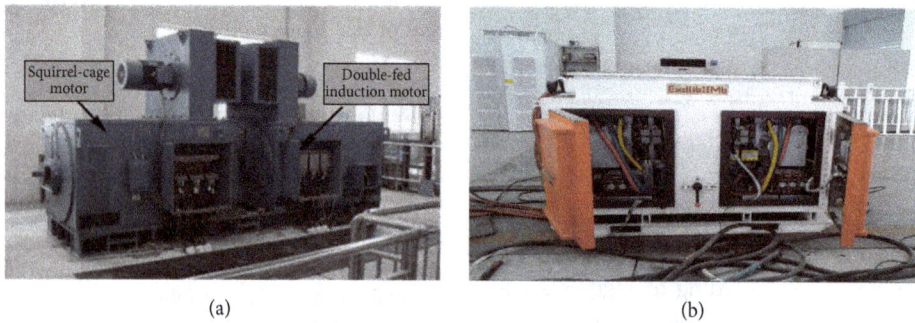

(a)   (b)

FIGURE 7: Experiment platform. (a) Motor part; (b) inverter part.

TABLE 3: Parameters of winding asynchronous motor.

Rated power P_d (kW)	475
Stator voltage U_s (V)	6000
Stator current I_d (A)	59
Rotor voltage U_r (V)	640
Rotor current I_r (A)	435
Rated speed (r/min)	735
Power factor	0.85

TABLE 4: Component parameters of inverter main circuit.

U_{dc}	1100 V
DC-link capacitor	1800 μF/1300 V
Power device parameters	Infineon, FF1400R17IE4
Switching frequency	2000 Hz

The well depth is 348 m. Meanwhile, the lifting conditions of one cycle are shown in Figure 8.

Simulation analysis of power losses in the same bridge $T1$ and its anti-parallel diode in NPC three-level explosion-proof inverter, based on two kinds of holographic equivalent mode, can be obtained in accordance with [14–17]. The simulation results are as shown in Figure 9.

It can be seen by analyzing the simulation results in Figures 9(a)–9(d) that power losses of $T1$ and its anti-parallel diode, based on simplified SVPWM mode, are much larger than the ones under quasi-square-wave modulation. In Figure 9(e), the power losses curves of power devices under dual-mode modulation were figured out. What is more, $T0$ in Figure 9(e) is the switching point between quasi-square-wave modulation and simplified SVPWM, and the quasi-square-wave modulation was used firstly at the start stage to reduce the power losses in case of overtemperature caused by high voltage and large current during overload period. Put the power loss values in Figure 9 into high-power explosion-proof inverter mode as heat sources, and the thermal simulation can be obtained in Figure 8.

Through the thermal analysis in Figure 10 based on two kinds of holographic equivalent modes, it can be found out that, in inverter system, temperature rises of heatsink and substrate using quasi-square-wave modulation mode have an average advantage over using simplified SVPWM mode by appropriate 15°C.

The electrical performance of the three-level explosion-proof inverter under normal condition in one cycle is shown in Figure 11, in which the mode switching point is the time $T0$. It can be seen from the Figure 11 that the inverter has not very good output properties during quasi-square-wave modulation compared with simplified SVPWM, because the torque current has a much higher harmonic content and it will bring pulse vibration into the motor. As shown in Figure 12, the harmonic characteristic based on the simplified SVPWM is a little better than the one based on the dual-mode modulation. But both of them have an acceptable harmonic performance and in this high-power application

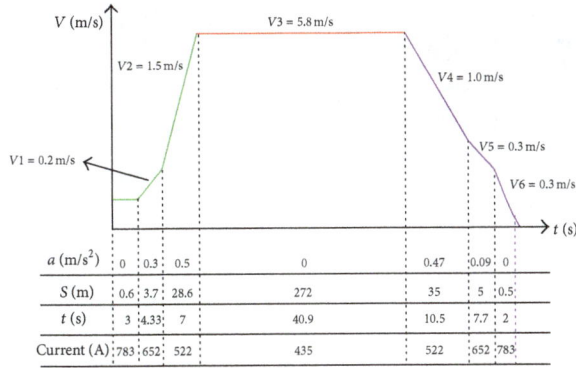

FIGURE 8: Lifting conditions of mine hoist in one cycle.

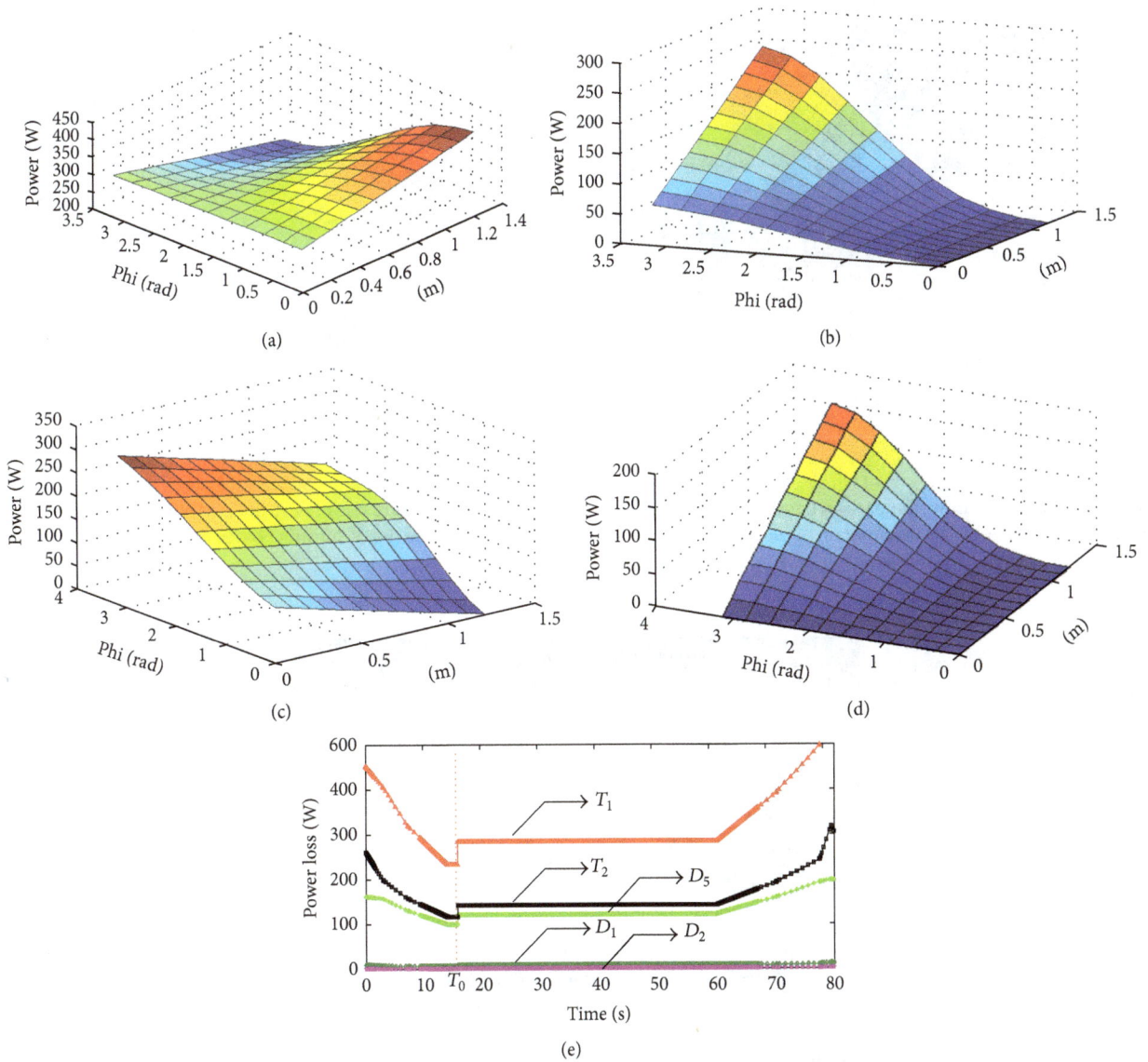

FIGURE 9: Power loss comparison between two equivalent modes. (a) Power loss of $T1$ based on simplified SVPWM; (b) power loss of $T1$ based on quasi-square-wave modulation; (c) power loss of diode based on simplified SVPWM; (d) power loss of diode based on quasi-square-wave modulation; (e) power loss curves of power devices under dual-mode modulation.

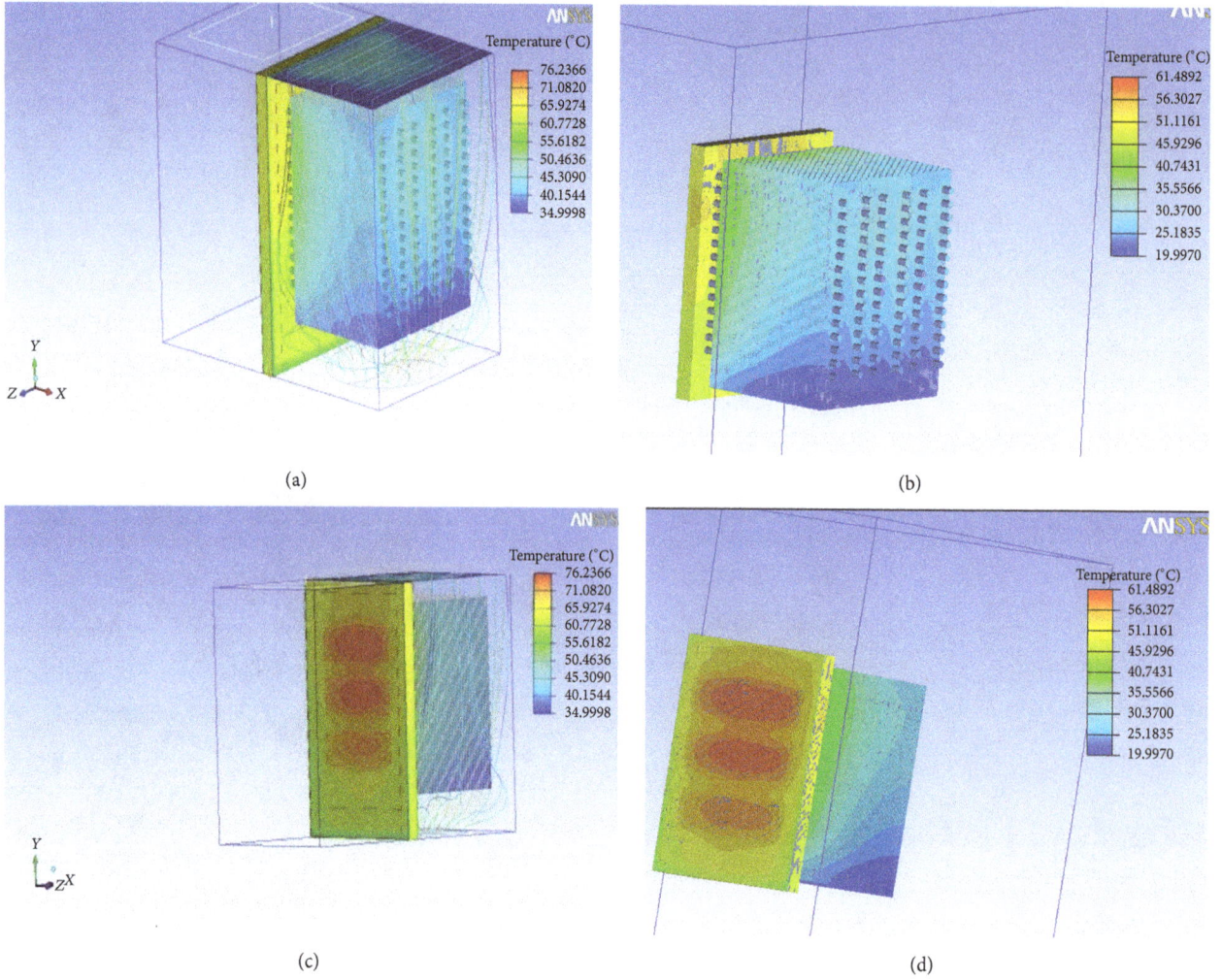

(a)

(b)

(c)

(d)

FIGURE 10: Thermal analysis comparison between the two equivalent modes. (a) Temperature rise of heatsink based on simplified SVPWM; (b) temperature rise of heatsink based on quasi-square-wave modulation; (c) temperature rise of substrate based on simplified SVPWM; (d) temperature rise of substrate based on quasi-square-wave modulation.

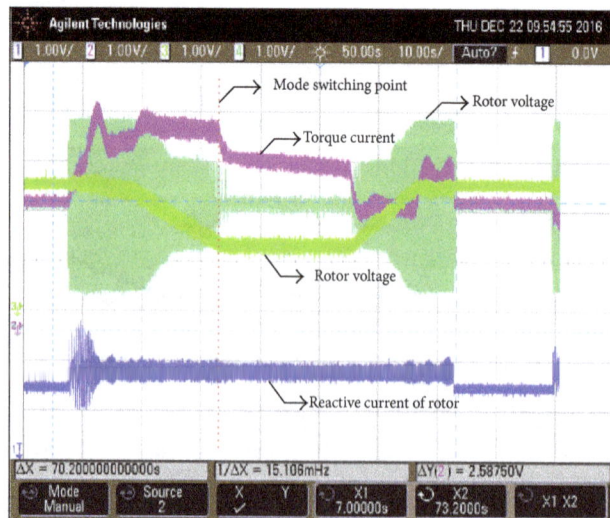

FIGURE 11: Electrical performance of inverter.

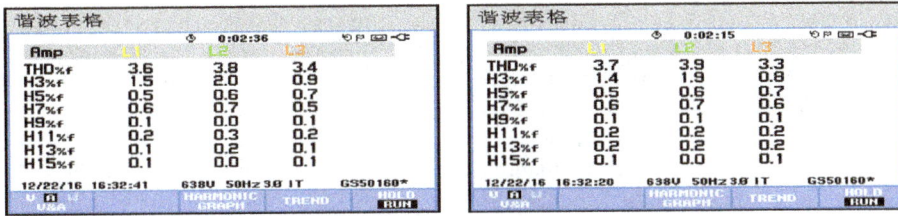

(a) Harmonic characteristic based on simplified SVPWM

(b) Harmonic characteristic based on dual-mode modulation

FIGURE 12: Harmonic characteristic comparison.

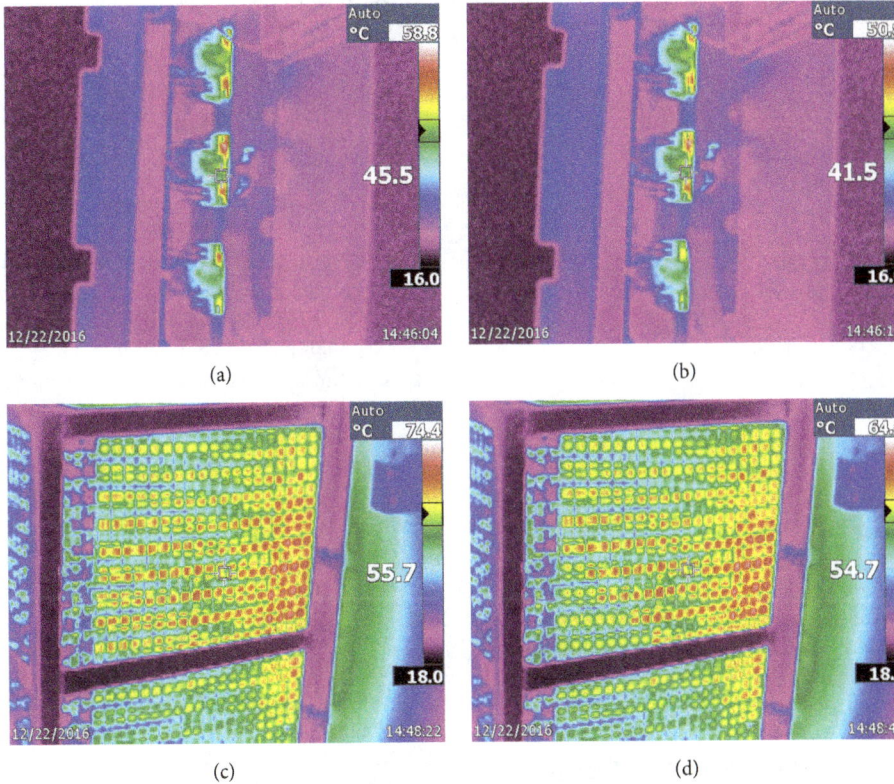

(a)

(b)

(c)

(d)

FIGURE 13: Surface and the substrate temperature rises of IGBT based on holographic equivalent dual mode. (a) Surface temperature rise of IGBT based on simplified SVPWM; (b) surface temperature rise of IGBT based on quasi-square-wave modulation; (c) substrate temperature rise of IGBT based on simplified SVPWM; (d) substrate temperature rise of IGBT based on quasi-square-wave modulation.

they almost are the same. Based on the experimental result comparison of power loss analysis, thermal analysis, and the temperature rise of the same IGBT $T1$ and its anti-parallel diode in Figure 13, it can be seen that the power loss and temperature rise using simplified SVPWM mode are greater than the ones using quasi-square-wave modulation mode by approximately 9°C. The thermal stability fault tolerant control based on holographic equivalent dual mode is to use this characteristic in different stages of the hoist's each operation cycle, according to the need of the controlled motor performance and thermal condition. In the start, the motor has a much larger overload and the inverter has a much larger output current as well, which is just stage asking for starting the given mode of quasi-square-wave modulation mode, to reduce the power losses of power devices effectively,

maintaining the stability of hoist running all the way. In constant speed stage, with a higher speed, it requires a much smaller motor vibration and a much higher running stability, and then the modulation switches into the simplified SVPWM, which can guarantee the high quality power of inverter output and good running characteristics of motor and also guarantee that the power loss of power devices is still controllable under the inverter cooling system in constant speed period to realize power device thermal fault tolerant control of the whole inverter system.

4. Conclusion

Based on high-power three-level explosion-proof inverter output characteristic and the special requirements of thermal stability, in this paper, we put forward a holographic equivalent dual-mode control algorithm based on the simplified SVPWM and quasi-square-wave modulation and calculate the power losses of power devices in three-level explosion-proof inverter on this basis and establish thermal models of power devices and cooling system for thermal simulation. The effectiveness of holographic equivalent dual-mode control on both the system output characteristics and thermal stability is verified by experimental analysis, so this mode plays an important role in realizing power device thermal fault tolerant control and improving the thermal stability of high-power explosion-proof inverter.

Conflicts of Interest

The authors declare that there are no conflicts of interest regarding the publication of this article.

Acknowledgments

The authors would like to thank the Doctoral Scientific Research Start-Up Foundation of Henan Normal University, Grant no. 5101239170001.

References

[1] Y. Firouz, M. T. Bina, and B. Eskandari, "Efficiency of three-level neutral-point clamped converters: Analysis and experimental validation of power losses, thermal modelling and lifetime prediction," *IET Power Electronics*, vol. 7, no. 1, pp. 209–219, 2014.

[2] O. S. Senturk, S. Munk-Nielsen, R. Teodorescu, L. Helle, and P. Rodriguez, "Converter structure-based power loss and static thermal modeling of the press-pack igbt-based three-level ANPC and HB VSCs applied to multi-MW wind turbines," in *Proceedings of the 2010 2nd IEEE Energy Conversion Congress and Exposition, (ECCE '10)*, pp. 2778–2785, IEEE, Atlanta, GA, USA, September 2010.

[3] R. Pittini, S. D'Arco, M. Hernes, and A. Petterteig, "Thermal stress analysis of IGBT modules in VSCs for PMSG in large offshore wind energy conversion systems," in *Proceedings of the 2011 14th European Conference on Power Electronics and Applications, (EPE '11)*, pp. 1–10, IEEE, Birmingham, UK, September 2011.

[4] A. V. Rocha, H. De Paula, M. E. Dos Santos, and B. J. Cardoso Filho, "A thermal management approach to fault-resilient design of three-level IGCT-based NPC converters," *IEEE Transactions on Industry Applications*, vol. 49, no. 6, pp. 2684–2691, 2013.

[5] D. Zhou, F. Blaabjerg, M. Lau, and M. Tonnes, "Thermal analysis of multi-MW two-level wind power converter," in *Proceedings of the 38th Annual Conference on IEEE Industrial Electronics Society, (IECON '12)*, pp. 5858–5864, IEEE, Montreal, QC, Canada, October 2012.

[6] K. Ma, F. Blaabjerg, and M. Liserre, "Thermal analysis of multilevel grid side converters for 10 MW wind turbines under low voltage ride through," in *Proceedings of the 3rd Annual IEEE Energy Conversion Congress and Exposition, (ECCE '11)*, pp. 2117–2124, IEEE, Phoenix, AZ, USA, September 2011.

[7] W. U. N. Fernando, L. Papini, and C. Gerada, "Converter temperature regulation with dual mode control of fault-tolerant permanent magnet motors," in *Proceedings of the 4th Annual IEEE Energy Conversion Congress and Exposition (ECCE '12)*, pp. 1902–1908, IEEE, Raleigh, NC, USA, September 2012.

[8] Z.-B. Ye, *Key Technology Research of High Power Three-level Double-fed Mine Hoist [Ph.D. dissertation]*, China University of Mining and Technology, Beijing Shi, China, 2010.

[9] U. R. Prasanna and A. K. Rathore, "Analysis, design, and experimental results of a novel soft-switching snubberless current-fed half-bridge front-end converter-based pv inverter," *IEEE Transactions on Power Electronics*, vol. 28, no. 7, pp. 3219–3230, 2013.

[10] F. Krismer and J. W. Kolar, "Accurate power loss model derivation of a high-current dual active bridge converter for an automotive application," *IEEE Transactions on Industrial Electronics*, vol. 57, no. 3, pp. 881–891, 2010.

[11] D. Xu, H. Lu, L. Huang, S. Azuma, M. Kimata, and R. Uchida, "Power loss and junction temperature analysis of power semiconductor devices," *IEEE Transactions on Industry Applications*, vol. 38, no. 5, pp. 1426–1431, 2002.

[12] S. Bernet, "Recent developments of high power converters for industry and traction applications," *IEEE Transactions on Power Electronics*, vol. 15, no. 6, pp. 1102–1117, 2000.

[13] Jing W., *Study on Power Device Losses of High-Power Three-Level Converter [Ph.D. dissertation]*, China University of Mining and Technology, Beijing Shi, China, 2010.

[14] S.-Z. Xu, Y.-F. Peng, and S.-Y. Li, "Application thermal research of forced-air cooling system in high-power NPC three-level inverter based on power module block," *Case Studies in Thermal Engineering*, vol. 8, pp. 387–397, 2016.

[15] S.-Z. Xu and F.-Y. He, "The optimized design of a NPC three-level inverter forced-air cooling system based on dynamic power-loss calculations of the maximum power-loss range," *Journal of Power Electronics*, vol. 16, no. 4, pp. 1598–1611, 2016.

[16] F.-Y. He, S.-Z. Xu, and C.-F. Geng, "Improvement on the laminated busbar of NPC three-level inverters based on a supersymmetric mirror circulation 3D cubical thermal model," *Journal of Power Electronics*, vol. 16, no. 6, Article ID JPE 16-6-10, pp. 2085–2098, 2016.

[17] N. N. Yang, C. J. Wu, R. Jia, and C. Liu, "Modeling and characteristics analysis for a buck-boost converter in pseudo-continuous conduction mode based on fractional calculus," *Mathematical Problems in Engineering*, vol. 2016, Article ID 6835910, pp. 1–13, 2016.

Permissions

List of Contributors

Muhammad Nawaz
ABB Corporate Research, Froskargränd 7, 724 78 Västerås, Sweden

Munir A. Al-Absi
EE Department, King Fahd University of Petroleum & Minerals, Dhahran, Saudi Arabia

Taoufik Benyetho, Jamal Zbitou and Larbi El Abdellaoui
LMEET, FST of Settat, Hassan 1st University, Settat, Morocco

Hamid Bennis
TIM Research Team, EST of Meknes, Moulay Ismail University, Meknes, Morocco

Abdelwahed Tribak
Microwave Team, INPT, Rabat, Morocco

Vinod Kumar Joshi and Chetana Nayak
Department of Electronics and Communication Engineering, Manipal Institute of Technology, Manipal Academy of Higher Education, Manipal 576104, India

Xiaomin Zheng, Yuejun Zhang, Jiaweng Zhang and Wenqi Hu
Institute of Circuits and Systems, Ningbo University, No. 818 Fenghua Road, Ningbo 315211, China

Xiuqin Xu, Hui Xu, Yongheng Shang, Zhiyu Wang, Yang Wang, Liping Wang, Hao Luo, Zhengliang Huang and Faxin Yu
School of Aeronautics and Astronautics, Zhejiang University, Hangzhou 310027, China

Hazem K. Khanfar
Department of Telecommunication Engineering, Arab-American University, Jenin, State of Palestine

A. F. Qasrawi and Yasmeen Kh. Ghannam
Department of Physics, Arab-American University, Jenin, State of Palestine

Yanbin Hou, Wanrong Sun, Aifeng Ren and Shuming Liu
School of Electronic Engineering, Xidian University, Xi'an 710071, China

Supachai Klungtong and Dusit Thanapatay
Department of Electrical Engineering, Faculty of Engineering, Kasetsart University, Bangkok 10900, Thailand

Winai Jaikla
Department of Engineering Education, Faculty of Industrial Education, King Mongkut's Institute of Technology Ladkrabang, Bangkok 10520, Thailand

Munir Al-Absi, Zainulabideen Khalifa and Alaa Hussein
Electrical Engineering Department, Faculty of Engineering, King Fahd University of Petroleum and Minerals, Dhahran, Saudi Arabia

Neelofer Afzal and Devesh Singh
Department of Electronics and Communication Engineering, Jamia Millia Islamia University, New Delhi 110025, India

Satyam Shukla, Sandeep Singh Gill and Navneet Kaur
Department of Electronic and Communication Engineering, Guru Nanak Dev Engineering College, Ludhiana 141006, India

Satyam Shukla, H. S. Jatana and Varun Nehru
Semi-Conductor Laboratory, Department of Space, Government of India, Mohali 160071, India

Rawid Banchuin
Faculty of Engineering and Graduated School of Information Technology, Siam University, Bangkok, Thailand

Weiping Li, Zongxi Tang and Xin Cao
School of Electronic Engineering, University of Electronic Science and Technology of China, No. 2006, Xiyuan Ave., West Hi-Tech Zone, Chengdu 611731, China

Weiping Li
School of Information Engineering, East China Jiaotong University, No. 88, Shuanggang Road, Nanchang 330013, China

Mohannad Jabbar Mnati, Dimitar V. Bozalakov and Alex Van den Bossche
Department of Electrical Energy, Metals, Mechanical Constructions and Systems, Ghent University, Technologiepark Zwijnaarde 913, B-9052 Zwijnaarde, Gent, Belgium

Mohannad Jabbar Mnati
Department of Electronic Technology, Institute of Technology Baghdad, Middle Technical University, Al-Za'franiya, 10074 Baghdad, Iraq

Praveen Kumar and Sajal Kumar Paul
Department of Electronics Engineering, Indian School of Mines, Dhanbad, India

Neeta Pandey
Department of Electronics and Communications, Delhi Technological University, Delhi, India

Yasuhisa Omura
ORDIST, Grad. School of Sci. & Eng., Kansai University, 3-3-35 Yamate-cho, Suita, Osaka 564-8680, Japan

Qifeng Pan and Qiao Liu
College of Big Data and Information Engineering, Guizhou University, Guiyang, Guizhou, China

Qifeng Pan, Yuanjiang Yang and Dongbin Tian
Xinyun Electronics Components Corporation, Guiyang, Guizhou, China

Mariagrazia Graziano, Malik Ashter Mehdy and Gianluca Piccinini
Electronics and Telecommunication Department of Politecnico di Torino, Torino, Italy

Ali Zahir
Department of Electrical Engineering of the COMSATS Institute of Information Technology, Abbottabad, Pakistan

Weiping Li, Zongxi Tang and Xin Cao
School of Electronic Engineering, University of Electronic Science and Technology of China, No. 2006 Xiyuan Ave, West Hi-Tech Zone, Chengdu 611731, China

Weiping Li
School of Information Engineering, East China Jiaotong University, No. 88 Shuanggang Road, Nanchang 330013, China

Liyun Shi and Jianjun Gao
School of Information and Science Technology, East China Normal University, 500 South Lianhua Road, Shanghai, China

Liyun Shi
School of Electronic Information and Electronical Engineering, Shanghai Jiao Tong University, 800 Dongchuan Road, Shanghai, China

Shi-Zhou Xu, Chun-jie Wang and Yu-feng Peng
College of Electronic and Electrical Engineering, Henan Normal University, Xinxiang, China

Index